ENVIRONMENTAL
SEPARATION
of HEAVY METALS

ENGINEERED PROCESSES

ENVIRONMENTAL SEPARATION
of HEAVY METALS

ENGINEERED PROCESSES

Edited by

Arup K. SenGupta

Professor of Environmental Engineering
Lehigh University

CRC Press
Taylor & Francis Group
Boca Raton London New York

CRC Press is an imprint of the
Taylor & Francis Group, an **informa** business

CRC Press
Taylor & Francis Group
6000 Broken Sound Parkway NW, Suite 300
Boca Raton, FL 33487-2742

First issued in paperback 2019

© 2002 by Taylor & Francis Group, LLC
CRC Press is an imprint of Taylor & Francis Group, an Informa business

No claim to original U.S. Government works

ISBN-13: 978-1-56676-884-9 (hbk)
ISBN-13: 978-0-367-39695-4 (pbk)

Visit the Taylor & Francis Web site at
http://www.taylorandfrancis.com

and the CRC Press Web site at
http://www.crcpress.com

*To All Our Teachers Who Inspired Us
to Be Creative without Sacrificing the Truth*

Table of Contents

Preface

WE all know that an unmixed blessing is non-existent in the real world. Since the advent of the industrial revolution, the beneficial use of metals has seen a tremendous growth around the world and essentially has reshaped our civilization. For a long time, a number of these metals, many of them toxic, were indiscriminately released into the environment. It required decades for the fate, transport, and deleterious effects of these metals to be comprehended. Gradually their impact on the quality of life on our planet surfaced as a matter of grave concern, ushering in a series of environmental regulations during the last thirty years. Metals that are potential carcinogens warranted closer scrutiny and were later included in the United States Environmental Protection Agency's (U.S.EPA's) list of priority pollutants. It would, however, be wrong to isolate industrial emissions and discharges as the sole source of environmental problems pertaining to toxic metals. In some instances, nature plays an unfriendly role in deteriorating environmental quality. Worldwide contamination of groundwater by arsenic, chromium, and selenium through geochemical soil leaching represents one example to this effect.

In order to be in regulatory compliance, industries and organizations in both the private and public sectors have sought new technologies for efficient heavy metals removal. Over the course of time, the emphasis has shifted from mere removal to recovery and reuse. Attaining "zero discharge" is today a common slogan in almost every industrial sector. Such cultural changes, along with new technologies, have warranted judicious use of efficient heavy-metal-separation technologies. This book seeks to provide comprehensive coverage of advanced technologies for heavy metals separation from groundwater, wastewater, contaminated soils, flue gases, industrial sludges and other contaminated sources. Separation of heavy metals is carried out to meet stringent environmental regulation with respect to priority pollutants, especially arsenic, mercury and lead, which are discussed in depth. Altogether the book contains ten chapters. The first chapter sets the stage by introducing heavy metals chemistry and the underlying principles for heavy metals separation. The remaining nine chapters pertain to technologies such as specialty sorption, magnetic separation, chelating ion exchange, polyelectrolyte-aided membrane separation, and advanced precipitation, among others. The book is designed to enable engineers and non-engineers alike to learn, select and adapt the best means to remove and, in certain instances recover, heavy metals through the use of appropriate technologies. The book is also intended to be of use to graduate students and professionals undertaking research and development work in heavy metals, remediation and related areas.

Before closing, I would like to sincerely thank the lead authors of the chapters and their co-authors for their time, effort and active cooperation during the last eighteen months. Ping Li's assistance in preparing several illustrations included in the book is gratefully acknowledged. Eleanor Nothelfer's editing of several manuscripts helped improve the final quality. Thanks are also due to Joe Eckenrode and Teresa Wiegand of Technomic Publishing Co., Inc. for their prompt help and quick turnaround time during the production stage of the book. Last but not the least, I would like to thank my wife, Susmita, for her encouragement and forbearance throughout this endeavor, which made working on it a pleasure.

ARUP K. SENGUPTA

Contributing Authors

Kashi Banerjee, USFilter Corporation, USA

Amjad Fataftah, ARCTECH, Inc., USA

Matthias Franzreb, Forschugxzentrum Karlsruhe GmbH, Germany

John E. Greenleaf, Lehigh University, USA

Roberto Passino, Instituto di Ricerca Sulle Acque, Italy

Domenico Petruzzelli, The Polytechnic University of Bari, Italy

Brian Reed, University of Missouri at Columbia, USA

Thomas W. Robinsson, Los Alamos National Laboratory, USA

H. G. Sanjay, ARCTECH, Inc., USA

Nancy N. Sauer, Los Alamos National Laboratory, USA

Arup K. SenGupta, Lehigh University, USA

Sukalyan Sengupta, University of Massachusetts at Dartmouth, USA

Barbara F. Smith, Los Alamos National Laboratory, USA

Giovanni Tiravanti, Instituto di Ricerca Sulle Acque, Italy

Radisav D. Vidic, University of Pittsburgh, USA

Daman Walia, ARCTECH, Inc., USA

James H. P. Watson, University of Southampton, UK

CHAPTER 1

Principles of Heavy Metals Separation: An Introduction

ARUP K. SENGUPTA[1]

HEAVY METALS: WHAT ARE THEY?

THE term "heavy metal," in spite of its widespread usage among professionals and lay-men, does not have a rigorous scientific basis or a chemical definition. Although many of the elements listed under "heavy metals" have specific gravities greater than five, major exceptions to this rule remain. In hindsight, this group should preferably have been referred to as "toxic elements," for they are all included in the United States Environmental Protection Agency's (USEPA's) list of priority pollutants. Figure 1.1 shows the periodic table containing the heavy metals that are of significant environmental concern. For comparison, commonly occurring light alkali and alkali-earth metals have also been included in the same figure. Strictly from a chemical viewpoint, heavy metals constitute transition and post-transition elements along with metalloids, namely, arsenic and selenium. They are indeed significantly heavier (i.e., higher specific gravities) than sodium, calcium and other light metals. These heavy metal elements often exist in different oxidation states in soil, water and air. The reactivities, ionic charges and solubilities of these metals in water vary widely. For their short- and long-term toxic effects, the maximum permissible concentrations of these heavy metals in drinking water as well as in municipal and industrial discharges are closely regulated through legislation. Nevertheless, barring the exceptions of cadmium, mercury and lead, heavy metals are also required micronutrients, i.e., essential ingredients for living cells. Toxicity effects of these elements are, thus, largely a function of concentration. These elements are beneficial and have nutritional values lower than some critical dosages but become inhibitory to toxic with an increase in concentration, as shown in Figure 1.2. The threshold toxic concentrations differ for each heavy metal and are governed primarily by the chemistry of each heavy metal in question and associated physiological effects. On the contrary, nonessential heavy metal elements are inhibitory at all concentrations.

Metal cycles on a regional and global basis have been profoundly modified by human ac-

[1]Department of Civil and Environmental Engineering, Lehigh University, 13 E. Packer Avenue, Bethlehem, PA 18015, U.S.A., aks0@lehigh.edu

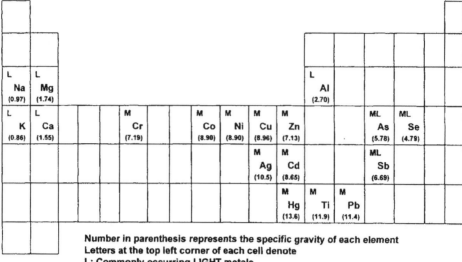

Figure 1.1 A modified periodic table showing commonly encountered regulated heavy metals, metalloids and unregulated light metals.

tivity and industrial development during the last fifty years. While mining, metallurgical, electroplating industries, etc., have greatly boosted the production and usage of heavy metals in our life cycles, the lowering of pH in rain and surface waters and the increased use of surfactants, which have greatly enhanced the mobility of heavy metals in the environment. Understandably, the presence of heavy metals in aquatic, terrestrial and atmospheric environment is of concern. In the aqueous phase, such heavy metals may exist as cations, anions, nonionized species and complex macromolecules. As most of the heavy metals and their compounds have extremely high boiling points, they are practically absent in the atmo-

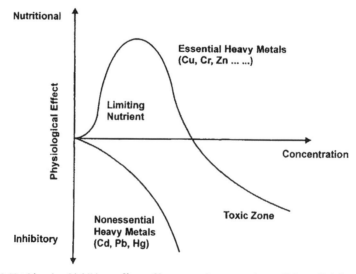

Figure 1.2 Nutritional and inhibitory effects of heavy metal concentration on living cells/microorganisms.

sphere under ambient conditions, with the glaring exception of elemental mercury. Flue gases from fossil fuel-fired steam generators and waste incinerators are major industrial sources of mercury emission into the atmosphere. Higher volatility and relative inertness compared to other heavy metals allow elemental mercury to persist in the environment for a prolonged period of time. Following the phaseout of leaded gasoline in industrial countries, the short- and long-term presence of lead in the atmosphere has greatly subsided. In the soil phase, heavy metals exist primarily as insoluble precipitates or as bound solutes on the surface sorption sites of microparticles. Mobility and fate of the heavy metals in the soil phase are often influenced by the chemical composition of the contacting liquid phase.

HEAVY METALS SEPARATION: UNDERLYING CHEMISTRY

Lewis Acid-Base Interaction

The speciation and fate of metals in the natural environment as well as their separation and/or control by engineered processes are ultimately governed by the electronic structures of the heavy metals. Such electronic structures also dictate the biochemical actions of metals as nutrients or toxicants. In order to develop an insight, let us consider the electronic configurations of a light metal cation (say Ca^{2+}) and a heavy metal cation (say Cu^{2+}) as shown below:

$$Ca^{2+}: 1s^22s^22p^63s^23p^6 \tag{1}$$

$$Cu^{2+}: 1s^22s^22p^63s^23p^63d^9 \tag{2}$$

Note that Ca^{2+} has the noble gas configuration of Krypton, i.e., its outermost electron shell is completely filled, and the octet formation is satisfied. Thus, Ca^{2+} is not a good electron acceptor and, hence, a poor Lewis acid. Ions like Ca^{2+} are not readily deformed by electric fields and have low polarizabilities. They are referred to as "hard" cations, and they form only outer sphere complexes with aqueous-phase ligands containing primarily oxygen donor atoms.

In contrast, the transition metal cation, Cu^{2+} or Cu(II), has an incomplete d-orbital and contains electron clouds more readily deformable by electric fields of other species. In general, these ions are fairly strong Lewis acids and tend to form inner sphere complexes with ligands in the aqueous phase. Electrostatically, Ca^{2+} and Cu^{2+} are identical, i.e., both Ca^{2+} and Cu^{2+} have two charges. However, Cu(II) is a stronger Lewis acid or electron acceptor and a relatively "soft" cation. Table 1.1 classifies several metal cations in three categories, namely, hard, borderline and soft [1,2]. Note that most of the heavy metals of interest fall under "borderline" and "soft." In general, the toxicity of metals increases as one moves from hard cations to borderline and then to soft. Relative affinities of these metal ions to form complexes with O-, N- and S-containing ligands, vary widely. While hard cations prefer oxygen-donating ligands (Lewis bases), borderline and soft cations exhibit higher affinities toward nitrogenous and sulfurous species. The soft cations thus bind strongly with sulfhydryl groups in proteins of the cells. Because sulfhydryl groups form active sites on proteins, their blockages through heavy metal binding result in severe toxic effects [3].

The foregoing phenomenon prompted Nierboer and Richardson to recommend that toxic metals be classified by their relative complex forming abilities with O-, N- and S-containing ligands, for such affinities are the primary determinants of physiological toxicity

TABLE 1.1. Classification of Selected Metal Cations.

Type	Name of Cations	Salient Properties
1. Hard cations	Na^+, K^+, Mg^{2+}, Ca^{2+}, Al^{3+}, Be^{2+}, etc.	Spherically symmetric and electronic configurations conform to inert gases; form only outer-sphere complexes with hard ligands containing oxygen donor atoms; weak affinity toward ligands with nitrogen and sulfur donor atoms; besides beryllium, most are non-toxic at low concentrations.
2. Borderline cations	Fe^{3+}, Cu^{2+}, Pb^{2+}, Fe^{2+}, Ni^{2+}, Zn^{2+}, Co^{2+}, Mn^{2+}	Spherically not symmetric, and electronic configurations do not conform to inert gases; form inner-sphere complexes with O- and N-atom-containing ligands; excepting iron and manganese, all are toxic.
3. Soft cations	Hg^+, Cu^+, Hg^{2+}, Ag^+, Cd^{2+}	Spherically not symmetric, and electronic configurations do not conform to inert gases; except high affinity toward S-atom-containing ligands; they are most toxic from a physiological viewpoint.

caused by the metals [4]. The fact that many heavy metals bind strongly onto proteins also suggests that these functional groups in proteins, when immobilized onto a solid phase, may selectively capture dissolved heavy metals from the aqueous phase. Widely used chelating exchangers essentially conform to this principle of separation. Tens of polymeric chelating exchangers have been synthesized to date, and are commercially available with various types of covalently attached functional groups. Physically, they are all the same, i.e., spherical beads with high mechanical strength and durability. Figure 1.3 illustrates several commercially available chelating exchangers with linear polymer chains, cross-linkings and a variety of covalently attached functional groups. Understandably, it is the Lewis acid-base interaction that governs the binding affinity of a heavy metal cation to a chelating exchanger. Such binding affinities (often expressed as separation factor values) are correlated to corresponding aqueous-phase stability constant values between the heavy metal ions and the representative ligands, and they can be modeled by the Linear Free Energy Relationship (LFER) [5]. Figure 1.4 shows the relationship between copper/calcium separation factor values for three commercial chelating exchangers and the corresponding aqueous-phase stability constant values for representative ligands [6]. Noteworthy is the fact that as the composition of the functional groups in Figure 1.4 changes from hard oxygen donor atoms (i.e., carboxylate) to relatively soft nitrogen donor atoms (bispicolylamine), the affinity of Cu(II), a borderline Lewis acid, is greatly enhanced over the hard cation, Ca^{2+}. Understandably, the composition of the functional groups in chelating exchangers can be judiciously tailored to improve specific affinities toward target metal ions. Chelating exchangers with S-containing thiol functional groups offer significantly higher selectivity for soft Hg(II) over Cu(II) and Zn(II). Along the same vein, Figure 1.5 shows the separation factor values of five different heavy metal cations for a weak-acid cation exchange resin with carboxylate functional groups (IRC DP-1, Rohm and Haas Co., Philadelphia, PA, USA). Note that the sequence and relative affinity of dissolved heavy metals are strongly correlated to their aqueous phase metal-acetate stability constant values [7].

Metals' sorption onto polymeric chelating exchangers is, however, kinetically slow, and

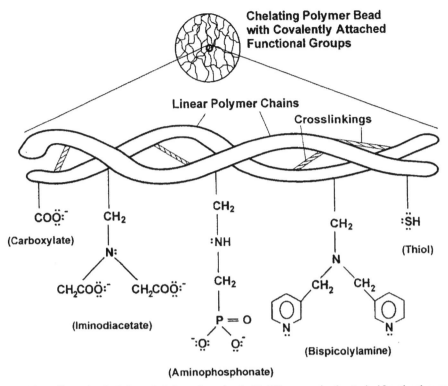

Figure 1.3 An illustration depicting a chelating polymer bead with different covalently attached functional groups.

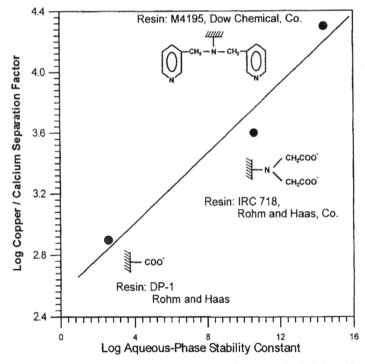

Figure 1.4 Relationship between copper/calcium separation factors for commercial chelating exchangers and corresponding aqueous-phase stability constant values with representative ligands.

intraparticle diffusion is often, if not always, the rate-limiting step. Due to the rigid structure and tortuous pathways, the effective intraparticle diffusivities for metals within chelating ion exchangers are several orders of magnitude lower than they are in the solvent phase. Solvent Impregnated Resins (SIRs) can greatly overcome this shortcoming. Even more important, SIRs do not require covalent attachment of organic functional groups onto the parent polymer beads and thus conveniently avoid major steps in chemical synthesis for their preparation. In SIR, an organophilic complexant is sorbed within macroporous copolymer beads, and the combined material serves as the metal-selective sorbent [8,9]. One critical disadvantage of SIRs is the gradual loss of complexant through aqueous-phase dissolution, which is a significant problem and precludes adaptation of SIRs in environmental application.

In order to eliminate the loss of complexant, a new SIR has been prepared wherein a thin coating is formed around each bead [10]. This coating is hydrophilic, thus preventing transport of the hydrophobic complexant out of the bead while permitting transport of the hydrophilic metal ion into the bead. Figure 1.6 illustrates the characteristic features of the modified SIR. Conceptually similar techniques have also evolved in membrane processes where hollow-fiber contained liquid membranes are used to selectively remove dissolved heavy metals from the aqueous phase with a minimum loss of the organic extractant in the aqueous phase [11,12].

Equipment configurations and physical arrangement of the processes for heavy metals removal often vary widely. Nevertheless, barring a few minor exceptions, Lewis acid-base interaction aided by precipitation, sorption, sieving, etc., constitutes the primary mechanism for heavy metals separation. Many biorenewable materials such as naturally occurring humus, dead bacterial and fungal cells and seaweeds, contain surface functional groups (carboxylate, carbonyl, phenolic) with moderate to high affinity toward heavy metals. Significant progress has been made in the recent past in modifying such materials into chemically stable, mechanically strong, durable sorbents [13,14]. As we lay an increased empha-

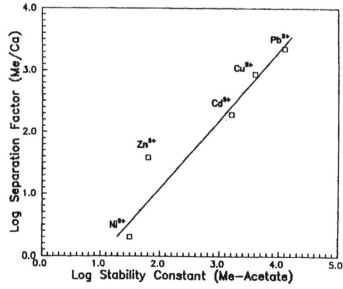

Figure 1.5 Relationship between experimentally determined metal/calcium separation factors for IRC DP-1 and aqueous-phase metal-acetate stability constant values.

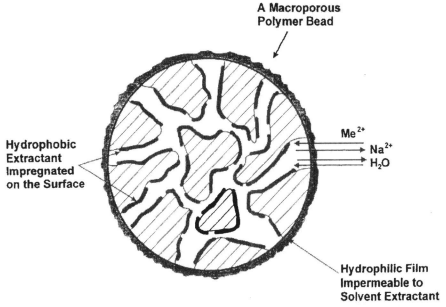

Figure 1.6 An illustration of the characteristic features of the modified solvent-impregnated resin (SIR).

sis on sustainable development, these sorbent materials are likely to be economically competitive, and large-scale commercial production will follow.

Redox Chemistry

Cycling of heavy metals in the environment as well as their removal by engineered processes are influenced by redox reactions. The following four examples to this effect cover a broad spectrum and provide an engineering perspective regarding how redox chemistry can be manipulated to achieve efficient separation of heavy metals.

$$CrO_4^{2-} + 5H^+ + 3e^- \xrightarrow{\text{Reduction}} Cr(OH)_{3(s)} \downarrow + H_2O \qquad (3)$$
$$\text{(Mobile)}$$

$$H_2AsO_4^- + 3H^+ + 2e^- \xleftarrow{\text{Oxidation}} H_3AsO_3 + H_2O \qquad (4)$$
$$\text{(Strongly Adsorbable)} \qquad\qquad \text{(Poorly Adsorbable)}$$

$$Cd^{2+} + SO_4^{2-} + 8H^+ + 8e^- \xrightarrow{\text{Reduction}} CdS_{(s)} \downarrow + 4H_2O \qquad (5)$$
$$\text{(Mobile)}$$

$$\begin{array}{l} Hg^\circ \xrightarrow{\text{Oxidation}} \\ S^\circ \xrightarrow{\text{Reduction}} \end{array} \longrightarrow HgS_{(s)} \quad \downarrow \qquad (6)$$

Each of the four foregoing reactions deserve some discussion to elucidate possible application of redox chemistry in achieving heavy metals separation.

(1) Toxicity, accompanied by widespread industrial applications and high mobility, has earned Cr(VI) an unusual notoriety in the area of environmental pollution. Contrary to Cr(VI) species or chromates, Cr(III) is less toxic and very insoluble at neutral to alkaline pH. As a result, chemical reduction of Cr(VI) to Cr(III), followed by precipitation as chromic hydroxide, $Cr(OH)_{3(s)}$, has been the traditional approach for treating Cr(VI)-laden wastewater [15,16]. For relatively low concentrations of Cr(VI) (i.e., in micrograms/liter), anion exchange process can be used to scavenge Cr(VI) using polymeric anion exchangers [17,18]. Subsequently, Cr(VI) can be concentrated through regeneration, and Cr(VI) in the spent regenerant can efficiently be reduced to relatively innocuous Cr(III) hydroxide.

(2) In many naturally contaminated groundwaters, dissolved As(III) compounds are significantly present along with As(V) species [19,20]. Sorption affinity of As(III) oxyacid (i.e., $HAsO_2$) onto aluminum oxide particles is poor [21]. That is why removal of As(III) through use of activated alumina adsorbent or alum coagulant is very inefficient [22]. Arsenic(V) oxyanions, on the contrary, are well adsorbable onto alumina particles. Thus, oxidation of As(III) to As(V) by chlorination or through use of manganese dioxide solids significantly improves overall arsenic separation [23,24].

(3) Changes in the oxidation states of accompanying species may also be utilized to remove heavy metals from the aqueous phase. Under anoxic or anaerobic conditions, sulfate in water is reduced by sulfate-reducing bacteria to sulfide, thus facilitating heavy metals precipitation due to very low solubility products of metal sulfides. Microbially mediated sulfate reduction is thus a viable mechanism for heavy metals separation. Many engineered biological processes and wetland systems judiciously utilize sulfate reduction as a means to reduce heavy metals concentrations from industrial wastewaters [25,26].

(4) Of all the heavy metals present in the environment, mercury is conspicuous due to its volatility in the elemental state; its ability to exist in all three phases, namely, air, soil and water, in different oxidation states; and its susceptibility to undergo biomethylation and concentrate in the food chain. Flue gases from industrial furnaces and incinerators are the primary contributors of mercury into the atmosphere. One viable process for removal of elemental mercury involves use of novel sulfur-impregnated porous adsorbents [27,28]. Selective mercury separation within this adsorbent is primarily a redox chemisorption process as shown in Equation (6), where elemental mercury is essentially immobilized within the porous adsorbent as highly insoluble mercuric sulfide, $HgS_{(s)}$, in +II oxidation state.

It is well recognized that in many hazardous waste sites, the chemical state and mobility of toxic metals are closely linked to biogeochemical redox reactions that occur as a result of organic carbon being degraded by different microorganisms using a series of terminal electron acceptors [29,30]. In such environments, trace heavy metals can be mobilized/immobilized via processes such as reduction/oxidation, sorption/desorption, precipitation/dissolution, and/or the formation of complex compounds. Figure 1.7 illustrates how the arsenic or chromium speciation changes with the changes in electron acceptors responsible for biological reactions. It is important to note that in the reduced state, chromium or Cr(III) becomes insoluble and, hence, immobile. In contrast, arsenic in the reduced state, i.e., As(III), is nonionized and, hence, more mobile. The underlying scientific principles governing adsorbability and solubility of heavy metals in the natural environment are often the same as those applied in engineered processes for achieving efficient metals separation.

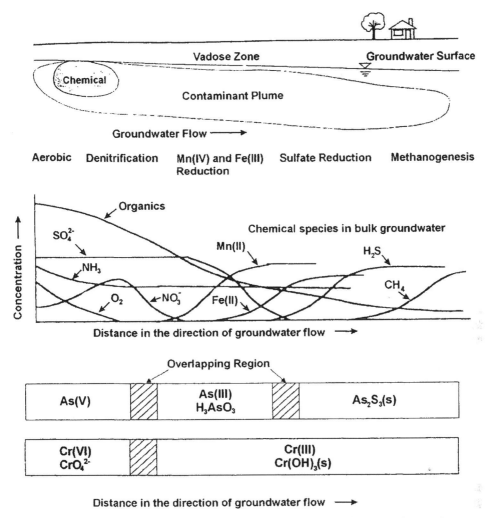

Figure 1.7 Change in arsenic and chromium speciation with the change in biogeochemistry of the subsurface environment (adapted from Bouwer and Zehnder, 1993, *Trends in Biotechnology*, 11, 360–367).

SEPARATION STRATEGIES AND CHAPTERS IN THE BOOK

Ionic charges, Lewis acidity/basicity, sorption affinity onto particulates containing surface functional groups, aqueous-phase solubility, physical sizes of the metal-ligand complexes, redox state, etc., can be manipulated to achieve efficient separation of heavy metals from the aqueous phase and other complex systems. Speciation of heavy metals in the dissolved states varies, and so do the sizes of these species. Table 1.2 provides the estimated sizes of divalent heavy metal cations, Me(II), in different physicochemical forms. Fate and transport behaviors of these heavy metals in a natural environment are also influenced by their relative sizes. Figure 1.8 shows a schematic illustrating a wide variety of strategies for heavy metals separation. Understandably, each of them has the potential to be a viable metal separation process under a specific set of conditions. In certain instances, combination of

TABLE 1.2. Size of a Heavy Metal Cation (Me^{2+}) in Water in Different Physicochemical States.

State Dissolved	Speciation	Approximate Diameter (Nanometer)
Water	H_2O	0.2
Hydrated free metal ion	$[Me(H_2O)_n]^{2+}$	Around 0.5
Inorganic complexes	$[Me(NH_3)_n]^{2+}$ $[MeOH]^+$ $[Me(OH)_2]^0$ $[MeCO_3]_2^{2-}$	Less than 1.5
Organic complexes	$[Me(COO)_2]^0$ $[Me(NH_3)_n]^{2+}$ $[Me(EDTA)]^{2-}$	1–5
Macromolecules/colloids	Me—Humate Complex Me—Fulvate Complex Me—NOM-coated silica	10–500
Surface binding onto microparticles	$\begin{array}{l}FeO^- \\ FeO^-\end{array}(Me^{2+})$	100–10,000
Precipitates	$Me(OH)_{2(s)}$ $MeCO_{3(s)}$	>500

more than one, i.e., a hybrid process, may be the most suitable. All such applications, however, tend to have one major drawback—they are unable to recover individual heavy metals with a high degree of purity and reuse it. With pollution prevention guidelines and the concept of industrial ecology in place, research and development works are under way to separate individual heavy metals and enhance their purities in recovered materials.

In addition to this concise introduction, the book contains nine chapters—all geared toward environmental separation of heavy metals. In choosing topics for inclusion in the book, the primary intent has been to provide comprehensive coverage to those recent developments that have potentials for applications in the near future. It is true that most of the chapters have very specific titles. However, their contents go well beyond the targeted goals. Underlying fundamentals and approaches elaborated in each chapter can be extended to understand and investigate other problems. The following provides a brief summary delineating the contents of each chapter.

The presence of elemental mercury in the flue gases of coal-fired utilities and waste incinerators has emerged as a major environmental concern. Chapter 2 provides a detailed account of equilibrium and kinetic properties of mercury removal using well-characterized carbon-based materials.

Underlying principles of chelating ion exchange and recent advances toward tailoring highly selective polymeric exchangers are elaborated in Chapter 3. Also included in this chapter are two challenging issues confronting heavy metals separation: first, selective removal of trace amounts of heavy metal precipitates from the background of innocuous sludge or residual solids; and second, separation of individual heavy metals.

Chapter 4 is dedicated toward the growing field of magnetic separation of heavy metals. In general, heavy metal precipitates have poor magnetic properties, i.e., they are dia- or weak paramagnetic particles. Co-precipitation with magnetically active iron oxide particles greatly enhances their specific magnetic susceptibilities. Subsequently, the resulting precipitates can be effectively removed using high gradient magnetic separation

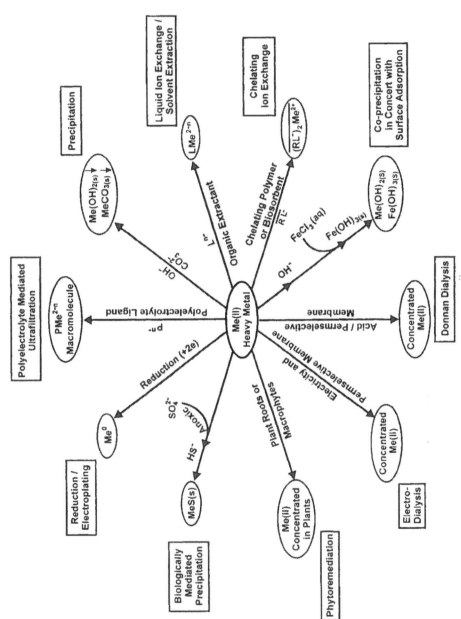

Figure 1.8 A schematic illustrating various engineered processes for heavy metals separation.

technique or other magnetic processes. The chapter also highlights how microbially mediated biological reactions can be integrated into magnetic separation of heavy metals.

Sizes of dissolved heavy metal ions are often less than 10 Angstroms (or one nanometer) and, hence, they are not amenable to separation by low pressure ultrafiltration processes. Heavy metal ions can, however, form macromolecules with tailored polyelectrolytes through formation of metal-ligand complexes. These soluble macromolecules are often two to three orders of magnitude greater in size than pure metal ions and are removable by low-pressure membrane processes. Chapter 5 provides a detailed account of polyelectrolyte-mediated membrane processes for metals removal from contaminated soils or sludges.

Precipitation alone is often inadequate to satisfy the stringent regulations in regard to residual dissolved metal concentrations in the liquid phase. Co-precipitation in concert with specific surface adsorption may greatly overcome such shortcomings in many real-life situations. Two detailed case histories to this effect are summarized in Chapter 6.

Activated carbon is an inexpensive and widely used adsorbent, but its application to date is limited to removing dissolved hydrophobic organic compounds. Through controlled oxidation, concentration of carboxylate and phenolic groups in activated carbon particles can be greatly enhanced, leading to high metal removal capacity. Chapter 7 discusses metals sorption by surface complexation, regenerability and possible reuse of activated carbons for a multiple number of cycles.

In January 2001, under a federal rule signed by the United States Environmental Protection Agency (USEPA) and the former US president, a revised arsenic Maximum Contaminant Level (MCL) to 10 µg/L was made effective five years from its promulgation; however, this ruling was later reversed in March 2001, by the current USEPA administration. Currently, the 60-year old arsenic MCL of 50 µg/L is the highest among all the developed countries in the world but a revised standard is scheduled to come into effect by January 2002. A world away, over seventy million people in Bangladesh and the Indian subcontinent are routinely exposed to arsenic poisoning through drinking water. Chapter 8 discusses salient aspects of geochemical contamination of groundwater by arsenic and underlying principles of various arsenic removal technologies.

Use of trivalent chromium or Cr(III) reagent has been widely practiced in leather or tanning industries around the world for centuries. It is well recognized that once discharged into the environment, Cr(III) is amenable to oxidation into more toxic and mobile Cr(VI). Chapter 9 provides information pertaining to recovery and reuse of Cr(III) in tanning industries.

Naturally occurring humus is a biorenewable material with an ability to complex heavy metals through its carboxylate, phenolate and similar other oxygen-containing functional groups. Chapter 10 discusses techniques used to chemically modify humus materials into viable heavy metal sorbents and presents experimental data to this effect.

As the editor of this treatise, I am quite optimistic about its value and usefulness to the professional community. The practicing engineers, scientists, and researchers who are directly involved in the application and research pertaining to heavy metals would very likely welcome this book as a ready reference for its thoroughness and up-to-date information on areas of current interest. Also, the book would serve the needs of those trying to explore and identify new technologies in the areas of heavy metals control and pollution prevention.

ACKNOWLEDGEMENT

I am immensely thankful to Dr. Ping Li for his assistance in preparing the illustrations included in this chapter.

REFERENCES

1 Pearson, R. (1968). Hard and Soft Acids and Bases, HSAB, Part 1. *Jour. Chem. Ed.,* 45, 9, 581–587.

2 Pearson, R. (1969). Hard and Soft Acids and Bases, HSAB, Part 2. *Jour. Chem. Ed.,* 45, 10, 643–648.

3 Forstner, U. and Wittman, G. (1979). *Metal Pollution in the Aquatic Environment.* Springer-Verlag, Berlin.

4 Nieboer, E. and Richardson, D. H. S. (1980). The Replacement of the Nondescript Term "Heavy Metals" by a Biologically and Chemically Significant Classification of Metal Ions. *Environ. Pollut. Ser. B.,* 1, 3.

5 Zhao, D. and SenGupta, A. K. (2000). Ligand Separation with a Copper(II)-Loaded Polymeric Ligand Exchanger. *Ind. Eng. Chem. Res.,* 39, 2, 455–462.

6 Zhu, Y. (1992). Ph.D. Dissertation. Department of Civil and Environmental Engineering, Lehigh University, Bethlehem, PA, USA.

7 Roy, T. K. (1989). MS Thesis. Department of Civil and Environmental Engineering, Lehigh University, Bethlehem, PA, USA.

8 Kabay, N., Demircioglu, M., Ekinci, H., Yuksei, M., Saglam, M. and Streat, M. (1998). Extraction of Cd(II) and Cu(II) from Phosphoric Acid Solutions by Solvent-Impregnated Resins (SIR) Containing Cyanex 302. *React. Func. Polym.,* 38, 219–226.

9 Warshawsky, A. (1981). In: J. A. Marinsky, Y. Marcus (Eds.), *Ion Exchange and Solvent Extraction,* Vol. 8, Marcel Dekker Inc., New York.

10 Alexandratos, S. D. and Ripperger, K. D. (1998). Synthesis and Characterization of High-Stability Solvent-Impregnated Resins. *Ind. Eng. Chem. Res.,* 37, 4756–4760.

11 Majumdar, S., Sirkar, K. K. and SenGupta, A. K. (1992). In: W. S. Ho and K. K. Sirkar (Eds.), *Membrane Handbook,* Van Nostrand Reinhold, New York.

12 Matsumoto, M., Shimauchi, H., Kondo, K. and Nakashio, F. (1987). Kinetics of Copper Extraction with Kelex 100 Using a Hollow Fiber Membrane Extractor. *Solvent Extr. Ion Exch.,* 5, 2, 301–323.

13 Volesky, B. (1990). *Biosorption of Heavy Metals.* CRC Press Inc., Boca Raton, FL.

14 Trujillo, E. M., Spinti, M. and Zhuang, H. (1995). In: A.K. SenGupta (Ed.), *Ion Exchange Technology: Advances in Pollution Control,* Technomic Publishing Co., Inc., Lancaster, PA.

15 Eary, L. E. and Rai, D. (1988). Chromate Removal from Aqueous Wastes by Reduction with Ferrous Ion. *Environ. Sci. Technol.,* 22, 972–977.

16 Kunz, R.G., Hess, T. C., Yen, A. F. and Arseneaux, A. A. (1980). Kinetic Model for Chromate Reduction in Cooling Tower Blowdown. *Jour. WPCF,* 52, 9, 2327.

17 SenGupta, A. K. and Clifford, D. (1986). Important Process Variables in Chromate Ion Exchange. *Envir. Sci. Technol.,* 20, 149.

18 SenGupta, A. K. and Lim, L. (1988). Modeling Chromate Ion Exchange Processes. *AIChE Jour.,* 34, 12, 2019.

19 Hering, J. G. and Chiu, Van. Q. (2000). Arsenic Occurrence and Speciation in Municipal Groundwater-Based Supply System. *Jour. Env. Engr. ASCE,* 126, 5, 471–474.

20 Safiullah, S., et al. (1998). *Proc. Int. Conf. on Arsenic* in Dhaka, Bangladesh, 8–12 February, 1998.

21 Clifford, D. (1999). Ion Exchange and Inorganic Adsorption. In R.D. Letterman (Ed.), *Water Quality and Treatment,* McGraw-Hill Inc., New York.

22 Hering, J. G. and Elimelech, M. (1996). Arsenic Removal by Enhanced Coagulation and Membrane Processes. *AWWARF Report,* Denver, CO.

23 Frank, P. and Clifford, D. (1986). As(III) Oxidation and Removal from Drinking Water. EPA Project Summary, Report No. EPA/600/S2-86/021. USEPA Office of Research and Development, Cincinnati, OH.

24 Bajpai, S. and Chaudhury, M. (1999). Removal of Arsenic from Manganese Dioxide Coated Sand. *Jour. Env. Engr. ASCE,* 125, 8, 782–784.

25 Reed, S.C., Middlebrooks, E.J. and Crites, R.W. (1988). *Natural Systems for Waste Management and Treatment.* McGraw-Hill Inc., New York.

26 Gersberg, R.M., Lyon, S.R., Elkins, B.V. and Goldman, C.R. (1985). Removal of Heavy Metals by Artificial Wetlands. *Proceedings American WaterWorks Assoc., Water Reuse Symp.,* Denver, CO., pp. 639–645.

27 Korpiel, J.A. and Vidic, R.D. (1997). Effect of Sulfur Impregnation Method on Activated Carbon Uptake of Gas-Phase Mercury. *Environ. Sci. Technol.,* 31, 8, 2319–2326.

28 Liu, W., Vidic, R.D. and Brown, J.D. (2000). Optimization of High Temperature Sulfur Impregnation on Activated Carbon for Permanent Sequestration of Elemental Mercury Vapors. *Environ. Sci. Technol.,* 34, 3, 483–488.

29 Smith, S.L. and Jaffe, P.R. (1998). Modeling the Transport and Reaction of Trace Metals in Water Saturated Soils and Sediments. *Water Resour. Res.,* 34, 3135–3147.

30 Massacheleyn, P.H., Delaune, R.D. and Patrick, W.H. (1991). Effect of Redox Potential and pH on Arsenic Speciation and Solubility in Contaminated Soil. *Envir. Sci. Technol.,* 25, 1414–1419.

CHAPTER 2

Adsorption of Elemental Mercury by Virgin and Impregnated Activated Carbon

RADISAV D. VIDIC[1]

INTRODUCTION

Mercury in the Environment

ENVIRONMENTAL control agencies and researchers have become increasingly concerned with the mobilization of trace elements to the environment from fossil fuel burning and solid waste incineration. Mercury is the trace element of particular concern, because during combustion, most of the mercury present in the feed stream is transferred into the vapor phase due to its high volatility. There is considerable evidence in the literature that currently used pollution abatement technologies (flue gas desulfurization, control of NO_x and SO_x emissions and particulate control devices) are not capable of controlling gas-phase mercury emissions [7–11,56].

Mercury is emitted into the atmosphere from various anthropogenic and natural sources. Natural sources of mercury include volatilization from soils, vegetation and the ocean. The global marine emissions of vapor-phase mercury are estimated to be about 2,000 ton/year [1]. Nriagu and Pacyna [2] estimated that the global anthropogenic emissions of vapor-phase mercury produced from human activities are about 1,000–6,000 ton/year, while Porcella [1] reported that about 2,000–3,400 ton/year of mercury emitted from anthropogenic sources accounts for about 30–55% of global atmospheric mercury emissions.

Physical forms of mercury in ambient air can be divided into two categories: vapor phase, which is dominant in the atmosphere, and particulate phase (associated with aerosols), which only comprises a few percent of total airborne mercury emissions [3]. Chemical forms determine the transport between different environmental media (air, water and soil) [1,4], and the mercuric compounds can be classified into elemental and divalent forms. The elemental form of mercury (Hg°) is the dominant form (>98%) of vapor-phase mercury

[1]Department of Civil and Environmental Engineering, 943 Benedum Hall, University of Pittsburgh, PA 15261, U.S.A., vidic@civeng1.civ.pitt.edu

in the atmosphere [1,3,5], and, following dissolution in cloud water or rainwater, is readily converted (oxidized) to more soluble mercury species [3]. Elemental mercury possesses relatively high vapor pressure and low solubility [6]. The former property leads to considerable mercury evaporation into the ambient air, while the latter makes it difficult for the existing air pollution control devices to remove mercury from the emission sources. Divalent mercury forms include inorganic (Hg^{2+}, HgO, $HgCl_2$) and organic oxidized forms (CH_3Hg, CH_3HgCl, CH_3HgCH_3)[1,3,6]. Divalent forms possess higher solubilities and readily combine with a variety of reactants, such as sulfite, chloride and hydroxide ions, in the aqueous phase [6].

Once discharged to the atmosphere, mercury persists in the environment and creates a long-term contamination problem. Furthermore, well-documented food chain transport and bioaccumulation of mercury, together with high toxicity to mammalians and severe health problems caused by the ingestion of mercury even at low levels, require strict control of mercury emissions from coal-fired power plants. Subsequently, the Clean Air Act Amendments of 1990 (Title III, Section 112[b][1]) require major sources to use maximum available control technology (MACT) to reduce mercury emissions.

Current State of Mercury Control Technologies

Many existing air pollution control technologies and several innovative methods have been evaluated for the control of vapor-phase mercury emissions from combustion processes. Sodium sulfide (Na_2S) has been used for vapor-phase mercury control in municipal solid waste combustors in Canada, Sweden, Germany and British Columbia. Sodium sulfide injection is usually combined with dry sorbent injection (DSI) and fabric filters (FFs) for acid gas and particulate matter (PM) control [7,8]. It has been reported that mercuric sulfide (HgS) is generated as a fine particulate in the process, which may prove difficult to capture in less efficient electrostatic precipitators (ESPs). Other potential problems for this process include corrosion, hydrogen sulfide formation and chemical storage and handling [9]. These problems, compounded by the lack of test data on full-scale coal-fired power plants [7,8], cloud the utility of sodium sulfide injection for the control of mercury emissions.

Wet scrubbers have been routinely used to remove hydrochloric acid and sulfur dioxide from the flue gases of industrial factories, coal-fired power plants and municipal waste combustors. Considerable interest in the use of wet scrubber systems to simultaneously remove sulfur dioxide and mercury has recently been expressed [7,9–11]. The removal of vapor-phase mercury in the wet scrubber system would also occur by absorption in the scrubbing slurry, whereby the mechanism of mercury removal depends on the solubility of mercury in the scrubbing slurry, contact time and solution chemistry. Elemental mercury is essentially insoluble in the wet scrubbing slurry, while some of the oxidized species, such as mercuric chloride, are highly soluble. Therefore, oxidized mercury can be easily absorbed with sufficient gas-liquid contact, while the removal of elemental mercury would remain limited [7–10]. Chang and Owens [12] reported that the treatment of a coal-fired power plant flue gas using only a wet scrubber allowed 70–75% of elemental mercury to be discharged into the atmosphere, while other studies reported 30–70% removal of elemental mercury by wet scrubbers [11].

In the past decade, spray dryer adsorption (SDA) systems have been applied to more than 17,000 MW of coal-fired boilers and several hundred municipal waste and hazardous waste incinerators [13,14]. The SDA process has been specified as one of the best available control technologies (BACT) to treat gas contaminants. The essence of the contaminant removal in the SDA process is the reaction with the alkali adsorbents. The alkali adsorbent can be lime (CaO), soda ash (Na_2CO_3), NaOH or $NaHCO_3$. Because of the nature of the waste product and its capability of meeting emission requirements, lime is the best choice for the adsorbent in most applications. Total mercury removal in the range of 3–50% can be achieved by the SDA process because only the oxidized forms of mercury can be captured, which elemental mercury will pass through the system [13,14]. For the effective control of vapor-phase mercury emission, the SDA system can be augmented with an activated carbon injection system that can increase the removal of mercury to more than 90% [13,14].

Adsorption-Based Technologies for Mercury Control

Activated carbon adsorption is the technology that offers great potential for the control of mercury concentrations in gas-phase emission. Otani et al. [15] studied the adsorption of mercury vapor on particles at room temperature using soot particles generated by incineration of sewage sludge and activated carbon particles. They observed that ash adsorbed very little mercury so that most mercury behaves as a vapor even in the presence of soot particles. Activated carbon, on the other hand, had an order of magnitude higher capacity that was accurately described by the Freundlich isotherm equation. Sinha and Walker [16] reported that sulfur-impregnated carbon exhibits faster initial breakthrough at room temperature than the virgin-activated carbon due to the reduction in the surface area induced by the impregnation process. However, at higher temperatures (150°C), the adsorptive capacity of sulfur-impregnated carbon greatly surpassed the capacity of virgin-activated carbon due to chemisorption of mercury and formation of mercuric sulfide. Furthermore, they reported that water vapor reduces adsorption of mercury for sulfur-impregnated carbon.

Matsumura [17] used steam-activated carbon from coconut shell in his studies of the effects of oxidation and iodization of activated carbon surface on the removal efficiency for mercury vapor. He concluded that oxidized or iodized activated carbon adsorbed mercury vapor 20–160 times more than untreated activated carbon when mercury vapor in concentrations of up to 40 mg/m^3 in a nitrogen stream at 30°C was brought into contact with these adsorbents. Oxidized carbons were successfully regenerated with hydrochloric acid. Iodized activated carbons were shown to be suitable adsorbents for mercury vapor though adsorbed mercury was not proportional to the amount of iodine adsorbed on the carbon.

Teller and Quimby [18] evaluated the performance of activated carbons impregnated with copper chloride or sulfur for the removal of mercury under the conditions representative of solid waste incinerators. They concluded that moisture content of the carrier gas and temperatures tested in their study had no effect on copper chloride-impregnated carbon capacity for mercury vapor. They also concluded that as the impregnate concentration increases (for copper chloride), mercury removal increases, but they were not able to correlate these two parameters. They observed that copper chloride-impregnated carbon exhibits as much as 300 times higher capacity for mercury removal as compared to untreated activated carbon. Sulfur-impregnated carbon exhibited only a 60% improvement in the breakthrough time. On the other hand, Henning et al. [19] examined the influence of potassium iodide and sulfur as impregnates for improving the ability of activated carbon to adsorb mercury. Their results indicated that 11 wt% sulfur addition increased the adsorption capacity by a factor of 400.

Direct injection of activated carbon into the flue gas stream has been proposed as a relatively simple approach for controlling mercury emissions [12]. Injected activated carbon binds the vapor-phase mercury through physical adsorption and/or chemisorption and is collected in downstream particulate collection devices, such as FFs or ESPs. Results from several tests indicated that effectiveness of activated carbon injection in removing mercury vapor depends on the type and composition of burned materials, flue gas composition and temperature, mercury speciation, activated carbon properties and injection rate, and operating conditions [5]. Lowering the flue gas temperature from 345 to 250°F with direct injection of virgin activated carbon improved mercury removal efficiency from 0% to 37%. Further tests showed that virgin activated carbon injection at 200°F resulted in greater than 90% mercury removal [13]. Because activated carbon can be collected effectively in the existing particulate control devices, direct activated carbon injection has several potential advantages over SDA and wet scrubbing processes: (1) simpler operation, (2) lower operational cost, (3) lower capital cost, (4) no wastewater problems, (5) simpler waste-disposal, (6) easier maintenance and repair and (7) greater efficiency for mercury removal.

Based on the success of field-scale trials conducted on a number of municipal solid waste (MSW) incinerators, it was anticipated that powdered activated carbon injection into the flue gas with subsequent collection at the fabric filters can also be used for the control of mercury emissions from coal-fired power plants. However, due to the fact that mercury concentrations in MSW incinerator flue gas (200–1,000 $\mu g/m^3$) are one to two orders of magnitude higher than for coal-fired power plants (2–20 $\mu g/m^3$), as well as other differences in process conditions (e.g., HCl content, flow rate, SO_2 and NO_x concentration), carbon to mercury ratios required for effective control of mercury in MSW flue gas are an order of magnitude lower than those necessary to achieve similar mercury removals in coal combustors [5]. Preliminary cost estimates for the control of mercury emissions from coal-fired power plants using powdered activated carbon injections range from $14,400 to $38,200 per pound of mercury removed [12]. Clearly, the use of virgin activated carbon is too costly and too low in efficiency to be practical. Therefore, further sorbent improvements must be made in order to facilitate commercial success of this promising technology.

MERCURY UPTAKE BY VIRGIN ACTIVATED CARBON

Experimental Approach

Detailed description of the experimental system and analytical methods used to evaluate the rate of mercury uptake by commercially available activated carbon (BPL, Calgon Carbon Corporation, Pittsburgh, PA, USA) is provided by Vidic et al. [20] and is briefly described here. Schematic representation of the experimental setup is shown in Figure 2.1. Differential-bed reactor was charged with 100 mg of activated carbon, placed in a laboratory oven and operated in a downflow mode. Six feet of Teflon tubing was placed in the oven upstream of the column to facilitate heating of the influent gas to a desired temperature. In order to prevent damage of the system components due to high effluent temperature in the recirculation loop, the recirculation loop was first passed through a condenser to cool the effluent gas coming out of the oven. Mercury laden gas was recirculated in the system at a flow rate of 75 L/min, which provided a gas turnaround time of 10 seconds. Mercury concentration in this closed system was continuously monitored using the atomic adsorption spectrophotometer equipped with a quartz cell and a hollow cathode lamp adjusted to a wavelength of 253.6 nm.

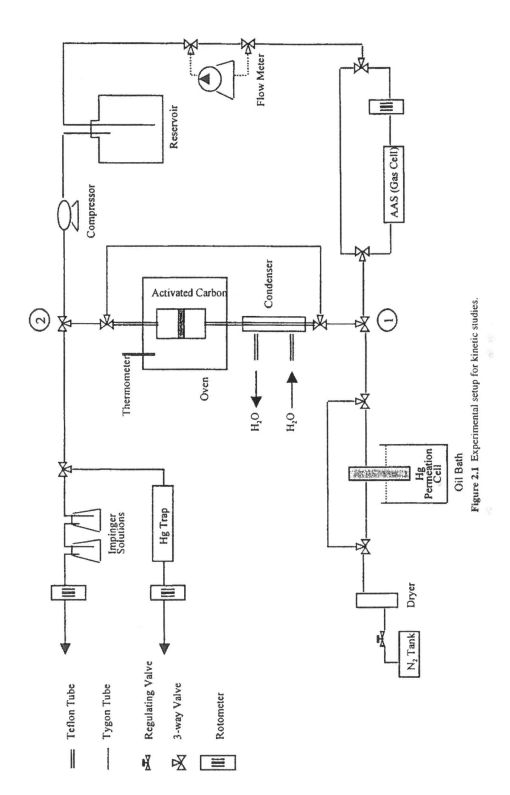

Figure 2.1 Experimental setup for kinetic studies.

19

Modeling Adsorption Kinetics

The homogeneous surface diffusion model (HSDM) was used to describe the kinetics of adsorption of compounds from the carbon surface into the particle. Because numerous articles have extensively described the theory behind the HSDM [21–23], only the primary assumptions are presented here. The HSDM was coupled to a model for a batch system to describe the uptake of Hg from the bulk phase onto the activated carbon. The key assumptions for the batch model and HSDM include the following: (a) removal of Hg from the bulk is due solely to adsorption by activated carbon; (b) the activated carbon particles are all spherical, uniform in size and well-dispersed in the gas phase; (c) a mass transfer boundary layer causes resistance to mass transfer from the bulk to the activated carbon surface; (d) adsorption equilibrium between the activated carbon surface and the gas phase is described by the Langmuir isotherm; (e) mass transfer within the activated carbon particle is controlled by surface diffusion; and (f) the activated carbon particle is isothermal.

A mass balance on Hg in the activated carbon particle and the bulk can be written as

$$\frac{\partial q}{\partial t} = \frac{D_s}{r^2} \frac{\partial}{\partial r} r^2 \frac{\partial}{\partial r} q \tag{1}$$

$$V \frac{dc_b}{dt} = -m_c \frac{dq_{ave}}{dt} \tag{2}$$

where

$$q_{ave} = \frac{3}{r_p^3} \int_0^{r_p} q(r,t) r^2 dr \tag{3}$$

Equations (1) and (2) are subject to the following initial and boundary conditions:

$$q(r, t = 0) = 0 \tag{4}$$

$$c_b(t = 0) = c_o \tag{5}$$

$$\frac{\partial q(r = 0, t > 0)}{\partial r} = 0 \tag{6}$$

$$\rho_p D_s \frac{\partial q(r = r_p, t > 0)}{\partial r} = k_f(c_b - c_s) \tag{7}$$

$$q(r = r_p, t > 0) = \frac{q_{max} b c_s}{1 + b c_s} \tag{8}$$

Using the following dimensionless variables,

$$C = \frac{c}{c_o} \quad Q = \frac{q}{q_o} \quad T = \frac{D_s t}{r^2} \quad R = \frac{r}{r_p} \quad Bi = \frac{k_f r_p c_o}{D_s \rho_p q_o} \quad \lambda_b = \frac{m_c q_o}{V c_o} \tag{9}$$

where

$$q_o = \frac{q_{max} b c_o}{1 + b c_o} \tag{10}$$

Equations (1) to (8) can be rewritten in the following dimensionless format:

$$\frac{\partial Q}{\partial T} = \frac{1}{R^2} \frac{\partial}{\partial R} R^2 \frac{\partial}{\partial R} Q \tag{11}$$

$$\frac{dC_b}{dT} = -\lambda \frac{dQ_{ave}}{dT} \tag{12}$$

$$Q_{ave} = 3 \int_0^1 Q(R,T) R^2 dR \tag{13}$$

$$Q(R, T = 0) = 0 \tag{14}$$

$$C_b(T = 0) = 1 \tag{15}$$

$$\frac{\partial Q(R = 0, T > 0)}{\partial R} = 0 \tag{16}$$

$$\frac{\partial Q(R = 1, T > 0)}{\partial R} = Bi(C_b - C_s) \tag{17}$$

$$Q(R = 1, T > 0) = \frac{1 + b c_o}{1 + b c_o C_s} \tag{18}$$

Equations (11) and (12) were reduced to ordinary differential equations using orthogonal collocation methods with 11 collocation points in the activated carbon particle [24]. The resulting system of equations was solved using DDASSL, a subroutine capable of simultaneously solving algebraic and ordinary differential equations [24].

The Langmuir isotherm constants (q_{max} and b), the film mass transfer coefficient (k_f), and the surface diffusion coefficient (D_s) were obtained by fitting the mathematical model simultaneously to several data sets obtained from batch experiments conducted at the same temperature using different initial mercury concentrations. The sum of the squares of the differences between the model output and the data sets (i.e., sum of squares of errors or SSE) was minimized by varying the four parameters using simulated annealing as a global minimization algorithm [24].

Table 2.1 shows the parameter values obtained at the two temperatures, and Figures 2.2(a) and 2.2(b) show the model fit and experimental data at 25°C and 140°C, respectively [24].

An increase in temperature results in a lower capacity of the carbon for Hg, which is characteristic of an exothermic adsorption process and is consistent with other studies [2,12,13]. The Langmuir coefficient, b, which can be conceptualized as the ratio of the kinetic coefficient for adsorption to the kinetic coefficient for desorption, increased with tem-

TABLE 2.1. Parameter Values Obtained from Model Fit (from Flora, Vidic, Liu and Thurnau, *J. AWMA*, 48:1051, 1998. With Permission from Air and Waste Management Association).

Parameter	Temperature	
	25°C	140°C
q_{max} (µg/g)	123	27.5
b (m^3/µg)	9.15×10^{-3}	2.06×10^{-2}
k_f (cm/s)	12.0	19.4
D_s (cm^2/s)	2.29×10^{-7}	1.21×10^{-6}

perature. If the effect of temperature on both kinetic coefficients can be described using an Arrhenius relationship, then b should not change with temperature. It is possible that with an increase in temperature, the kinetic coefficient for adsorption increased, but the kinetic coefficient for desorption did not increase with the same magnitude. A shift in the adsorption mechanism from physical to chemical at a higher temperature could result in a proportionately lower increase in the desorption kinetic coefficient [12].

Model Predictions

Because this research focuses on the removal of Hg under conditions that may be encountered in the flue gases of coal-fired power plants, the isotherm parameters and D_s obtained from the batch experiments at 140°C were used for the simulations. Because the process configuration for the laboratory tests was different from what would be encountered in a flue gas stream, k_f was estimated using an empirical correlation for forced convection around a solid sphere as described by Flora et al. [24].

The model developed for a batch system was adapted to estimate the performance of virgin activated carbon in a flue gas stream, where steady-state, plug-flow conditions were assumed. Figure 2.3 shows the impact of particle radius on the fractional removal of Hg from the influent of flue gases as a function of retention time in the flue gas stack for a C/Hg ratio of 10^6. An initial mercury concentration of 20 µg/m^3 was used in all calculations. The fractional removal of Hg increases with longer retention times, with the fractional removal asymptotically reaching a maximum value. At this maximum value, the carbon particles are at equilibrium with the bulk Hg concentration, and further Hg removal cannot be realized. Longer residence time is required for particles with larger radii to reach equilibrium because of the longer distance that the Hg has to diffuse into the carbon. Assuming the average residence time of activated carbon injected into the flue gas of 2 seconds, it can be seen from Figure 2.3 that the equilibrium is reached for particles with radius below 30 µm, with larger particles having lower mercury removal.

One strategy to increase mercury removal from flue gases is to use carbon particles with a higher adsorptive capacity. To analyze the impact of carbon capacity, simulations were performed with varying q_{max} for various particle radii, while keeping other parameters constant. Figure 2.4 shows the required C/Hg ratio to achieve 90% removal of Hg for various carbon capacities. Because equilibrium is achieved for small particle radii, the required C/Hg ratio decreases linearly with an increase in capacity. This linear relationship holds for a fixed removal efficiency if all other parameters remain constant. For large particle radii, the required C/Hg ratio is less sensitive to the carbon capacity because the system is mass-transfer limited. This is particularly evident for particles with a radius greater than 3×10^{-3} cm (30 m) and for q_{max} greater than 200 µg/g. Under these circumstances, Hg removal

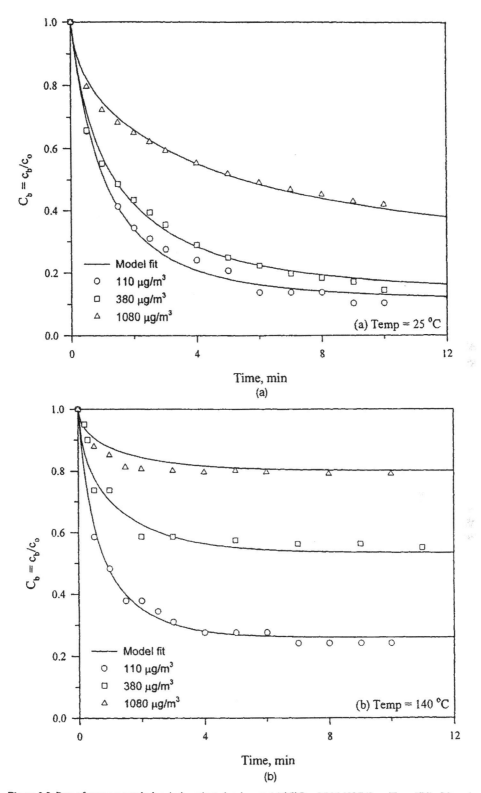

Figure 2.2 Rate of mercury uptake by virgin activated carbon at: (a) 25°C and (b) 140°C (from Flora, Vidic, Liu and Thurnau, *J. AWMA*, 48:1051, 1998. With permission from Air and Waste Management Association).

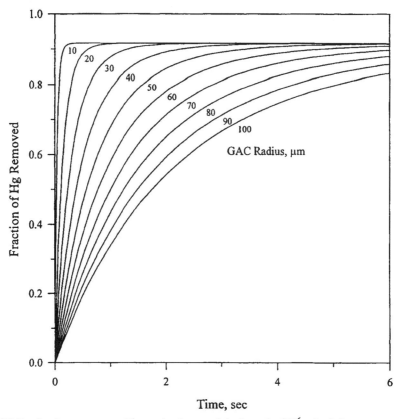

Figure 2.3 Fractional mercury removal for a carbon/mercury injection ratio of 10^6 and an influent mercury concentration of 20 mg/m^3 (from Flora, Vidic, Liu and Thurnau, *J. AWMA,* 48:1051, 1998. With permission from Air and Waste Management Association).

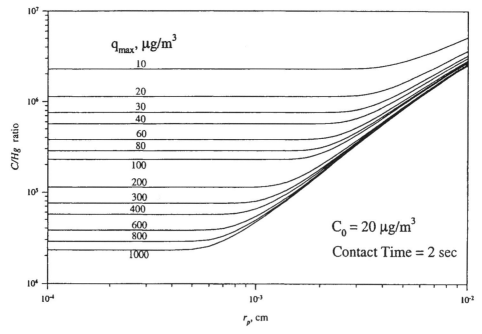

Figure 2.4 Impact of adsorptive capacity on carbon/mercury injection ratio required for 90% mercury removal.

is primarily dependent on the amount of Hg that can diffuse from the bulk into the carbon particle. Using a smaller particle size would provide more efficient use of the carbon for Hg removal.

MERCURY UPTAKE BY IMPREGNATED ACTIVATED CARBONS

Sulfur-Impregnated Activated Carbons

Initial studies on the effectiveness of sulfur impregnation on mercury uptake were performed using a commercially available sulfur-impregnated activated carbon (HGR, Calgon Carbon Co., Pittsburgh, PA, USA) and activated carbon produced by the reaction with elemental sulfur at 600°C and denoted as BPL-S [25]. This temperature was selected based on an earlier report that the chemisorption of sulfur onto activated carbon is maximized at 600°C [24]. Although the two different impregnation methods used in the production of these sorbents yielded similar sulfur contents (Table 2.2), the BET specific surface area of BPL carbon that was used as a starting material for both sorbents was reduced by 53 and 20% for HGR and BPL-S, respectively (Table 2.2). Due to the higher impregnation temperature, the sulfur in BPL-S carbon is suspected to be more evenly distributed in the pore structure, occupying the deeper, narrower pores. The sulfur in HGR carbon, on the other hand, is most likely condensed at the external surface of the carbon particle, blocking the access to the narrower high-energy pores [25].

Thermogravimetric analysis (TGA) was conducted by heating samples of BPL, HGR and BPL-S carbons up to 400°C in an argon atmosphere. Both BPL and BPL-S carbons underwent negligible decreases in weight, while the weight of the HGR carbon sample decreased by 8.5%. Because both HGR and BPL-S carbons are manufactured by impregnating BPL carbon with sulfur, this outcome implies that BPL-S carbon lost a negligible amount of its impregnated sulfur, while HGR carbon lost 88% of its sulfur content.

Sulfur exists in several allotropes, including S_λ (S8 rings), S_π (S8 chains) and S_μ (chains of variable length), with S8 rings as the only form at room temperature [27–29]. Because the sulfur vapor at 200°C is in the form of S_8 (76.5%) and S_6 (23.5%) rings [27,29], it is reasonable to assume that HGR carbon contains sulfur predominantly in the form of voluminous S8 rings. At 600°C, sulfur vapor possesses a significant fraction of S_6 (58.8%) and S_2 (16.4%) molecules [29], which are less voluminous and more reactive because they possess a greater fraction of sulfur terminal atoms [30]. Therefore, the smaller S_2 and S_6 chains can more easily migrate into the narrower pores of the carbon matrix and, as the carbon cools to room temperature at the completion of the impregnation process, steric hindrance impedes reformation of the more voluminous S_8 rings from the other two allotropes [30–33].

Based on the discussion presented above, it is believed that the sulfur in HGR carbon is predominantly in the form of S_8 rings and is weakly bonded to the carbon surface in the macroporous region of the carbon particle. On the other hand, the sulfur in BPL-S carbon is

TABLE 2.2. Comparison of GAC Types (from Korpiel and Vidic, *Environ. Sci. Technol.,* 31(8):2319, 1997. With Permission from American Chemical Society).

GAC Type	Sulfur Content (wt%)	BET A_{sp} (m^2/g)
BPL	0.7	1026
HGR	9.7	482
BPL-S	10.0	824

strongly bonded to the carbon surface in the microporous region of the carbon particle, requiring mercury to diffuse a longer distance for chemisorption to occur. A much larger proportion of the sulfur in BPL-S carbon is believed to be in the form of S_2 and S_6 chains, which are more reactive than the S_8 rings present in HGR carbon.

In order to compare the effect of temperature on the performance of HGR and BPL-S carbons, adsorber experiments using 200 mg of 60×80 U.S. Mesh carbon were conducted for a duration of 10 hours at the low influent mercury concentration of 55 $\mu g/m^3$, a flow rate of 1.0 L/min, and at temperatures of 25, 90 and 140°C using the experimental setup similar to that depicted in Figure 2.1, except that the effluent from the adsorber was not recirculated back into the system. The breakthrough curves illustrated in Figure 2.5 show that HGR and BPL-S carbons performed similarly in the uptake of mercury vapor at 25 and 90°C. However, when the temperature was increased to 140°C, the performance of BPL-S carbon improved slightly while HGR carbon exhibited significant deterioration in the ability to remove mercury from the feed stream. This may be due to the fact that 140°C is above the melting point of sulfur (115.2°C), which induces the sulfur that is weakly bonded to the surface of HGR carbon to melt and agglomerate as a liquid in the form of long polymer chains [27] and decreases the sulfur surface area available for contact with the incoming mercury molecules. Thus, the performance of HGR carbon at 140°C may be limited by the slow diffusion of mercury through the liquid state sulfur. The stronger bonding and more uniform distribution of sulfur in BPL-S carbon prevented the sulfur from agglomerating, which ensured that the performance of BPL-S carbon did not deteriorate at higher temperatures.

An adsorber experiment in which the mercury-laden gas was periodically diverted around the adsorber (while the temperature of the adsorber was maintained constant) was conducted at 140°C using the influent mercury concentration of 55 $\mu g/m^3$, a flow rate of 1.0 L/min and 100 mg of 60×80 U.S. Mesh HGR carbon. As shown by the breakthrough curve in Figure 2.6, four mercury loading steps were performed over a period of eight days. Note that in the first two loading steps, after 100% breakthrough was reached and the adsorber

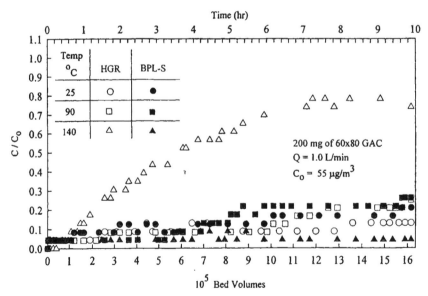

Figure 2.5 Effect of temperature on mercury breakthrough for HGR and BPL-S adsorbers (from Korpiel and Vidic, *Environ. Sci. Technol.,* 31(8):2319, 1997. With permission from American Chemical Society).

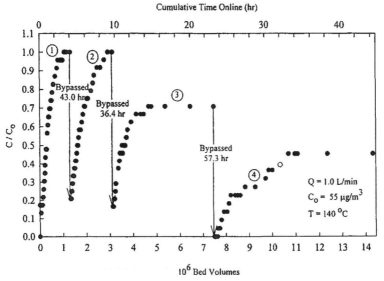

Figure 2.6 Mercury breakthrough from HGR adsorber in on-line/off-line operation (from Korpiel and Vidic, *Environ. Sci. Technol.*, 31(8):2319, 1997. With permission from American Chemical Society).

was bypassed for over a day, the adsorber exhibited additional capacity after it was placed back on-line. The performance of HGR in each mercury loading step is summarized in Table 2.3.

The following is a possible explanation for the behavior illustrated in Figure 2.6 [25]: by the time 100% breakthrough is reached in the first loading step, the available surface area of the sulfur agglomerates is completely covered by a monolayer of HgS. The diffusion of HgS through liquid state bulk sulfur is slow, and therefore, the HgS monolayer blocks the incoming Hg molecules from reacting with the bulk sulfur that is still available in the carbon pores. During the period that the adsorber was bypassed, the HgS molecules had sufficient time to diffuse into the sulfur liquid phase, breaking up sulfur chains to provide additional sulfur terminal atoms to react with mercury molecules in a subsequent loading step. The observation that the mercury removal performance actually improved in the third and fourth loading steps may be due to the fact that the sulfur polymer chains began to break apart into smaller chains as more and more mercury reacted with sulfur, thereby increasing the fraction of sulfur terminal atoms which led to an increase in the rate of HgS formation in the later part of the column experiment.

TABLE 2.3. Summary of HGR Performance in Each Mercury Loading Step (from Korpiel and Vidic, *Environ. Sci. Technol.*, 31(8):2319, 1997. With Permission from the American Chemical Society).

Mercury Loading Step	Duration (hr)	Hg Uptake (μg Hg/g)	Cumulative Hg Uptake (μg Hg/g)
1	4.1	34.9	34.9
2	5.6	45.4	80.3
3	13.5	155.0	235.3
4	21.1	263.0	498.3

OPTIMIZATION OF SULFUR IMPREGNATION

Three different temperature settings were selected for the impregnation process, namely 250, 400 and 600°C. The designation of newly derived sorbents was based on the starting material, the initial sulfur to carbon ratio (SCR) and the impregnation temperature. For example, BPL-S-4/1-600 denotes a BPL carbon that was impregnated with sulfur at SCR of 4:1 and temperature of 600°C.

Newly developed sulfur-impregnated carbons used in this study were evaluated for their efficiency of vapor-phase elemental mercury removal in the fixed-bed reactor. The empty bed contact time was about 0.01 seconds. A detailed description of the testing protocol and analytical procedures can be found elsewhere [34].

Because the performance of sulfur-impregnated activated carbons should be a function of both SCR and impregnation temperature, it was decided to divide sorbents into two groups when evaluating their performance. Three carbons prepared with the SCR of 4:1 and impregnation temperatures of 250, 400 and 600°C constituted Group A. Four carbons produced at the impregnation temperature of 600°C using the SCR of 4:1, 2:1, 1:1 and 1:2 constituted Group B.

To facilitate direct comparison of these mercury sorbents, experimental conditions for each column run were always kept the same: 100 mg of sulfur-impregnated carbon was placed in the fixed bed reactor operated at 140°C using the influent mercury concentration of 55 μg/m^3 and a N_2 flow rate of 1.0 L/min. The normalized effluent concentration as a function of time (breakthrough curve) was generated for each adsorbent, and the amount of mercury adsorbed in each run was calculated by integrating the area above the breakthrough curve.

Table 2.4 summarizes mercury removal for the carbons in Groups A and B. As can be seen from this table, mercury uptake is strongly related to the impregnation temperature. As the impregnation temperature increased, the capacity for mercury removal also increased. For example, the maximum mercury uptake for BPL-S-4/1-250 is about 550 μg Hg/g carbon, while the capacity for BPL-S-4/1-600 is about 2,200 μg Hg/g carbon. When SCR changed from 4:1 to 2:1, the capacity for mercury removal did not experience a significant decrease. Major loss in capacity was observed when SCR was reduced from 2:1 to 1:1. As shown in Table 2.4, both BPL-S-4/1-600 and BPL-S-2/1-600 exhibited high capacities for mercury uptake, while the capacities of BPL-S-1/1-600 and BPL-S-1/2-600 are approximately 40 to 50% lower.

A previous study [16] stipulated that the mechanism for mercury removal by sulfur-im-

TABLE 2.4. Performance of Sulfur-Impregnated Carbons.

Sample Type	Sulfur Content (wt%)	Specific Surface Area (m^2/g)	Dynamic Mercury Uptake (μg/g)
BPL-S-4/1-250	36.2–38.5	164.4–170.6	560
BPL-S-4/1-400	10.27–10.65	628.7–634.4	1400
BPL-S-4/1-600	10.04–10.18	823.7–845.7	2200
BPL-S-2/1-600	9.12–9.21	897.6–909.5	2000
BPL-S-1/1-600	8.31–8.39	847.6–860.6	1260
BPL-S-1/2-600	7.11–7.23	859.0–888.6	1070
BPL	0.51–0.73	987.7–1026.0	0.7
BPL (heated to 600°C)	0.45–0.51	960.7–989.5	NA
HGR	9.66–10.01	782.6–823.1	38

TABLE 2.5. Dominant Sulfur Allotropes at Different Temperatures (from Liu, Vidic, and Brown, *Environ. Sci. Technol.,* 32(4):531, 1998. With Permission from the American Chemical Society).

Temperature (°C)	S_8	S_7	S_6	S_5	S_2	Vapor Pressure* (atm)
180	65%	15%	15%			0.0027
250	49%	27%	21%			0.017
400	22%	36%	28%	5%	2%	0.496
600			59%		16.4%	>3.817

*Calculated based on Reference [30].

pregnated carbons was primarily through the formation of mercuric sulfide (HgS). Therefore, the physical and chemical properties of sulfur, the predominant form of sulfur, the available surface area and the pore size distribution are essential factors that influence the performance of these sorbents.

As discussed previously, the presence of various sulfur allotropes on the carbon surface depends on their distribution in the vapor phase, which is a strong function of ambient temperature and is shown in Table 2.5 [33]. The difference between various sulfur allotropes in terms of the availability of active terminal sulfur atoms to react with mercury and associated differences in the performance of sulfur-impregnated carbons is discussed by Korpiel and Vidic [25].

The last column of Table 2.5 shows the saturated sulfur vapor pressures at different temperatures, which were calculated based on the empirical equations given by Hampel [28]. When the temperature increases, the vapor pressure of sulfur increases rapidly, indicating that sulfur molecules are present mainly in the vapor phase, not in the liquid phase. Lower amounts of sulfur could be attached to the carbon surface at higher impregnation temperatures because nitrogen gas was continuously flushing the quartz cell during the impregnation process. As can be seen from Table 2.4, which also provides key properties of sulfur-impregnated sorbents utilized in this study, the sulfur content decreased from about 37% to about 10% as the impregnation temperature increased from 250°C to 400°C. When impregnation temperature was further increased to 600°C, the effect of saturated vapor pressure became less significant because the sulfur content of BPL-S-4/1-600 was only slightly lower than that of BPL-S-4/1-400.

The data in Table 2.4 also show that the sulfur content in the BPL-S-600 series decreased from about 10% to about 7% as the SCR decreased from 4:1 to 1:2. This change in SCR not only decreased the absolute amount of sulfur on the carbon surface, but also reduced the number of terminal active sulfur atoms. The direct consequence was that the mercury removal capacity became smaller.

As discussed above, the mechanism for mercury adsorption is governed by the reaction between active sulfur atoms and vapor phase mercury molecules. Although the impregnation temperature was the most important factor influencing the efficiency of these new adsorbents, lower SCR values also decreased their efficiency for mercury removal. However, SCR did not exhibit as strong an influence on the performance of these novel carbons as did the impregnation temperature. Therefore, the assessment of the performance of sulfur-impregnated carbon for mercury removal cannot be based solely on the measured sulfur content of the sorbent.

The classical nitrogen BET method was used to study the microstructure of these new activated carbons. The specific surface area for each adsorbent is listed in Table 2.4. For the BPL-S-4/1 series, the specific surface area decreased with a decrease in the impregnation temperature. For instance, the specific surface area of BPL-S-4/1-250 was only 20% of the BPL-S-4/1-600 surface area. Comparing the surface area of virgin and heated (at 600°C without the addition of sulfur) BPL carbons, it can be seen that the surface area does not change significantly as a result of exposure to 600°C (Table 2.4). This indicates that the major reason for the loss of surface area for all of these carbons is the impregnation with sulfur.

FATE OF SPENT SORBENT

In order to determine the fate of mercury adsorbed on activated carbon-based adsorbents, a modified TCLP test was conducted by placing 0.1 g of spent adsorbent in a glass vial with 20 ml of Extraction Fluid 1 [35]. The volume of the vial was also 20 mL so that there would be no headspace above the liquid. By doing so, the possibility of losing any mercury into the vapor phase can be eliminated. All the sample vials were capped and sealed tightly by Teflon tapes. The vials were placed in a 30 rpm tumbler for 18 to 20 hours of the extraction process.

The supernatant was separated by pressurized filtration, and the liquid mercury analysis method [25] was used to measure mercury concentration in the leachate. The solid adsorbent was collected to verify the remaining mercury concentration by combusting the adsorbent in oxygen atmosphere at 800°C. The combustion off-gas was collected in impingers (15 g of $KMnO_4$ in 1 L 10% H_2SO_4) that were analyzed using the liquid mercury analysis method.

Table 2.6 shows the mercury concentrations measured in extracts of different spent adsorbents. It can be seen from this table that the amount of mercury extracted by acidic solution is below the detection limit of the analytical method used (about 0.1–0.2 µg/L). The sample storage time varied from 2 to 330 days, and the adsorbents included all types of sulfur-impregnated carbons [25,34,35] (i.e., HGR and BPL-S series). The last column in Table 2.6 shows the mercury loading ratio that represents a ratio of the expected mercury mass on adsorbents after breakthrough tests and the amount of mercury measured by combusting the adsorbents after the TCLP test. These results revealed that the amount of mercury remaining on sulfur-impregnated adsorbents after prolonged storage and after the TCLP test was equal to the amount of mercury adsorbed during the breakthrough test.

The test results for BPL carbons loaded with mercury and stored for 2 and 310 days revealed different behavior. Although there was no mercury detected in the TCLP leachate, the combustion tests showed that the amount of mercury present on this adsorbent after 310

TABLE 2.6. TCLP Results for Different Carbon Samples (from Liu, Vidic, and Brown, *Environ. Sci. Technol.*, 34(3):483, 2000. With Permission from the American Chemical Society).

Sample	Storage Time (days)	Hg in Leachate (µg/L)	Mercury Loading Ratio
BPL-S-4/1-600	30	Not detected	2287:2170
BPL-S-4/1-600	330	Not detected	2312:2421
BPL-S-2/1-600	240	Not detected	2010:1907
BPL-S-1/1-600	200	Not detected	1256:1302
BPL-S-4/1-400	260	Not detected	1448:1331
BPL-S-4/1-250	180	Not detected	550:603
HGR	260	Not detected	38:36
HGR	60	Not detected	36.9:38
BPL	310	<0.2	0.65:0.42
BPL	2	<0.2	0.68:0.71

days of storage was much lower than the expected value, while the amount of mercury present on BPL carbon after two days of storage was within the experimental error of the expected value.

The results described above are in agreement with the hypothesis that mercury is chemisorbed by sulfur-impregnated carbons as HgS. HgS is a very stable chemical with a high sublimation point (583.5°C), and it is not soluble in water and weak acids [36].

Although TCLP results showed that the amounts of mercury extracted from carbons were far below the RCRA limit (200 µg/L) [37,38], the stability of mercury on the adsorbent surface is strongly related to the adsorption mechanism and predominant mercury forms. Apparently, mercury adsorbed by sulfur-impregnated carbons is more stable than that adsorbed by virgin carbons. It is obvious that converting elemental mercury to HgS would be beneficial in terms of handling the spent adsorbents. Another important conclusion is that although TCLP analysis of spent virgin carbon may pass the regulatory limit, the elemental mercury could re-enter the vapor phase during a long storage time. This will certainly present a threat to the environment, and the spent virgin carbon may have to be treated as hazardous waste regardless of the TCLP results.

IMPACT OF MAJOR FLUE GAS CONSTITUENTS ON MERCURY UPTAKE BY SULFUR-IMPREGNATED CARBONS

Impact of Carbon Dioxide

Effect of CO_2 on the performance of sulfur-impregnated activated carbon (SIAC) was tested at two different concentrations, namely 5% and 15% (both tests were performed in duplicate), while the remainder was pure N_2 gas [39]. Identical breakthrough curves obtained in all tests indicated that CO_2 behaves like an inert gas and does not affect the performance of SIAC.

Impact of Oxygen

The concentration of O_2 in the carrier gas was varied from 0 to 9%, and the resulting mercury uptake is shown in Figure 2.7 [39]. When the concentration of O_2 was increased from 0% to 3%, the mercury uptake capacity of SIAC remained almost unchanged. The overall mercury removal capacity increased by 16 and 33% as the O_2 concentration increased to 6% and 9%, respectively.

Because oxygen is readily chemisorbed by activated carbons to form carbon-oxygen complexes that are important in determining surface reactions and adsorptive behavior [26], it was necessary to study the possibility of carbon-oxygen complexes formation during the column tests conducted in the presence of oxygen and their impact on mercury removal.

Both virgin carbon (BPL) and SIAC were tested for the formation of acidic oxygen complexes upon exposure to oxygen using the procedure described by Tessmer et al. [40]. Pre-oxidized samples were prepared by contacting 2 grams of the unoxidized samples with a stream of air (flow rate of 1.0 L/min) in a ceramic boat at 140°C for a period of seven days. Total acidic surface oxygen content of virgin BPL increased from 445 ± 15 to 578 ± 8 µeq/g (based on triplicate measurements) as a result of air treatment, while that increase was more pronounced for SIAC (from 130 ± 20 to 620 ± 5 µeq/g). Lower acidic surface functional group content of unoxidized SIAC compared to BPL can be explained by outgassing effects

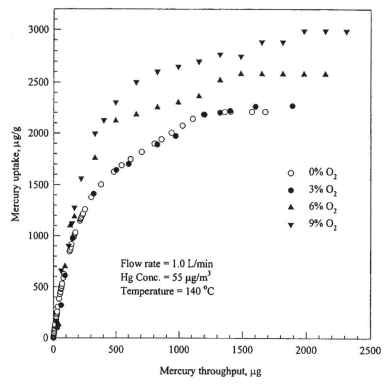

Figure 2.7 Effect of oxygen on mercury uptake by SIAC (from Liu, Vidic, and Brown, *Environ. Sci. Technol.*, 34(1):154, 2000. With permission from the American Chemical Society).

during exposure of BPL carbon to a temperature of 600°C [40]. The finding that carbon-oxygen complexes formed on SIAC after exposure to oxygen are equal to those formed on pre-oxidized BPL carbon suggests that a significant portion of the original surface area of BPL carbon remained reactive even after impregnation with sulfur. Sulfur analysis of the SIAC showed that a negligible amount of sulfur (less than 1% of impregnated sulfur) was lost during seven days of contact with an airstream at 140°C.

Mercury breakthrough tests were performed using these unoxidized and pre-oxidized carbons with pure N_2 as a carrier gas [39]. The breakthrough curves clearly showed that air pretreatment had no impact on the performance of SIAC. Identical observation was made in the case of BPL carbon. In addition, 1% sulfur loss did not reduce the adsorptive capacity of SIAC. These results indicate that air can oxidize a carbon surface and increase its acidic surface functional group content, but these changes have no impact on the performance of activated carbon for mercury removal.

It is reasonable to assume that the reaction between O_2 and mercury that is catalyzed by activated carbon surface [41] is responsible for the enhanced performance of SIAC in the presence of oxygen (Figure 2.7), because oxidized mercury is much more adsorbable than elemental mercury. Virgin BPL was tested for mercury uptake using a carrier gas containing 9% oxygen to test this hypothesis. Figure 2.8 illustrates that virgin carbon showed a much higher capacity for mercury uptake (20 µg Hg/g carbon) after four hours of contact in 9% O_2 than in the absence of oxygen (1.5 µg Hg/g carbon at saturation). Because virgin BPL can only remove 1.5 µg Hg/g carbon in the absence of oxygen, the additional capacity for mercury removal in the presence of oxygen can only be explained by the conversion of mercury to mercuric oxide, as there was no reaction between oxygen and mercury in the absence of an activated carbon surface.

Although, the virgin activated carbon acted as a catalyst for mercury oxidation, it is possible that its catalytic effect could decrease due to the loss of surface area resulting from sulfur impregnation. Surface area of activated carbon covered with sulfur molecules for SIAC would be about 7% of the initial surface area of virgin BPL of 900 m^2/g, if it is assumed that sulfur molecules (diameter, $D = 2.08 \times 10^{-10}$ m) [42] are impregnated onto the carbon surface in a monolayer. The actual available activated carbon surface area should be higher because the sulfur molecules may form multiple layers on the carbon surface, and it is reasonable to assume that SIAC still has enough active sites to catalyze the reaction between mercury and oxygen as seen for virgin activated carbon.

Impact of Moisture

Breakthrough tests conducted in the presence of varying amounts of moisture in the carrier gas showed that 5% moisture had virtually no impact on the performance of SIAC for mercury uptake (based on duplicate tests) [39]. However, carbon adsorptive capacity decreased as much as 25% when the moisture content increased to 10% (based on duplicate tests). Because the adsorptive capacity of carbon did not change at low moisture content, it

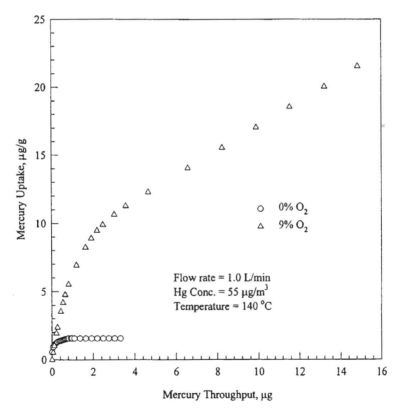

Figure 2.8 Catalysis of mercury and oxygen reaction by BPL (from Liu, Vidic, and Brown, *Environ. Sci. Technol.*, 34(1):154, 2000. With permission from the American Chemical Society).

can be concluded that moisture does not affect the reaction between sulfur and mercury. Therefore, it is postulated that the effect of moisture is related to the adsorption of water by the carbon surface. The carbon surface can bind water molecules to form hydrogen bonds with other molecules [43]. Higher vapor pressure will increase the amount of adsorbed water. For the 5% moisture in the carrier gas, the capillary condensation may be the dominant process. As the water vapor pressure increased to 10%, water molecules were able to fill the micropores so that isolated water zones merged to block access to some active sites on the carbon surface and active sulfur molecules, thereby creating additional mass transfer resistance for the adsorption of elemental mercury.

Another factor that could influence the carbon performance is hydrogen formation due to the dissociation of water induced by carbon. It was reported that hydrogen and CO can be formed if sufficient water vapor pressure is present above the carbon surface in the temperature range used in this study and that hydrogen is preferentially retained by the carbon [26]. Due to their extremely small size, hydrogen molecules can easily reach small carbon pores to form strong hydrogen-carbon complexes. As a result, the available surface area in the meso- and micro-porous region is decreased, and mercury is unable to react with sulfur retained in those pores.

Impact of Sulfur Dioxide

Figure 2.9 compares mercury uptake by SIAC in pure N_2 and in 1,600 ppm SO_2/N_2 mixture [39]. This test indicated that the mercury uptake capacity of SIAC was not affected by the presence of SO_2. Identical observation was made for virgin BPL (data not shown).

It is well known that SO_2 could interact with activated carbon via physisorption or chemisorption at relatively high temperatures (>500°C) and high concentrations [44–46] according to the following overall reaction: $SO_2 + C \rightarrow CO_2 + 1/2\ S_2$.

Thermogravimetric analysis (TGA) used to investigate the potential impact of SO_2 on virgin BPL and SIAC revealed that the weight of BPL and SIAC samples remained unchanged even after five hours of exposure to 3,000 ppm SO_2 at 140°C (data not shown). From the overall reaction, every mole of SO_2 would consume 1 mole of carbon and deposit 0.5 moles of S_2, thereby increasing the net weight of carbon by about 20 g per mole of SO_2 reacted. Based on the experimental conditions used in the TGA test (100 ml/min of 3,000 ppm SO_2 in N_2 at 140°C), the weight increase that could have been observed for the five-hour period was estimated at 3.54 mg. TGA instrumentation used in this study would have registered even if only 1/3,500 of SO_2 reacted with carbon, because the detection limit is 0.001 mg. The results of TGA tests suggest that SO_2 did not react with the virgin carbon surface or with impregnated sulfur. This behavior is most likely due to low SO_2 concentration and low reaction temperature that retarded possible reactions.

Impact of Nitric Oxide

Figure 2.9 indicates that the performance of SIAC did not exhibit a significant change as the carrier gas was switched from pure N_2 to NO/N_2 mixture. Identical observation was made for virgin BPL carbon (data not shown). TGA tests showed that the weight of both BPL and SIAC samples remained unchanged upon exposure to 100 ml/min of 400 ppm NO in N_2 after five hours of exposure at 140°C (data not shown).

Based on the experimental conditions used in TGA tests, the amount of NO passed through the system in five hours was about 11.59 mg. Because the detection limit of the

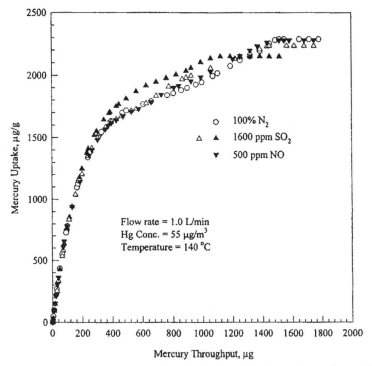

Figure 2.9 Effect of SO₂ and NO on mercury uptake by SIAC (from Liu, Vidic, and Brown, *Environ. Sci. Technol.*, 34(1):154, 2000. With permission from the American Chemical Society).

TGA instrument used in this study is 0.001 mg, it would have registered if only 1/10,000 of NO had been adsorbed by the carbon. This was not observed, and it is postulated that NO was not adsorbed by the carbon at the test conditions employed. Several studies found that NO could react with carbon to form various gas products [47,48] based on the following reactions:

$$C + 2NO \rightarrow CO_2 + N_2$$
$$C + NO \rightarrow CO + 1/2N_2$$
$$CO + NO \rightarrow CO_2 + 1/2N_2$$

According to these reactions, carbon weight loss could occur due to gasification. However, this was not observed under the experimental conditions used in this study. One possibility is that the concentration of NO in this study was extremely low (0.0004%) compared to 1 to 10% NO used in other studies. Low temperature was another reason for the lack of gasification that was shown to occur at 500–800°C [47,48]. Because the SIAC did not experience any weight loss in the presence of NO, it can be concluded that there was no reaction between NO and elemental sulfur.

Impact of Gas Mixture

Preliminary studies were performed to investigate the combined effects of SO₂ and water vapor and NO and water vapor [39]. These tests showed that the performance of SIAC in the presence of SO_2 2 (1,600 ppm) and 10% moisture or NO (300 ppm) and 10% moisture did not differ very much from the tests conducted in the presence of 10% moisture alone [41]. Even if weak acids (H_2SO_3 or HNO_2) formed on the carbon surface, the low concentration of the acid resulting from the low concentration of SO_2 and NO did not have a major impact on the performance of SIAC, and moisture was the dominant factor influencing the adsorptive capacity of this sorbent.

METHODS OF SULFUR IMPREGNATION

Sulfur impregnation onto activated carbon with elemental sulfur was compared to another method for sulfur impregnation onto BPL carbon that involved hydrogen sulfide oxidation with oxygen at relatively low temperatures (below 200°C) that is catalyzed by the substrate surface [49]. Sulfuration according to this new approach was carried out in a 0.25-inch inner diameter stainless steel fixed-bed reactor charged with 250 mg of BPL and placed in a laboratory oven to maintain constant temperature during impregnation. A clean nitrogen gas was used as an inert carrier gas while oxygen and H_2S were supplied from separate tanks as pure streams. The total gas flow rate was 0.2 L/min, and the gas mixture was adjusted at O_2:H_2S:N_2 ratio of 1:1:2, respectively. The reaction temperature was ramped at 7–8°C/min to 150°C where it was maintained for a predetermined period of time to promote effective sulfur deposition onto the substrate surface, while reducing the formation of undesired compounds, such as SO_2 [50]. The impregnation of BPL was performed for 0.25, 0.5, 1 and 2 hours, and these sorbents are denoted as BPL-HS-0.25, BPL-HS-0.5, BPL-HS-1 and BPL-HS-2, respectively.

Table 2.7 shows that sulfur loading does not increase linearly with impregnation time. While impregnation may be carried out for two hours, over 70% of sulfur will be impregnated within the first 30 minutes. As can be seen in Table 2.7, 90% reduction in surface area of BPL carbon due to sulfur deposition is achieved within 30 minutes, resulting in a significant decrease in sulfur deposition rate due to the lack of active catalytic surface and pore volume for sulfur deposition.

Comparison of mercury breakthrough profiles for different BPL-HS sorbents revealed that BPL-HS-0.25 (15-minute impregnation) had the highest efficiency for mercury removal. It is, therefore, important to note that the performance of BPL-HS sorbents cannot be compared based on the sulfur content alone. The relationship between the sulfur content and adsorptive capacity is highly nonlinear, as shown in Figure 2.10, and further testing and optimization is needed to find the sulfur content and impregnation protocol that would result in optimal performance. Liu et al. [34] already showed that the predominant sulfur form and pore size distribution are more important for effective mercury uptake by activated carbons impregnated with elemental sulfur at elevated temperatures than the total sulfur content. These results suggest that total sulfur content is also not a key factor governing the per-

TABLE 2.7. Sulfur Content and BET Surface Area of BPL-Based Adsorbents.

Adsorbent	Reaction Time (hr)	Surface Area (m²/g)	Sulfur Content (wt%)
BPL	—	1020	0.1
BPL-S	—	820	10
BPL-HS-0.25	0.25	570	12.7
BPL-HS-0.5	0.5	89	35.2
BPL-HS-1	1	<50	44.3
BPL-HS-2	2	<50	50.8

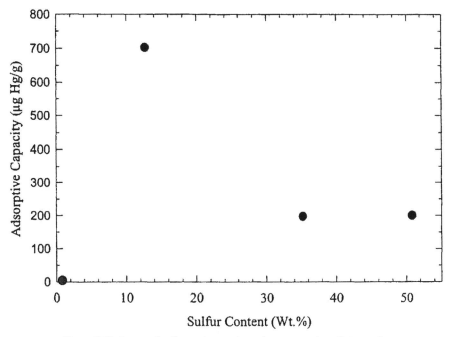

Figure 2.10 Impact of sulfur content on dynamic mercury adsorptive capacity.

formance of BPL-HS in a fixed-bed adsorber operated with a short EBCT. Explanation for such behavior in terms of pore blockage with excess sulfur is discussed above, while the arguments of Korpiel and Vidic [25] and Liu et al. [34] about the importance of predominant sulfur forms are also valid here.

Breakthrough tests with BPL-S (10 wt% sulfur), and BPL-HS-0.25 (12.7 wt% sulfur) revealed superior performance of BPL-S sorbent. While BPL-HS-0.25 reached 100% breakthrough within three days, it took almost 18 days for BPL-S to reach 100% breakthrough despite lower sulfur content. BPL-S was produced by impregnating BPL with elemental sulfur at 600°C, which is a much higher temperature than that used for impregnation with H_2S. Consequently, the sulfur in BPL-S is more evenly distributed in pore structure, occupying deeper, narrower pores [25,34]. On the other hand, the sulfur in BPL-HS carbons is most likely located close to the external surface of BPL particles as evidenced by the result of BET surface area analyses shown in Table 2.7. Sulfuration of BPL for 15 minutes caused a decrease in the specific surface area of about 45%, while impregnation with elemental sulfur at 600°C caused a decrease in surface area of only 20%.

Thermogravimetric analysis (TGA) was conducted on BPL-S and BPL-HS-2 carbons in order to determine the effect of impregnation method on thermal stability of sulfur deposited on an activated carbon surface [49]. Each sample was heated to 400°C in nitrogen atmosphere, and the percent decrease in weight of the samples was recorded as a function of time. As shown in Figure 2.11, weight loss for BPL and BPL-S carbons was almost negligible, while BPL-HS-2 underwent a 46% reduction in weight. If the weight loss was due to sulfur loss, it can be concluded that over 95% of sulfur on BPL-HS-2 was lost by heating to 400°C. This outcome implies that the bonding of sulfur molecules to the carbon matrix is much stronger for BPL-S carbon, making it more suitable for fixed-bed applications. However, this apparent sulfur loss of BPL-HS after exposure to high temperature for a long period of time might not be relevant for powder sorbent injection systems where a contact time of 1–2

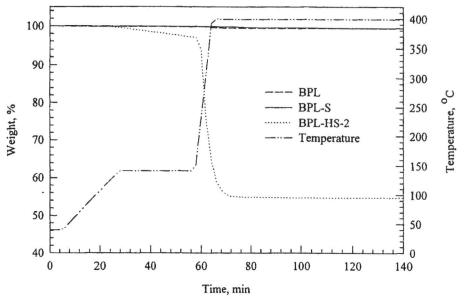

Figure 2.11 Thermogravimetric analyses of activated carbon-based sorbents.

seconds is typical. In addition, BPL-HS sorbents were prepared using H_2S, which is considered a waste product for a number of industries, and the use of BPL-HS sorbents for mercury control would have a positive impact on pollution prevention efforts by these industries.

Chloride-Impregnated Activated Carbon

CHLORIDE IMPREGNATION PROTOCOL

The chloride-impregnated carbon (BPL-C) was produced by the method of incipient wetness, whereby the virgin 60 × 80 BPL was contacted with a solution of copper chloride dissolved in 6 N nitric acid [51]. The volume of the copper chloride solution used to impregnate the carbon was based on the estimated pore volume of the carbon (0.55 ml/g) and the concentration of the copper chloride in the solution was dependent on the desired mass percentage of the chloride on the carbon. After impregnation, the carbon was dried in an oven at 90°C and placed in a desiccator until needed.

The chloride content of activated carbon was determined indirectly by measuring the copper content of the impregnated sample. The impregnated carbon was combusted, and the ash was dissolved in aqua regia and analyzed for copper using an atomic absorption spectrophotometer.

MERCURY UPTAKE BY CHLORIDE-IMPREGNATED ACTIVATED CARBON

Figure 2.12 illustrates the mercury breakthrough curve for BPL-C-1 (BPL carbon impregnated with 1 wt% chloride) at empty bed contact time (EBCT) of 0.017 seconds, while Figure 2.13 shows the impact of operating temperature on mercury uptake by BPL-C-1 [51].

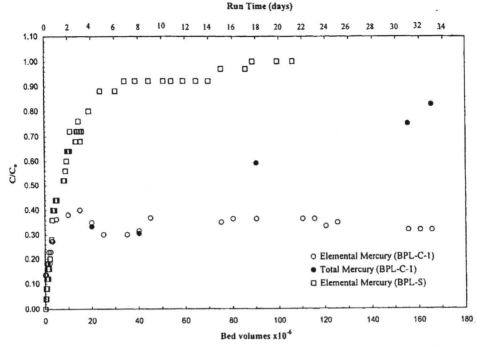

Figure 2.12 Mercury breakthrough for BPL-C-1 and BPL-S sorbents (EBCT = 0.017 s) (from Vidic and Siler, *Carbon*, 39, 3, 2001. With permission from Elsevier Science, Ltd.).

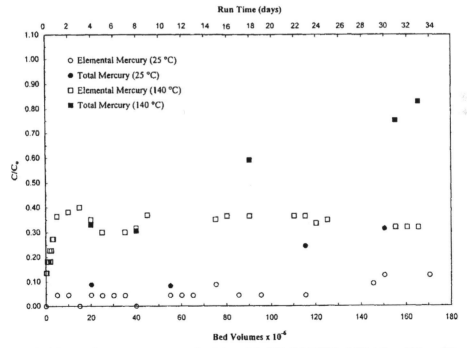

Figure 2.13 Impact of temperature on mercury breakthrough for BPL-C-1 (EBCT = 0.017 s) (from Vidic and Siler, *Carbon*, 39, 3, 2001. With permission from Elsevier Science, Ltd.).

Immediately evident in both figures is the presence of other species of mercury in the effluent besides the elemental form. This indicates that a reaction may be taking place between the copper chloride on the carbon surface and the influent mercury, with a percentage of the reacted mercury not being retained by the carbon. Such behavior suggests that the adsorption of elemental mercury onto the chloride-impregnated carbon is not as stable as that previously observed for sulfur-impregnated carbons [25,34]. Release of oxidized mercury forms into the effluent can be the result of either a weak carbon-chloride bond or a weak chloride-mercury bond or both. Because the Gibbs free energy for the vapor-phase reaction between mercury and chloride $(Hg^+_{(g)} + Cl^-_{(g)} \rightarrow HgCl_{(g)})$ at both 25 and 140°C is highly negative [52–54], any release of chloride from the carbon surface would result in the formation of mercuric chloride, which would account for the difference in column effluent concentrations between the direct AAS measurements of elemental mercury and the total mercury measurements obtained using the impinger method.

Additional observation from these figures shows that none of the chloride-impregnated carbons reached 100% elemental mercury breakthrough after five weeks of column operation, which could lead to conclusions about the superiority of chloride-impregnated carbons as mercury sorbents [18], without regard for the possibility of oxidized mercury release into the environment.

Impact of Empty Bed Contact Time (EBCT)

The performance of chloride-impregnated BPL carbons was examined using three different EBCTs, namely, 0.017, 0.034 and 0.068 seconds to determine the degree to which contact time affects the dynamic adsorption capacity of the impregnated carbon [51]. The temperature and influent mercury concentrations for all three reactors were kept constant at 140°C and 55 $\mu g/m^3$, respectively. Each test was performed for over 30 days without reaching 100% breakthrough. The dynamic adsorption capacity at 30% total mercury breakthrough, or total mass of mercury adsorbed per gram of carbon after 30% breakthrough (M30), during each test is listed in Table 2.8. As can be seen from Table 2.8, the M_{30} for the BPL-C-1 carbon operated at EBCT of 0.017 seconds is similar to that of the BPL-S [27,36].

Previous research by Korpiel [55] on the adsorption of mercury by virgin BPL found that increasing the EBCT from 0.011 to 0.055 seconds at the operating temperature of 140°C had minor impact on the dynamic adsorption capacity. Such a finding indicates that the EBCT of 0.011 second is sufficient to satisfy diffusional limitations of mercury adsorption on virgin carbon in a fixed-bed system. However, the comparison of dynamic adsorption capacities of BPL-C-1 adsorbers operated at EBCT of 0.017 and 0.034 seconds shows almost a thirteenfold increase in the capacity with only a twofold increase in the EBCT. These experimental findings with BPL and BPL-C-1 imply that mercury uptake by chloride-im-

TABLE 2.8. Impact of EBCT on Dynamic Adsorption Capacity of BPL-S and BPL-C-1 (from Vidic and Siler, *Carbon*, 39, 3, 2001. With Permission from Elsevier Science, Ltd.).

Sorbent Type	BPL-S	BPL-C-1		
Carbon mass (mg)	100	100	200	200
N_2 flow rate (L/min)	1.0	1.0	1.0	0.5
EBCT (s)	0.017	0.017	0.034	0.068
M_{30} (μg Hg/g carbon)	44	25	337	515

pregnated carbon is limited by the rate of chemical reaction between chloride and mercury rather than by the rate of mass transfer of mercury onto the carbon surface.

Further doubling of the EBCT from 0.034 seconds to 0.068 seconds at a constant temperature of 140°C and influent mercury concentration of 55 $\mu g/m^3$ resulted in an increase in the dynamic adsorption capacity from 337 $\mu g/m^3$ to 515 $\mu g/m^3$, or only about a 50% increase. This difference in the degree of sensitivity to flow rate is another indication that the interaction of mercury and chloride is the rate controlling factor. Each breakthrough experiment demonstrated a rapid breakthrough followed by a period of approximately 10 to 15 days where the concentration of elemental mercury in the effluent remained relatively constant, while the concentration of oxidized mercury slowly increased. This plateau in the elemental mercury breakthrough decreased as the EBCT increased (from approximately 30–40% at EBCT of 0.017 seconds to 15–20% at EBCT of 0.034 seconds and 5–15% at EBCT of 0.068 seconds), indicating that longer EBCTs allow for more elemental mercury to react and bond with the chloride impregnate, while the rate of oxidized mercury species released into the atmosphere remains unaffected.

Impact of Temperature

Elemental and total mercury breakthrough curves from a fixed-bed reactor charged with 100 mg of 60 × 80 U.S. Mesh BPL-C-1 and operated at 25 and 140°C showed that mercury uptake at 30% total mercury breakthrough decreased from 367 µg Hg/g carbon at 25°C to 25 µg Hg/g carbon at 140°C [51]. The greater dynamic adsorption capacity of mercury at the lower column temperature is consistent with similar column experiments using chloride-impregnated carbon [18] and sulfur-impregnated carbon [25] and can easily be explained by the exothermic nature of mercury reaction with chloride [52–54] or elemental sulfur [25,34].

Effect of Chloride Content

Virgin BPL was impregnated with approximately 1 and 5% chloride by weight to investigate how the dynamic adsorption capacity is affected by the amount of chloride available on a carbon surface to react with mercury [51]. Performance data summarized in Table 2.9 show that BPL carbon impregnated with 1% chloride has similar capacity at 30% total mercury breakthrough with BPL-S carbon, while the impregnation with 5% chloride yielded significantly higher adsorption capacity. The dynamic adsorption capacity of BPL-C-5 at 30% total mercury breakthrough was approximately an order of magnitude higher than that of BPL-C-1. It is also important to note that a much better utilization of chloride on the carbon surface was observed for BPL-C-5 than BPL-C-1 (Table 2.9). Although both values are far below the stoichiometric ratio, three times more chloride reacted with mercury in the case of BPL-C-5 than BPL-C-1 (0.69% versus 0.23%).

Teller and Quimby [18] observed a similar relationship between the dynamic adsorption capacity and the copper chloride content on the carbon despite the fact that those observations were made using an EBCT approximately 17 times greater than those listed in Table 2.9 and an order of magnitude greater influent mercury concentration. However, the dynamic adsorption capacities after 20% breakthrough reported by Teller and Quimby were approximately 75 to 370 times greater than those observed in this study. The difference in the influent mercury concentration and EBCT between this study and that of Teller and Quimby may account for the dissimilarity in adsorption capacities in which a longer contact

TABLE 2.9. Impact of Chloride Content on Dynamic Adsorption Capacity (from Vidic and Siler, *Carbon*, 39, 3, 2001. With Permission from Elsevier Science, Ltd.).

Carbon Type	BPL-S[27]	BPL-C-1	BPL-C-5
Cl content (wt%)	N/A	1.07	4.79
Carbon mass (mg)	100	100	100
EBCT (s)	0.017	0.017	0.017
M_{30} (μg Hg/g carbon)	44	25	332
M_{30}/Cl (%)	N/A	0.23	0.69
Hg_{ox}/Hg_{total} (%)	N/A	15	26

time and higher mercury concentration would allow for a more extensive reaction between mercury and chloride.

Another observation that can be made from Table 2.9 is that the total contribution of oxidized mercury species (Hg_{ox}) in the effluent increased significantly with the increase in chloride content. During the 20 days of column operation, the total mass of oxidized mercury species emitted in the effluent increased from 15 to 26% of total mercury emissions from the reactor, despite the fact that better chloride utilization and lower elemental mercury emissions were observed for the high chloride content carbon. These observations further underscore limited utility of chloride-impregnated activated carbons for the control of mercury emissions from the combustion process using a fixed-bed adsorption system.

REFERENCES

1 Porcella, D. B, Ed. *Proceedings of the 1993 International Conference on Managing Hazardous Air Pollution*, Washington, DC, July 13–15, 1993, pp. IV-33–IV-41.

2 Nriagu, J. O. and Pacyna, J. M., *Nature*, 1988, 333, 12, 134.

3 Schroeder, W. H., Yarwood, G., and Niki, Y. *Water, Air, and Soil Pollution*, 1991, 56, 653.

4 Chu, P. and Schmidt, C. *Proceedings of the 1994 Pittsburgh Coal Conference*, Volume 1, Pittsburgh, PA, Sept. 12–16, 1994, pp. 551–556.

5 Chang, R. and Offen, G. *Power Engineering*, Nov. 1995, 99(11), 51.

6 Seigneur, C., Wrobel, J., and Constantinou, E. *Proceedings of the 1993 International Conference on Managing Hazardous Air Pollutions*, Washington, DC, July 13–15, 1993, pp. III-39–III-57.

7 White, D. M., Nebel, K. L., and Johnson. M. G. *Municipal Waste Combustion: Papers and Abstracts from the Second Annual International Conference*, Tampa, FL, April 15–19, 1991, p. 652.

8 Donnelly, J. R. *Municipal Waste Combustion: Papers and Abstracts from the Second Annual International Conference*, Tampa, FL, April 15–19, 1991, p. 125.

9 Krivanek, C. S. *Proceedings of the 1993 International Municipal Waste Combustion Conference*, Williamsburg, Virginia, March, 1993, p. 824.

10 Meij, R. *Water, Air, and Soil Pollution*, 1991, 56, p. 21.

11 Brna, T. G. *Municipal Waste Combustion: Papers and Abstracts from the Second Annual International Conference*, Tampa, FL, April 15–19, 1991, p. 145.

12 Chang, R. and Owens, D. *EPRI Journal*, 1994, July/August, 46.

13 Felsvang, K., Gleiser, R., Juip, G., and Nielsen, K. K. *Proceedings of the 1993 International Conference on Managing Hazardous Air Pollutions*, Washington, DC, July 13–15, 1993, pp. VI-1–VI-7.

14 Felsvang, K., Gleiser, R., Juip, G., and Nielsen, K. K. *Fuel Processing Technology*, 1994, 39, p. 417.

15 Otani, Y., Emi, H., Kanaoka, C., Uchijima, I., and Nishino, H. *Environmental Science and Technology*, 1988, 22, p. 708.

16 Sinha, R. K. and Walker, P. L. *Carbon,* 1972, 10, p. 754.

17 Matsumura, Y. *Atmosph. Environ.,* 1974, 8, 1321.

18 Teller, A. J. and Quimby, J. M. *Proceedings of the AWMA Meeting,* Vancouver, BC, June 1991.

19 Henning, K. D., Keldenich, K., Knoblauch, K. and Degel, J. *Gas Separation & Purification,* 1988, 2, 20.

20 Vidic, R. D., Chang, M. T., and Thurnau, R. C. *J. Air and Waste Management Association,* 1998, 48, 247.

21 Traegner, U. K. and Suidan, M. T. *ASCE J. of Environmental Engineering,* 1989, 115(1), 109.

22 Traegner, U. K. and Suidan, M. T. *Water Research,* 1989, 23(3), 267.

23 Hand, D. H., Crittenden, J. C., and Thacker, W. E. *ASCE J. of Environmental Engineering,* 1983, 109(1), 82.

24 Flora, J. R. V., Vidic, R. D., Liu W., and Thurnau, R. C. *J. Air and Waste Management Association,* 1998, 48, 1051.

25 Korpiel, J. A. and Vidic, R. D. "Effect of Sulfur Impregnation Method on Activated Carbon Uptake of Gas-Phase Mercury." *Environmental Science & Technology,* 1997, 31(8), 2319–2326.

26 Puri, B. R. In *Chemistry and Physics of Carbon,* Walker, P. L. Jr., Ed.; Marcel Dekker Inc.: New York, 1970; Vol. 6, pp. 264–282.

27 Tuller, W. N., Ed. *The Sulfur Data Book.* McGraw-Hill Book Company, Inc.: New York, 1954.

28 Hampel, C. A., Ed. *The Encyclopedia of the Chemical Elements.* Reinhold Book Corporation: New York, 1968.

29 Pryor, W. A. *Mechanisms of Sulfur Reactions.* McGraw-Hill Book Company, Inc.: New York, 1962.

30 Daza, L., Mendioroz, S., and Pajares, J. A. *Clays and Clay Minerals,* 1991, 39, 14.

31 Daza, L., Mendioroz, S., and Pajares, J. A. *Solid State Ionics,* 1990, 42, 167.

32 Daza, L., Mendioroz, S., and Pajares, J. A. *Applied Catalysis B: Environmental,* 1993, 2, 277.

33 Meyer, B. *Chem. Rev.,* 1964, 64, 429–451.

34 Liu, W., Vidic, R. D., and Brown, T. D. *Environmental Science & Technology,* 1998, 32(4), 531.

35 Liu, W., Vidic, R. D., and Brown, T. D. *Environmental Science & Technology,* 2000, 34(3), 483–488.

36 Lide, D. R., Ed., *Handbook of Chemistry and Physics,* 73rd ed. CRC Press: Boca Raton, FL, 1992, pp. 4–74.

37 U.S.EPA. *Mercury and Arsenic Wastes—Removal, Recovery, Treatment, and Disposal.* Noyes Data Corporation: Park Ridge, NJ, 1993, p. 14.

38 Alloway, B. J., Ed., *Heavy Metals in Soils,* 2nd Edition. Chapman and Hall: New York, 1995, p. 245.

39 Liu, W., Vidic, R. D., and Brown, T. D. *Environmental Science & Technology,* 2000, 34(1), 154–159.

40 Tessmer, C. H., Vidic, R. D., and Uranowski, L. *J. Environ. Sci. Technol.* 1997, 31, 1872.

41 Hall, B., Schager, P., and Lindqvist, O. *Water, Air, and Soil Pollut.,* 1991, 56, 3.

42 Nebergall, W. H., Schmidt, F. C., and Holtzclaw, H. F. Jr. *College Chemistry,* 5th ed. D. C. Heath and Company: Lexington, 1976, post-face.

43 Dubinin, M. M. *Carbon* 1980, 18, 355.

44 Thrower, P. A., Ed. *Chemistry and Physics of Carbon,* Vol. 21. Marcel Dekker, Inc.: New York, 1989, p. 147.

45 Stacy, W. O., Vastola, F. J., and Walker, P. L. Jr. *Carbon,* 1968, 6, 917.

46 Puri, B. R. and Hazra, R. S. *Carbon,* 1971, 9, 123.

47 Teng, H., Suuberg, E. M., and Calo, J. M. *Energy and Fuels,* 1992, 6, 398.

48 Illan-Gomez, M. J., Linares-Solano, A., Salinas-Martinez de Lecea, C., and Calo, J. M. *Energy and Fuels,* 1993, 7, 146.

49 Kwon, S.-J. "Evaluation of Sulfur Impregnation Processes on Different Substrates for the Production of Elemental Mercury Sorbents" M.S. Thesis, University of Pittsburgh, 1996.

50 Coskun, I. and Tollefson, E. L. *The Canadian Journal of Chemical Engineering,* 1980, 58, 72.

51 Vidic, R. D. and Siler, D. P. *Carbon,* 2001, 39, 3.

52 Van Wylen, G. J., Sonntag, R. E., and Borgnakke, C. *Fundamentals of Classical Thermodynamics,* 4th ed. John Wiley and Sons, Inc.: New York, 1994, p. 555.

53 Smith, J. M. and Van Ness, H. C. *Introduction to Chemical Engineering Thermodynamics,* 4th Ed., McGraw-Hill, Inc., New York, 1987, p. 105.

54 Chase, M. W., Jr., Davies, C. A., Downey, J. R. Jr., Frurip, D. J., McDonald, R. A., and Syverd, A. N. *Journal of Physical and Chemical Reference Data, Supplement 1,* 1985, 14, 1.

55 Korpiel, J. A. "Effect of Sulfur Impregnation Method on Activated Carbon Adsorption of Vapor-Phase Mercury," M.S. Thesis, University of Pittsburgh, 1996.

56 Brown, T. D., Smith, D. N., Hargis, R. A., and O'Dowd, W. J. *J. Air & Waste Management,* Critical Review, June 1999, 1.

CHAPTER 3

Trace Heavy Metal Separation
by Chelating Ion Exchangers

SUKALYAN SENGUPTA[1]
ARUP K. SenGupta[2]

INTRODUCTION

VERY often, removal of heavy metal cations from water and wastewater streams involves a scenario where only a trace concentration of the heavy metal ion (typically in the range of ppm) needs to be selectively removed in the presence of other competing nontoxic, nonregulated ions (e.g., Ca^{2+}, Mg^{2+}, Na^+, etc.). The case for selectivity is strong; removing only the trace contaminant constitutes an efficient and economic solution, but it is also a challenging proposition. Use of treatment processes that remove all ions—heavy metals and nontoxic (e.g., solidification/stabilization, membrane filtration, precipitation, etc.)—invariably involve difficulties in justifying the associated costs. An additional problem relates to handling the copious quantities of sludge produced, disposal of which is again an environmental issue and an economic issue. Organic ion exchange polymers allow the benefit of selective removal of heavy metal ions and can overcome the limitations described above. This method also opens the possibility of reusing the heavy metal ions.

From a generic viewpoint, an organic-base ion exchanger may be defined as a solid phase where a number of functional groups with positive or negative charges are covalently attached to a polymer matrix that can be permeated by ions of opposite charge from the mobile liquid phase. Thus, for an anion exchanger, functional groups are positively charged (say quaternary ammonium group, R_4N^+) and immobilized in the polymer phase; electroneutrality is maintained by permeation of anions from the liquid phase. A cation exchanger is just the opposite; fixed functional groups are negatively charged (say sulfonic acid, RSO_3^-) and electrically balanced by cations permeating from the liquid phase. The process of ion exchange has been known for decades in naturally occurring soil and zeolites; organic-base exchangers are, in many ways, tailored to duplicate similar behavior with a view to achieving the intended goal with greater efficiency. It is essentially a hetero-

[1]Department of Civil and Environmental Engineering, University of Massachusetts Dartmouth, 285 Old Westport Road, North Dartmouth, MA 02747-2300, U.S.A., ssengupta@umassd.edu

[2]Department of Civil and Environmental Engineering, Lehigh University, 13 E. Packer Ave., Bethlehem, PA 18015, U.S.A., aks0@lehigh.edu

geneous process. The ions in the aqueous phase with the same sign as the fixed charge in the polymer phase are called "co-ions." Ions with the opposite charge are referred to as "counter-ions." During ion exchange, the counter-ions (cations for cation exchangers and anions for anion exchangers) are sorbed and can permeate in and out of the polymer, while the co-ions are excluded from the polymer phase by a phenomenon known as "Donnan Co-ion Exclusion Principle."

Ion-exchanging polymers are normally synthesized as spherical beads. For a fixed-bed process, the bead size normally varies from 0.3–1.2 mm. Figure 3.1 shows photomicrographs of two widely used ion exchangers that are very similar except for the fact that one (IRA-900) has a hydrophobic (polystyrene) matrix, whereas the other (IRA-958) has a hydrophilic (polyacrylate) matrix. One may note that in IRA-900, the resin beads are more spatially separated because of the hydrophobicity of polystyrene. Water molecules are always present in the interior of the ion exchanger beads. Because polyacrylate is more hydrophilic, IRA-958 beads display this tendency to form clusters, as can be clearly observed from Figure 3.1.

A polymeric ion exchanger, regardless of its type of application, is typically characterized by the following composition variables:

a. Type of polymer matrix
b. Type and degree of crosslinking
c. Type of functional group
d. Concentration of functional group or charge density
e. Porosity

The polymer matrix and cross-linking influence the hydrophobic/hydrophilic nature of the ion exchanger, including its ability to swell or shrink in a solvent, while functional groups and charge density are responsible for Coulombic, Lewis acid-base and other spe-

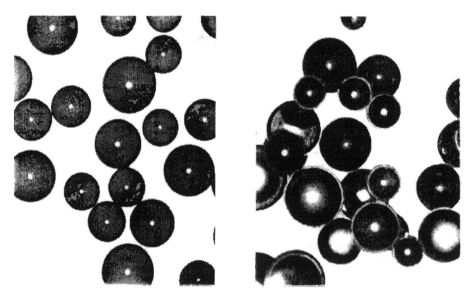

Figure 3.1 Photograph showing two macroreticular and macroporous anion exchangers synthesized by Rohm and Haas Co.: IRA-900 on left (polystyrene, hydrophobic matrix) and IRA-958 on right (polyacrylate, hydrophilic matrix).

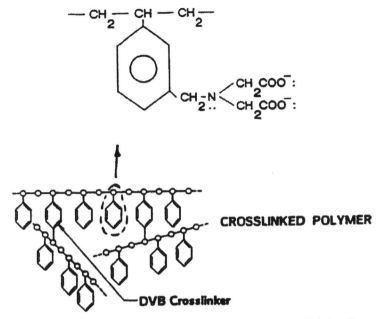

Figure 3.2 Schematic presentation of three-dimensional cross-linked polystyrene with imino-diacetate functional group.

cific types of interactions between the liquid-phase ions and the exchanger. For inorganic ions, pore size has minimal effect on selectivity but often influences diffusion-controlled rate processes significantly. Figure 3.2 is a chemical representation of a cation exchange resin where imino-diacetate-type functionality has been introduced into a polystyrene matrix. It may be observed that the linear polystyrene chain becomes three-dimensional due to incorporation of divinylbenzene cross-linker. This three-dimensional network imparts strength, insolubility and rigidity to the polymeric ion exchangers.

In a typical ion exchange reaction (cationic or anionic), the type of interaction between the functional groups on the polymer and exchangeable liquid-phase ions is primarily Coulombic (electrostatic), and the selectivity of the ion exchanger for a target ion in such cases is governed by:

a. Charge and hydrated ionic radii of the target ions compared to other competing ions

b. Hydrophobic/hydrophilic nature of the polymer matrix

c. Charge density and steric property of the functional groups

Following the syntheses of organic-base ion exchangers about 60 years ago [1,2], a number of investigations [3–7] have been carried out to study the effects of foregoing variables on an exchanger's selectivity toward a specific counter-ion; they will not be revisited here.

HEAVY METAL REMOVAL FROM AQUEOUS PHASE

Limitations of Polymeric Ion Exchangers

As mentioned before, in heavy metal contaminated aqueous solutions, the concentration of background, nontoxic cations is much higher than the target heavy metal cations. In a conventional ion exchange process, the law of mass action adversely affects its efficiency in such a scenario. For example, let us consider a typical cation exchange reaction between Ni^{2+} and Ca^{2+}:

$$\overline{R_2Ca} + Ni^{2+} \leftrightarrow \overline{R_2Ni} + Ca^{2+} \tag{1}$$

Here, the overbar represents the exchanger phase, and R denotes the polymer with a cation-exchanging functionality. Assuming ideality, the equilibrium constant or selectivity coefficient of Equation (1) is as follows:

$$K_1 = \frac{\overline{[R_2Ni]}\,[Ca^{2+}]}{[Ni^{2+}]\overline{[R_2Ca]}} \tag{2}$$

The ratio of Ni^{2+} to Ca^{2+} in the aqueous phase is, in general, much less than unity. Thus, in order for the Ni^{2+} removal process to be selective (i.e., high $\overline{R_2Ni}$), the equilibrium constant, K_1, needs to be extremely high. Conventional cation exchangers with Coulombic (electrostatic) type interaction are unable to attain such high selectivity (charge of calcium and nickel ion is the same). Table 3.1 shows relative selectivities of various cations with respect to H^+ for a strong-acid cation exchanger with sulfonic acid functionality and 8% divinylbenzene cross-linking [8]. Note the following:

a. Ca^{2+} is preferred over Zn^{2+}, Co^{2+}, Cu^{2+} and Ni^{2+}

b. Hg^{2+} and Pb^{2+} are preferred marginally over Ca^{2+}

TABLE 3.1. Relative Selectivity of Various Counter-Ions for a Strong-Acid Cation Exchanger.

Counter-Ion	Relative Selectivity for AG 50W-X8 Resin
H^+	1.0
Na^+	1.5
NH_4^+	1.95
K^+	2.5
Cu^+	5.3
Ag^+	7.6
Mn^{2+}	2.35
Mg^{2+}	2.5
Fe^{2+}	2.55
Zn^{2+}	2.7
Co^{2+}	2.8
Cu^{2+}	2.9
Cd^{2+}	2.95
Ni^{2+}	3.0
Ca^{2+}	3.9
Sr^{2+}	4.95
Hg^{2+}	7.2
Pb^{2+}	7.5
Ba^{2+}	8.7

AG 50W-X8 is a strong-acid cation exchanger with sulfonic acid functional group attached to a styrene-divinylbenzene copolymer matrix with 8% divinylbenzene and is available from Biorad Inc., CA, USA.

Such a low selectivity toward target heavy metal ions renders these commonly used cation exchangers uneconomical and, hence, unacceptable for commercial applications requiring specific removal of heavy metals.

A second problem faced by any ion exchange process stems from the fixed capacity (1.0–4.0 eq./L) of polymeric ion exchangers. When the capacity is exhausted, the ion exchangers need to be regenerated with an acid/alkali/salt. Frequent exhaustion and shorter run length during the fixed bed process render it less viable if the target metal is present at a concentration in the range of 100 ppm and higher. Other metal-ion removal processes, e.g., precipitation, membrane filtration, etc., outperform the ion exchange process economically under these circumstances. Application of ion exchange processes is, therefore, primarily limited as a selective polishing step to ensure high quality of the treated water, as often demanded by regulatory agencies. However, in certain situations, innovative process system design may help accomplish removal and recovery of fairly high concentrations of heavy metal ions by ion exchange processes. For details, please refer to References [9–12].

Emergence of Metal-Selective Chelating Exchangers

Most of the heavy metal cations of interest, such as Cu^{2+}, Hg^{2+}, Pb^{2+}, Ni^{2+}, Cd^{2+}, Zn^{2+}, etc., are transition-metal cations and exhibit Lewis-acid characteristics (electron acceptors). With organic and inorganic ligands (Lewis bases), all these heavy metal cations form fairly strong complexes. Most of the complexes of these metal cations, depending on their coordination number, have regular or slightly distorted tetrahedral, octahedral, or square pyramid structures [13]. Because Ca^{2+}, Mg^{2+}, and Na^+—the most commonly encountered competing nontoxic cations in water and wastewater—do not undergo such strong complexation, incorporating organic ligands as functional groups into the polymer matrix of the ion exchanger through covalent bonding was a natural progression of ideas to improve the exchanger's selectivity toward the toxic metal ions. These functionalized polymers are often referred to as chelating polymers, coordinating polymers or metal selective ion exchange resins. Figure 3.2 shows the chelating exchanger with imino-diacetate functionality, the workhorse of metal-selective exchangers, that are available from every major resin manufacturer around the world. Figure 3.3 shows two commonly used routes for synthesizing this chelating polymer from styrene monomers (cross-linking not shown) [14]. A typical ion exchange reaction between a metal ion, Me^{2+}, and Na^+ for this resin may be presented in the following way:

$$\overline{RN(CH_2COONa)_2} + Me^{2+} \leftrightarrow \overline{RN(CH_2COO)_2Me} + 2Na^+ \qquad (3)$$

Equation (3) however, fails to reveal the Lewis acid-base type interaction between the metal ion and the imino-diacetate functionality. Assuming the water molecule has four coordinated water molecules in the aqueous phase, $[Me(H_2O)_4]^{2+}$, the overall exchange involves the following:

$$\tag{4}$$

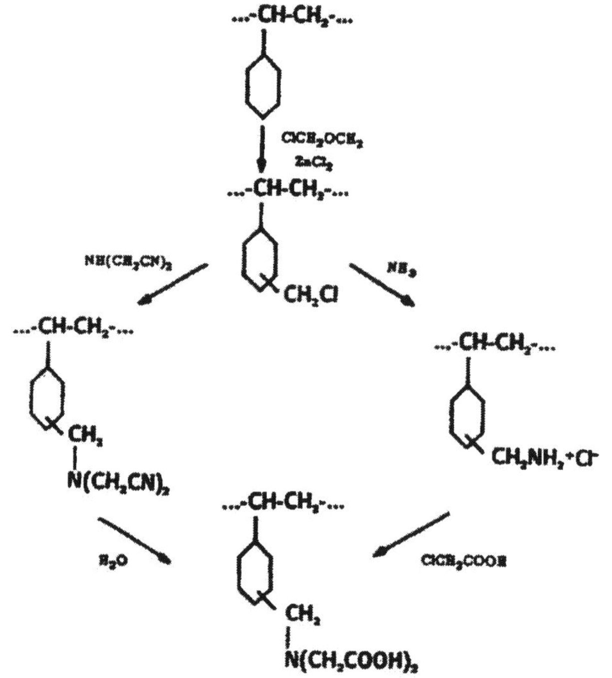

Figure 3.3 Two parallel routes for synthesis of imino-diacetate resins from polystyrene (cross-linking not shown).

Note that three water molecules (ligands) from the coordination sphere of the metal ion are replaced by one nitrogen and two oxygen donor atoms in the imino-diacetate functionality. The arrows indicate the metal-ligand or Lewis acid-base (LAB) interaction, and the high metal-ion selectivity for this type of functional group is often attributed to the accompanying coordination reaction in conjunction with exchange of ions. Figure 3.4 provides experimentally determined Me^{2+}/Ca^{2+} separation factors for three commercial ion exchange resins (imino-diacetate functionality—IRC-718, thiol functionality—GT-73, and picolylamine functionality—XFS-4195) for various heavy metals at varying pH values, and

$$\text{Separation Factor, } \alpha_{Me/Ca} = \frac{\overline{[RMe]}[Ca^{2+}]}{\overline{[RCa]}[Me^{2+}]} \tag{5}$$

the high selectivity of the metal ions can be readily noted [15]. Separation factor is a dimensionless measure of relative selectivity between two competing ions, and in this case, is equal to the ratio of the distribution coefficient of the metal ion concentration between exchanger and aqueous phases to that of calcium ion. Like Ca^{2+}, these metal ions have a charge 2+ and, therefore, only Coulombic/electrostatic interaction cannot be the reason for such a high Me^{2+}/Ca^{2+} separation factor. On the contrary, such a high selectivity is always attributed to the relatively strong Lewis acid characteristic of the toxic-metal cations, favoring their selective uptake through coordination reactions.

In order to characterize high metal-ion selectivity for chelating ion exchangers from a thermodynamic perspective, the metal ion uptake can be divided into two consecutive steps—ion exchange (IX) followed by Lewis acid-base (LAB) interaction, i.e.,

$$\text{Me}^{2+}(\text{aq}) \xrightarrow{\text{IX (Step I)}} \overline{\text{RMe}} \xrightarrow{\text{LAB (Step II)}} \overline{\text{RMe}}$$

At the standard state, the overall free energy at equilibrium between $\text{Me}^{2+}(\text{aq})$ and $\overline{\text{RMe}}$ is given by the following:

$$\Delta G^0_{\text{overall}} = \Delta G^0_{\text{IX}} + \Delta G^0_{\text{LAB}} \tag{6}$$

or $\quad -\text{RT} \ln K_{\text{overall}} = -\text{RT} \ln K_{\text{IX}} - \text{RT} \ln K_{\text{LAB}}$

or $\quad K_{\text{overall}} = K_{\text{IX}} * K_{\text{LAB}} \tag{7}$

In general, for ion exchangers with chelating functionality, K_{LAB} is very high for most of the heavy-metal ions of interest due to their Lewis acid characteristics. Therefore, the overall equilibrium constants, according to Equation (7), are also very high. For sodium, Na^+, Lewis acid-base (LAB) interaction (step II) is practically absent and, hence,

$$K_{\text{overall}} = K_{\text{IX}} \tag{8}$$

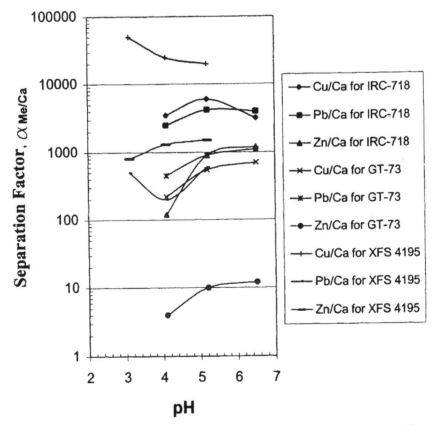

Figure 3.4 Experimentally determined Me(II)—Calcium separation factor values as a function of pH for various resins: IRC-718 = imino-diacetate functionality, GT-73 = thiol functionality and XFS 4195 = picolylamine functionality.

For Ca^{2+}, however, LAB interaction is present but is much weaker compared to most of the heavy metal cations, and thus as a general rule, the selectivity sequence for chelating exchangers may be written as follows:

$$K_{overall} \text{ (heavy metal)} \gg K_{overall} \text{ (calcium)} \gg K_{overall} \text{ (sodium)}$$

Such a high metal-ion selectivity and more stringent environmental regulations have aroused high interest in the application of these chelating polymers for removal, separation and purification of metal ions from heavy-metal contaminated water and wastewater streams [16–24].

Salient Properties of Chelating Exchangers

Many chelating exchangers have been synthesized during the last 25 years [25–27]. Table 3.2 provides the composition of several commercially available chelating exchangers and other relevant information. These exchangers, as indicated before, contain chelating functionalities with one or more donor atoms that can form coordinate bonds with metal ions. Depending on the number of donor atoms present in a repeating functionality, these exchangers are often referred to as mono-, bi- or polydentate. Figure 3.4 includes Me(II)/Ca separation factors for some of these resins at varying pH, and high metal-ion selectivity values are readily observed [15]. Obviously, for a particular metal ion, such high selectivity values depend on the composition of the chelating exchange, viz., polymer matrix, cross-linking, charge density and, above all, the functional group. Notwithstanding the varying effects of different chelating exchangers on the metal-ion selectivity, there remain some underlying commonalties; the next section provides some generic properties of this group of polymeric exchangers.

TABLE 3.2. Background Information on Some Chelating Polymers.

Functionality	Electron Donor Atoms (Double Strikethrough)	Matrix, Cross-linker	Nature of Chelating Group
®—N with CH₂COO⁻ / CH₂COO⁻	1 Nitrogen and 2 Oxygen atoms	Polystyrene, divinylbenzene	Imino-diacetate
®—COO⁻	1 Oxygen atom	Polymethacrylate	Carboxylate
®—CH₂—NH—CH₂—P(=O)(O⁻)(O⁻)	1 Nitrogen and 2 Oxygen atoms	Polystyrene, divinylbenzene	Aminophosphonate
pyridine—CH₂—N(®)—CH₂—pyridine	3 Nitrogen atoms	Polystyrene, divinylbenzene	Picolylamine-based
pyridine—CH₂—N(®)—CH₂—CHOH—CH₃	2 Nitrogen atoms	Polystyrene, divinylbenzene	Pyridine-based
®—S⁻	1 Sulfur atom	Polystyrene, divinylbenzene	Thiol

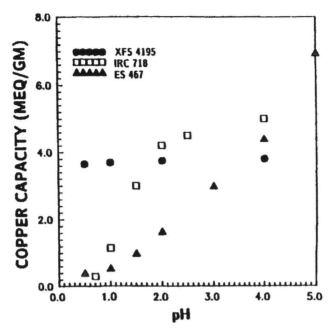

Figure 3.5 Comparison of copper uptake as a function of pH for three chelating exchangers in equilibrium with 300 mg/L Cu^{2+}. Data taken from Melling and West, Reference [28].

Characteristics of the Functional Groups

In spite of wide variation in the composition of chelating functionalities, nitrogen, oxygen and sulfur are the donor atoms in almost every chelating exchanger synthesized to date. Identifying the active donor atoms for a given application may provide useful clues to assess metal-ion selectivity and other related properties for a given chelating polymer. These donor atoms form only a part of the complete chelating functionality, which is essentially either weak-acid (say carboxylate, diacetate, thiol, etc.) or weak-base (tertiary amine, pyridine, etc.). Due to the weak-acid or weak-base characteristic, these chelating functionalities exhibit high affinity toward the hydrogen ion. As a result, selective uptake of heavy metal cations by chelating exchangers under highly acidic conditions (pH < 2.0) is adversely affected due to strong competition from H^+. On the other hand, at neutral to alkaline pH, heavy metal cations are quite insoluble because of low solubility product values for their hydroxides, carbonates, sulfides, etc. For effective heavy metals removal, the optimum pH range for most of the chelating polymers is, thus, often limited to 2.0–7.0. Figure 3.5 shows total copper uptake for different chelating exchangers as a function of pH [28]. Notice how metal removal capacity is essentially lost for IRC-718 (imino-diacetate functionality) and ES-467 (amino-phosphonate functionality) at pH < 1.5. Notice also the unusual behavior of XFS 4195; there is practically no reduction in copper uptake capacity even at a pH as low as 1.5. This is due to its unique metal-ion-binding mechanism that will be discussed later.

Solid chelating exchangers can be titrated in a manner similar to other aqueous-phase weak acids and weak bases. For example, strong-acid cation exchangers show an abrupt change in pH, whereas weak-acid cation exchangers show gradual increase in pH—indicative of the latter's high affinity for H^+—as may be seen from Figure 3.6. Titration curves

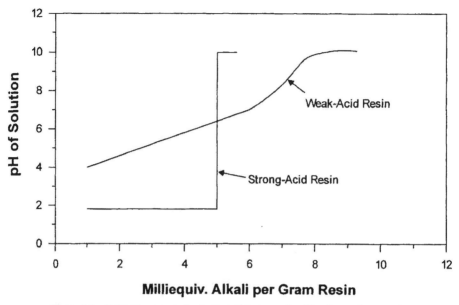

Figure 3.6 pH Titration curve of a strong-acid and a weak-acid cation-exchange resin.

are, however, dependent on the aqueous-phase electrolyte concentration; see discussions of this subject by Marinsky [29] and Helfferich [30]. A relationship used to compute the apparent dissociation constant, K_a, of a weak-acid exchanger as a function of pH and aqueous-phase electrolyte concentration has been proposed by Helfferich [30] and may be stated as follows:

$$pK_a = pH - \log[A^-] + \log\left[\frac{\bar{X}}{2}\right] \tag{9}$$

where $[A^-]$ is the aqueous-phase anion concentration and $[\bar{X}] = [\overline{RH}] + [\overline{R}]$ is the total concentration of (dissociated and undissociated) ionogenic groups in the exchanger. Note that the apparent pK_a decreases at a constant pH with an increase in electrolyte concentration (i.e., A^-), implying enhanced ionization of the exchanger.

Chelating exchangers' high preference for H^+ is often viewed as a shortcoming for heavy metal removal under highly acidic conditions, but it offers an excellent regeneration of metal-loaded chelating polymers with moderately concentrated (2–10%) mineral acid. From a practical viewpoint, high regeneration efficiency of a chelating exchanger is just as desirable as high metal-ion affinity. Figure 3.7 shows the regeneration of copper-loaded IRC-718 (imino-diacetate functionality) with 2% HCl [31]. As expected, copper desorption/elution was very sharp, with copper concentration in the spent regenerant as high as 15,000 mg/L, indicating efficient regeneration in accordance with the principles of displacement chromatography [32]. Incidentally, several toxic metals, viz., Cu, Pb, Hg, Cd, Zn and Ni, are included in EPA's list of priority pollutants [33], and one of the major challenges is to minimize the volume of these metal-contaminated wastes. High metal-ion selectivity of the chelating exchangers accompanied by excellent acid-regeneration efficiency offers opportunities to concentrate and reduce the volume of metal-laden dilute wastewater streams, often by over 1,000 times [34].

Nature of Metal-Ion Binding and Selectivity Sequence

In general, metal(II)-ligand complexes in the aqueous phase have regular or slightly distorted tetrahedral, octahedral or square pyramid structures depending on the metal ion's coordination number that, in turn, is related to the number of nonbonding electrons in its d-orbital [13,35]. It is true that in a chelating polymer with multiple binding sites, the metal ions try to reproduce their aqueous-phase stereochemistry. However, the functionality in a chelating polymer is often rigidly bound to a repeating monomer (e.g., styrene) that, again, is fixed as part of a three-dimensional network cross-linked through divinylbenzene. As a result, the donor atoms (N, O or S) in the polymer phase will experience considerable strain to orient themselves spatially around the receptor metal ions. This strain, which may be viewed as an extra thermodynamic parameter, may not allow the individual functionality in the polymer phase to reproduce its aqueous-phase metal-ligand configuration.

Experimental results are available for more widely used chelating exchangers with imino-diacetate, amino-phosphonate and carboxylate functionalities. In this context, the general consensus is that a metal(II) ion can bind at most with one nitrogen and two oxygen atoms for imino-diacetate or amino-phosphonate exchangers and with two oxygen atoms from two neighboring carboxylate groups for carboxylate-type exchangers [36,37]. Figure 3.8 shows a general schematic for these binding mechanisms where imino-diacetate and amino-phosphonate exchangers act as bi-dentate ligands for a metal(II) ion with a coordination number of 4 or 6. Also note that in all three cases, the 2+ charge of the metal ion is neutralized by the fixed negative charge in the polymer phase. This is the reason why these binding mechanisms are viewed as cation exchange accompanied by chelation, and the anions in the aqueous phase are excluded from the polymeric exchangers due to Donnan

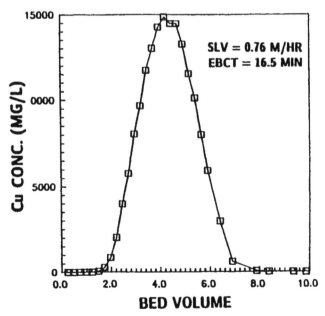

Figure 3.7 Demonstration of high regeneration efficiency of copper-loaded IRC-718 (imino-diacetate functionality) with 2% HCl. Reprinted from *Reactive Polymers*, 13, 241–253, Zhu, Millan, and Sengupta. Toward separation of toxic metal cations by chelating polymers: some noteworthy observations, 1990, with permission from Elsevier Science.

Functionality	Formula	Donor Atoms
Aminophosphonate	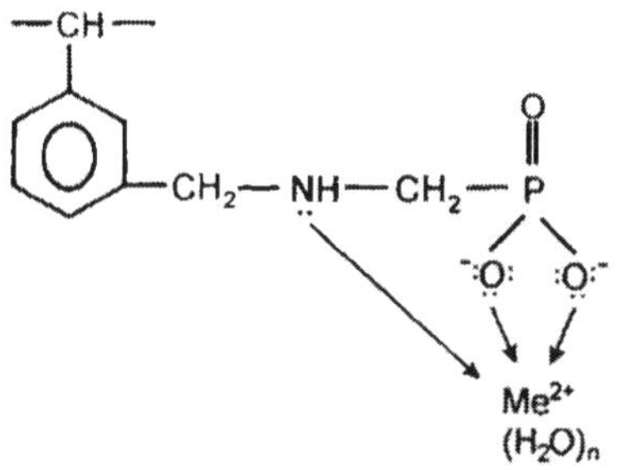	1 Nitrogen and 2 Oxygen
Iminodiacetate	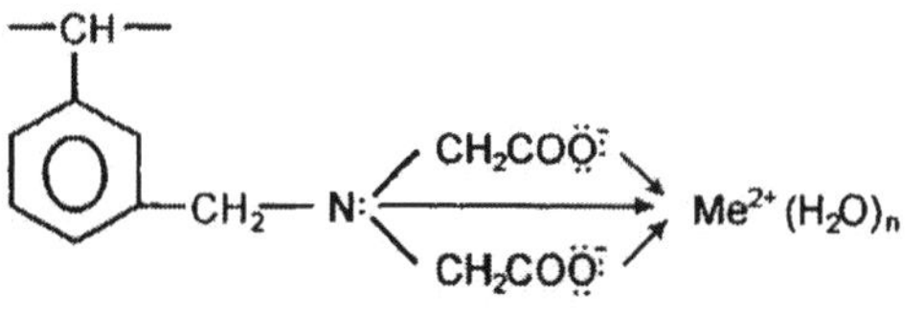	1 Nitrogen and 2 Oxygen
Carboxylate	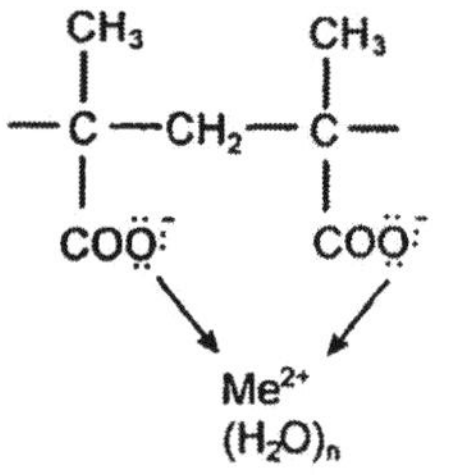	2 Oxygen

Figure 3.8 A schematic representation of the nature of metal-ion-binding (cation exchange followed by chelation) for aminophosphonate, imino-diacetate and carboxylate functional groups.

co-ion exclusion effect [30]. Such a model (cation exchange followed by chelation) can quantitatively explain the equilibrium behavior of these resins quite satisfactorily. For several chelating exchangers with bi- or polydentate functionality, however, individual donor atoms have been reported [27] to be binding metal ions independently on a molar basis. Such an independent-type binding, in some cases, may lead to quite unintuitive results; one such case for chelating exchangers with multiple donor atoms will be presented later.

Close proximity of the functional groups in a chelating polymer and their weak-acid or weak-base characteristics often lend themselves to special interactions through hydrogen bonding and/or Coulombic effects. For carboxylate functionality, Figure 3.9 shows open and closed forms of hydrogen bonding among neighboring functional-group atoms [15,38]. Such interaction would obviously enhance the exchanger's affinity toward H^+ and cause a reduction in metal-ion uptake, particularly at acidic pH. Reference [39] postulates that for imino-diacetate functionality, one of the two acetate groups in every functionality cannot be protonated even at pH as low as 1.0 because of its strong ion-pair formation with the neighboring positively charged, protonated nitrogen site. Figure 3.10 shows a schematic representation of the proposed interaction. One conspicuous corollary of such a postulate is

R — C
O
O
H
O
R — C
OH
O - - - - H — O
R — C
C — R
O — H - - - - O

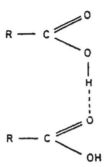

Figure 3.9 Open- and closed-form interaction between neighboring carboxylate functional groups through hydrogen bonding.

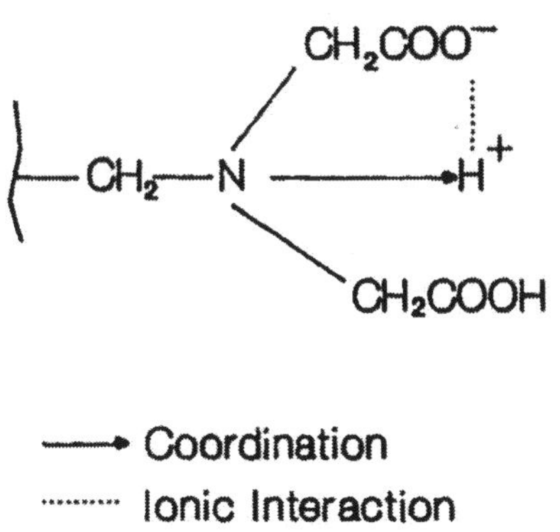

CH₂COO⁻
CH₂—N
H⁺
CH₂COOH
Coordination
Ionic Interaction

Figure 3.10 Proposed interaction within an imino-diacetate functional group at an acidic pH showing absence of

that, in spite of having tertiary amine functionality, the exchanger will have virtually no anion exchange capacity. Experiments carried out under highly acidic conditions with IRC-718 (imino-diacetate functionality) have confirmed this prediction, i.e., anion exchange capacity was practically absent.

Notwithstanding the complexities of metal-ion-binding mechanisms and interaction among neighboring functional groups, the relative selectivity values of various heavy-metal cations for a given chelating exchanger can often be qualitatively predicted based on their aqueous-phase stability constants with a chemically similar ligand. Consider the following metal-ligand reaction in the aqueous phase:

$$Me^{2+} + 2CH_3COO^- \leftrightarrow Me(CH_3COO)_2 \tag{10}$$

Relative values of stability constants in Equation (10) for various metal(II) cations are in the following order [40,41]:

$$Pb^{2+} > Cu^{2+} > Cd^{2+} > Zn^{2+} > Ni^{2+}$$

Metal-ion binding by chelating exchangers with carboxylate functionality (say DP-1) involves similar metal-ligand interaction in the polymer phase. At pH 6.5, IRC DP-1 is mostly deprotonated. It has been observed [15] that at pH 6.5, for example, Pb^{2+} is the most preferred ion followed by Cu^{2+}, Cd^{2+}, Zn^{2+} and Ni^{2+}. Quantitative prediction of the exchanger's metal-ion selectivity based on the corresponding aqueous phase metal ligand stability constants is often inaccurate. Such a simplified model does not account for other associated effects caused by hydrophobicity of the polymer matrix, steric property of the functional group, cross-linking, hydrated ionic radii of the counter-ions, nature of metal-ion binding and extent of hydrogen bonding, which again, is dependent on pH. In many cases, however, a model based on a linear free-energy relationship (LFER) can provide acceptable correlation for metal-ion selectivity at near neutral pH for exchangers with chelating functionality and the aqueous-phase metal-ligand equilibrium constant. For a given metal-ligand reaction, according to LFER, the aqueous-phase free energy at equilibrium is linearly related to the free-energy change in the polymer phase, i.e.,

$$\Delta G_L^0 = m\Delta G_R^0 \tag{11}$$

where subscripts L and R represent the liquid (water) and the resin phase, respectively, and m is the proportionality constant. An extension of the relationship in Equation (11) provides the following:

$$\log K_L = m * \log K_R + C \tag{12}$$

where K_L and K_R are the equilibrium constants for the reaction in the liquid and resin phase, respectively, and when plotted (K_L vs. K_R) in a log-log scale, should result in a straight line with a slope equal to m.

To test the predictive ability of Equation (12) in regard to metal(II) ion selectivity values for IRC DP-1 (carboxylate functionality), K_L or aqueous-phase stability constants for various metal ions in accordance with Equation (10) were obtained from Morel [40]. K_R can be replaced with experimentally determined Me^{2+}/Ca^{2+} separation factor values at pH 6.5 from Reference [15] without losing the generality of Equation (12). Figure 3.11 shows a

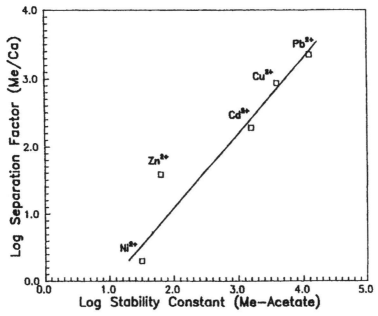

Figure 3.11 Relationship between experimentally determined metal/calcium separation factor for IRC DP-1 and aqueous-phase metal-acetate stability constant values.

plot of $\log K_L$ vs. $\log K_R$; the linearity of the plot provides a good starting point for estimating the selectivity of a given metal(II) ion toward IRC DP-1. At lower pH, however, such a simplified model should be used with caution and may need to be corrected for competing effects of H^+, ionic interactions and possible hydrogen bonding.

Kinetics and Rate Processes

In spite of their high affinity toward most of the heavy-metal cations and favorable thermodynamics over a wide pH range, the metal-selective chelating exchangers are often criticized because of their slow kinetics that are primarily diffusion-controlled rate-limiting processes [42–44]. Several factors influence kinetics, including degree of cross-linking and accompanying rigidity, steric property of the functional group, charge density, nature of polymer matrix, e.g., gel, macroporous or pellicular, mechanism of metal-ion binding, particle size and other hydrodynamic conditions. Although no two cases are likely to be identical, ion exchange kinetics for chelating polymers is improved by decreasing the amount of cross-linking, by decreasing the hydrophobic nature of the polymer matrix, and above all, by reducing the size of the polymer beads. Metal-ion diffusivity in the exchanger is strongly dependent on the cross-linking because it influences the swelling characteristic and water content of the polymer phase. Boyd and Soldanho [45] determined diffusivity of zinc ion for strong-acid cation exchangers with varying degrees of divinylbenzene (DVB) cross-linking, as reproduced in Table 3.3. Note that effective zinc-ion diffusivity drops by almost two orders of magnitude as the DVB increases from 5% to 23.8%.

In general, metal ion exchange by a chelating polymer is intra-particle diffusion controlled (ipdc) accompanied by chemical reaction (chelation) [42–49]. Both Fick's law and Nernst-Planck Equation (46) have been used to describe ion exchange kinetics by intra-par-

TABLE 3.3. Effective Diffusion Coefficients for Zinc Ion in
Sulfonic Acid Cation Resins of Different Cross-linking
(after Boyd and Soldanho, Reference [45]).

Crosslinking (% DVB̄)	D_{Zn} (cm^2/sec)
5	3.7×10^{-7}
10	3.8×10^{-8}
16.2	8.1×10^{-9}
23.8	3.2×10^{-9}

ticle diffusion. For ion exchange beads with spherical geometry and constant
exchanger-phase diffusion coefficient, the mathematical representation may be given as
follows:

$$\frac{\partial \bar{C}_i}{\partial t} = \bar{D}\left(\frac{\partial^2 \bar{C}_i}{\partial r^2} + \frac{2}{r}\frac{\partial \bar{C}_i}{\partial r}\right) \tag{13}$$

where r = radial space coordinate, i.e., distance from bead center.

This equation is to be solved under the appropriate initial and boundary conditions. For
the simple case of infinite solution volume (ISV), where the aqueous-phase concentration
does not change due to sorption or desorption, fractional attainment of equilibrium, F, is
given by

$$F \approx \left[1 - \exp\left(-\frac{\bar{D}t\pi^2}{r_0^2}\right)\right]^{1/2} \tag{14}$$

where $\bar{D}$ is intra-particle self-diffusion coefficient of the exchanging ion, t is the time since
the beginning of run and r_0 is the radius of the spherical ion exchange bead. The half-time,
$t_{1/2}$, for attainment of 50% equilibrium capacity (i.e., $F = 0.5$) can be computed from Equation (14) to give the following relationship:

$$t_{1/2} = \text{constant}\,\frac{r_0^2}{\bar{D}} \tag{15}$$

The relative rate (inverse of $t_{1/2}$) is thus proportional to the diffusion coefficient in the
exchanger and inversely proportional to the square of the bead radius. Note that $t_{1/2}$, according to ipdc, is independent of any concentration term.

The above rate expression assumes homogeneous exchanger phase where transport of
ions takes place only in the solid phase. Such an assumption is valid for a gel-type resin but
is truly debatable for macroporous exchangers. Figure 3.12 provides a clearer distinction
between the morphology of a gel and a macroporous resin. Note that for a macroporous
exchanger, solid phase and macropores (on the order of 500 Å) coexist in the interior of the
resin and, therefore, parallel diffusion of counter-ions in the solid phase and also in the
macropore is a strong possibility. Effective diffusivity in such cases tends to be concentration dependent [48,50,51]. Streat and his coworkers [50] studied this aspect with respect to
Ni^{2+}-Na^+ exchange for a macroporous chelating exchanger (ES-467, Duolite) with
aminophosphonate functionality. Some of the salient findings of the study are as follows:

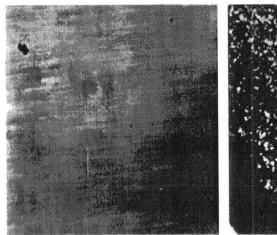 

Gel Type Ion Exchanger   **Macroporous Ion Exchanger**

Figure 3.12 Transmission electron microphotograph (TEM) of gel and macroporous ion exchangers ($\times$ 50,000 magnification) obtained through the courtesy of R. L. Albright.

- Experiments carried out at different temperatures (5–55°C) and different stirring speeds (200–3000 rpm) clearly indicated that the kinetics is intra-particle diffusion controlled. This was further confirmed by plotting $t_{1/2}$ versus ion exchanger bead radius, r_0; $t_{1/2}$ varied with r_0^2 in agreement with Equation (15).
- Aqueous-phase Ni concentration, however, was found to have a pronounced effect on ion exchange kinetics, which is not consistent with the prediction from an intra-particle diffusion controlled model based on the Nernst-Planck equation. Experimental results could, however, be accounted for with the aid of a macroporous model, originally developed by Yoshida and Kataoka [51] based on the assumption of parallel diffusion of counter-ions in the solid gel phase and in the macropores. This model predicts the concentration dependence of the effective self-diffusion coefficient, D_e, of counter-ion "i" according to the following equation:

$$D_e = \frac{D_p \varepsilon_p + D_g (1 - \varepsilon_p) Q / Z C_0}{\varepsilon_p + (1 - \varepsilon_p) Q / Z C_0} \tag{16}$$

where

D_p = diffusivity in the pore space
D_g = diffusivity in the gel phase
ε_p = void fraction of pores
Q = total exchange capacity per unit volume
Z = charge of counter-ion "i"
C_0 = concentration of "i" in the aqueous phase

An increased counter-ion concentration in the aqueous phase has a favorable effect on kinetics for macroporous resins, i.e., $t_{1/2}$ decreases with an increase in aqueous-phase nickel concentration. This feature helps macroporous resins during regeneration because the concentration of the regenerant (acid or alkali or salt) is usually high (2–10%).

For gel-type exchangers, macropores are absent, and intra-particle diffusion through the gel phase is the only mechanism for transport of the counter-ions. Hoell [44] provided an optical verification to this effect for a gel resin with carboxylate functionality (IRC-84, Rohm and Haas Co., Philadelphia, PA, USA) using Ca^{2+}-H^+ exchange. The exchanger was originally loaded with calcium ions and then placed in an acid solution; Ca^{2+} ions from the gel phase were slowly eluted by hydrogen ions, progressing from the circumference toward the center. With time, the calcium-loaded core gradually shrank and eventually disappeared. Figure 3.13 provides an excellent photographic testimony of this hypothesis [44,52].

Helfferich [53] discussed four scenarios of ion exchange kinetics and highlighted how

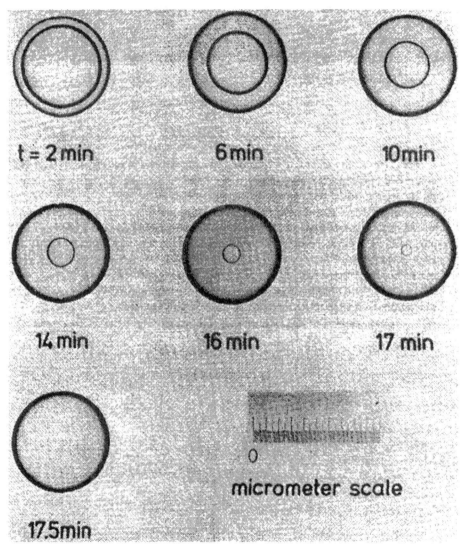

Figure 3.13 Development of the Ca^{2+}-H^+ exchange for Amberlite IRC-84 (weak-acid catio-exchange) resin in a 1 M HNO_3 solution. Reprinted from *Reactive Polymers*, 2, 93–101, Hoell. Optical verification of ion exchange mechanisms in weak electrolyte resins, 1989, with permission from Elsevier Science.

the premises of some mathematical models are incompatible with facets of physical reality. The specific cases considered were as follows:

(1) Nonlinear concentration gradients in Nernst-Planck film model
(2) Shell-core model in the absence of a mechanism that can produce shell-core behavior
(3) Reaction-controlled shell-core model
(4) Model for macroporous beads where diffusion in both the gel-phase and the macropores is comparable

The guidelines proposed in this communication [53] may serve as a basis to verify the applicability of any mathematical model under a given operating condition.

Metal-Ion Affinity in the Presence of Ligands

Although removal of heavy metals by chelating exchangers is a selective process due to the exchanger's high affinity toward these metal cations, specificity of the removal process is often adversely affected in the presence of ligands in the liquid phase. Such a reduction in metal affinity is caused by parallel, competing metal-ligand interactions that may be schematically represented as follows:

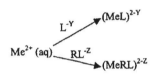

where L is the ligand (organic or inorganic) present in the aqueous phase, while RL is the solid, chelating polymer. Reduction in metal uptake for a given ion exchanger under such conditions is dependent on the following:

* concentration of ligand in the aqueous phase
* metal-ligand stability constants
* pH when the ligand has weak-acid or weak-base characteristics

The presence of inorganic (chloride, sulfate, ammonia, etc.) and organic (molecules with carboxylate, acetate, pyridine and amino groups, naturally occurring humates and fulvates) ligands is fairly common in many wastewater streams, and this may be viewed as an obstacle against selective removal of metal cations by ion exchange processes.

Dorner [54] performed soil-column tests to demonstrate how the presence of chloride ions causes early breakthrough of Cu(II), Ni(II) and Cd(II), as shown in Figure 3.14. Soil used in the column was Hanford sandy loam (typical Xerothent) with a cation exchange capacity of 0.05 meq/gm. The salient points of Dorner's study are as follows:

(1) For all the soil-column runs, competing Na^+ concentration was 0.5 M, while the metal(II) concentration in the influent was 10 mg/L.
(2) All other conditions remaining unchanged, as the conjugate anion was changed from ClO_4^- to Cl^-, the soil column was exhausted earlier, i.e., there was a significant reduction in the metal removal capacity in the soil column. Pore volume is a measure of the volume of water fed to the column, and "I" denotes the ionic strength.

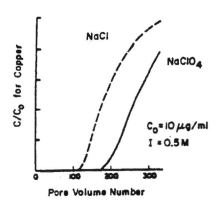

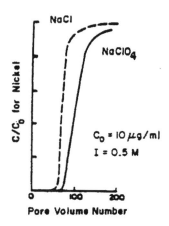

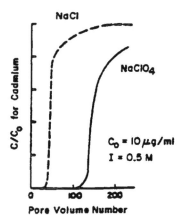

Figure 3.14 Relative effects of NaCl and NaClO4 on the mobility of metal cations in soil columns. Reprinted from *Soil Sci. Soc. Am. J.*, 42, 882–885, Donner. Chloride as a factor in mobilities of Ni(II), Cu(II), and Cd(II) in soil 1978, with permission from Soil Science Society of America.

(3) Perchlorate ion, ClO_4^-, is an inert ligand and does not form any complex with metal ions, while Cl^- forms complexes as shown below:

$$Me^{2+} + nCl^- \leftrightarrow MeCl^{2-n} \tag{17}$$

where n is the number of Cl^- ions complexed with the metal(II) ion. Cumulative stability constants for the complexation reactions are provided in Table 3.4 [41].

(4) Note that the relative decrease in the number of pore volumes treated in the presence of Cl^- is the greatest for cadmium and lowest for nickel, and this is in agreement with the descending order of first stability constants, i.e., highest for Cd(II) followed by Cu(II) followed by Ni(II).

Ion exchange capacity of the soil (Hanford sandy loam) is contributed by inorganic functional groups (alumino silicates), and the nature of interaction between heavy metal cations and soil is primarily Coulombic. As a result, the metal-ion affinity is relatively low, and the

TABLE 3.4. Cumulative Stability Constants for Cl⁻ Complexes to Ni(II), Cu(II) and Cd(II).

Metal	Cumulative Stability Constants, log			
	β_1	β_2	β_3	β_4
Ni(II)	−0.43	0.53	0.71	0.16
Cu(II)	0.21	−0.18	−1.96	−5.91
Cd(II)	2.00	2.70	2.11	1.7

presence of anionic ligands like Cl⁻ significantly reduces the metal removal capacity of the soil. Quantitative relationship to this effect—including the effects of total ligand concentration and metal-ligand stability constant—has been provided by SenGupta [55].

The metal-ion affinity for chelating exchangers is, however, far greater than the soil and the competing effects of weak inorganic ligands like Cl⁻ and SO_4^{2-}, even at extremely high concentration. These synthetic exchangers have been used successfully to concentrate various metal ions from seawater because they can break the chloro-complexes of the metal ions in the aqueous phase and selectively sorb the free metal ions, as shown below for a chelating exchanger in sodium form:

$$MeCl^+ + \overline{2RNa} \Leftrightarrow \overline{R_2Me} + 2Na^+ + Cl^- \qquad (18)$$

In the presence of polydentate organic ligands, common in many wastewater streams, the metal-removal capacity of chelating exchangers can be seriously impaired. An experimental study [56] was undertaken to investigate this aspect, using ethylenediamine, EN, a bidentate, weakly basic ligand with high affinity toward Me(II) cations. Figure 3.15 in-

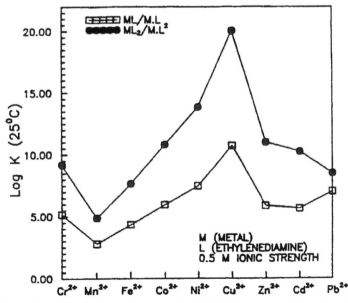

Figure 3.15 First and second stability constant values for Me(II)-EN complexes. Data taken from Smith and Martell, Reference [41].

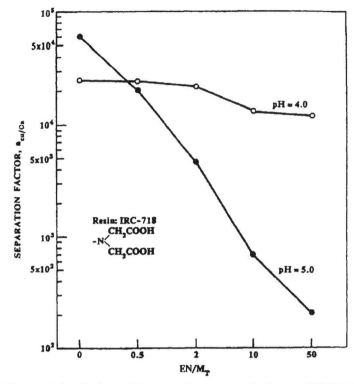

Figure 3.16 Effects of ethylenediamine and pH on copper/calcium separation factor for IRC-718 (imino-diacetate functionality).

cludes various Me(II)-EN stability constants [41], and it can be seen that they are, in general, orders of magnitude greater than those of chloride ions. Equilibrium tests were carried out at a constant pH (4.0 and 5.0) for a chelating exchanger with iminodiacetate functionality (IRC-718) at Cu(II):EN (molar), where Ca(II) is the competing cation.

Figure 3.16 shows how the Cu/Ca separation factor is influenced by EN/M_T ratio and pH, where M_T corresponds to concentration of the heavy metal cation. Note that at a pH of 4.0, an increase in EN/M_T had a negligible effect on α Cu/Ca, but at a pH of 5.0, the drop in (Cu/Ca was drastic. Quantitative prediction of such effects is provided in Reference [56]. Very simply, the governing factor here is complexation of EN with H^+; at a pH of 4.0, EN is highly protonated and does not like to complex with Cu(II). This example helps illustrate the complexity of a real system where multiple ligands, with varying acid-dissociation constant values, are present simultaneously. In a later section, how this concept of multiple-ligand interaction can be harnessed to remove heavy metal cations from a solid phase by artificially introducing an aqueous-phase ligand in the solid matrix will be presented.

DEVELOPMENTS IN OVERCOMING SHORTCOMINGS OF CONVENTIONAL ION EXCHANGE PROCESSES

Particle Size Effects and Recoflo® Short-Bed Process

Like other heterogeneous processes using small (<1 mm) particles, various reactor configurations have evolved for ion exchange, viz., packed-bed, continuous counter-current, fluidized bed, contact followed by sedimentation, etc. However, the packed-bed or fixed-bed process where the mobile liquid passes through stationary ion exchange beads in a column is by far the most popular unit operation due to its simplicity of construction and operation. This method is routinely used for metal-ion removal, water softening and water demineralization. The following steps are followed in a fixed-bed process:

- exhaustion or removal of the contaminant
- backdating
- in-situ regeneration
- rinsing
- return to exhaustion

Timewise, "exhaustion" is much longer than all the other steps combined and is practically equal to cycle time, t_{cycle}. Particle diameter in a packed bed influences pressure drop and kinetics of ion exchange. For rigid, spherical particles like ion exchange beads, the pressure drop under laminar conditions can be estimated from the following equation [57]:

$$\Delta P = \frac{\mu v L}{K_p d_p^2} \tag{19}$$

where K_p is the permeability, μ is the viscosity of the liquid stream, L is the length of the packed bed and v is the superficial liquid-phase velocity. Note that the pressure drop is inversely proportional to d_p^2. This is why smaller diameter particles cannot be used in a packed bed. Assuming μ, v and K_p to remain constant, the pressure drop remains the same if L/d_p^2 is kept constant. If, in addition, L/t_{cycle} is kept constant, the capacity of two columns of different height will be the same even though the shorter column will exhaust faster [58].

Poor kinetics is one of the major limitations of metal-selective ion exchange resins, and as pointed out earlier, intra-particle diffusion is often the rate-limiting step. During the exhaustion cycle, there are three specific zones for a solute in a fixed bed, viz., saturated, unused and mass-transfer zone (MTZ), as shown in Figure 3.17. As the cycle advances, MTZ moves from the inlet toward the exit of the column. For a favorable isotherm [59,60] L_{MTZ} does not change during its passage inside the column but is dependent on the size of the adsorbent particle. For intra-particle diffusion, L_{MTZ} is proportional to d_p^2. Improving kinetics in a fixed-bed process essentially means reducing L_{MTZ}. Thus, hydrodynamic (high-pressure drop) and mass transfer (intra-particle diffusion) limitations in a fixed-bed process can be overcome by reducing the size of the polymer beads in accordance with the following guidelines:

a. If d_p is reduced in a way that L/d_p^2 remains constant, pressure drop across the bed does not change. In addition, if L/t_{cycle} is kept constant, the column will have the same capacity as any larger column.

b. With a reduction in particle size, d_p, mass transfer rate by intra-particle diffusion will be enhanced, and the length of the mass-transfer zone, L_{MTZ}, will be reduced so that L_{MTZ}/L remains constant for a favorable isotherm, i.e., the fractional utilization of the column per cycle is unchanged.

The above theoretical principles form the basis of "Recoflo® Short Bed IX Process" of ECO-TEC Limited, Ontario, Canada [61,62]. Recoflo® uses a much finer particle size

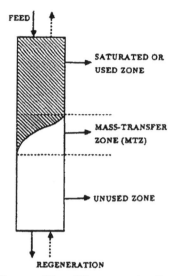

Figure 3.17 Presence of three different zones in a fixed-bed column.

(100–200 mesh) than normally used in industrial ion exchange processes (20–40 mesh). Depending on the application, total cycle time can vary from as low as two minutes to one hour. Consequently, the resin inventory for this process is low, and that helps reduce the capital cost. Upon exhaustion, the short beds are regenerated in a counter-current mode for maximum efficiency. The entire Recoflo® system is assembled in a compact, skid-mounted unit (Figure 3.18). Poor kinetics and the relatively high price of chelating exchangers tend to limit their use for metals recovery. The improved kinetics (due to smaller particle size) and short-bed height characteristic of the Recoflo® process greatly enhance the performance and economic viability of the chelating exchangers. They are now being used to recover metals from electroplating and hydrometallurgical wastestreams [63]. Grinstead [34] has also used this principle for removing heavy metal cations—particularly from dilute wastewater streams—and concentrating the heavy metals in the spent regenerant by over 1,000 times.

Figure 3.18 A skid-mounted short-bed Recoflo® unit. Source: Eco-Tec Ltd., Ontario, Canada.

Heavy Metal Removal under Highly Acidic Conditions with Specialty Chelating Polymers

One of the major shortcomings of most of the commonly used chelating exchangers is their drastic reduction in metal-ion uptake under highly acidic conditions (pH < 2.0), a situation not uncommon for many wastewater streams. Raising the pH by addition of a base is operationally easy but results in a substantial increase in operating expenses. Chelating functionalities are weakly acidic or weakly basic and, therefore, have a strong affinity toward H^+. Fierce competition from H^+ is the sole reason why metal-ion removal by these exchangers is not a viable process at pH < 2.0. SenGupta et al. [39] have shown, however, that chelating polymers with only nitrogen donor atoms perform at extremely low pH without any loss of efficiency, although they have weak-base functional groups (pyridine and tertiary amine). Figure 3.19 shows the chemical composition of two such exchangers. This truly unusual property of XFS 4195 makes it a strong candidate for metal-recovery application under highly acidic conditions.

Another interesting observation regarding XFS 4195 is the unusual effect of ionic strength on metal-ion uptake. Figure 3.20 [64] shows nickel uptake as a function of aqueous-phase nitrate concentration for batch tests at a constant pH of 3.0 and a constant aqueous-phase Ni^{2+} concentration of 2.0 ppm but varying concentrations from 50 to 1,000 mg/L (by addition of required amounts of sodium nitrate). Note that Ni^{2+} uptake increased with an increase in concentration, although both H^+ and Ni^{2+} concentrations in the aqueous phase remained constant. This observation suggests that electrolyte concentration or aqueous-phase ionic strength can be used as a process variable to improve the metal-ion uptake under highly acidic conditions. Other chelating exchangers with carboxylate, iminodiacetate and aminophosphonate functionalities do not exhibit such favorable effects of ionic-strength on metal-ion uptake; on the contrary, they show adverse effects from competition of the counter-ion (Na^+ in this case) [39].

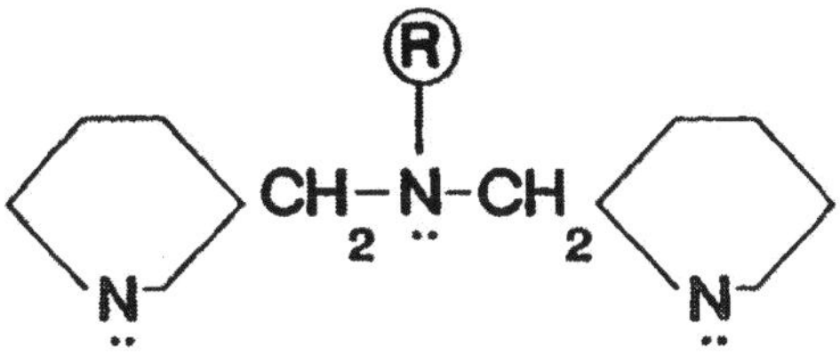

XFS 4195

(3 N-Donor Atoms per Functional Group)

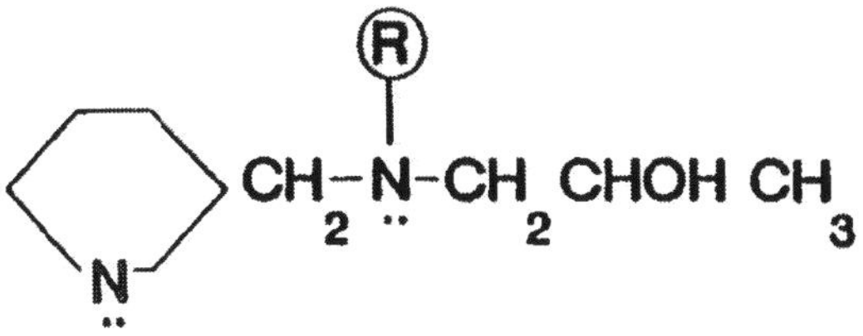

XFS 43084

(2 N-Donor Atoms per Functional Group)

Figure 3.19 Chemical composition of two chelating exchangers with multiple nitrogen donor atoms.

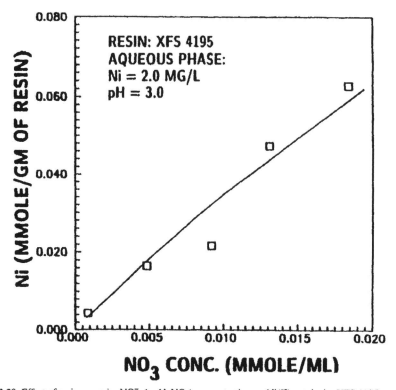

Figure 3.20 Effect of an increase in NO_3^- (as $NaNO_3$) concentration on Ni(II) uptake by XFS 4195 at a constant pH of 3.0 and constant aqueous-phase Ni(II) concentration. Reprinted with permission from Reference [39], SenGupta, Zhu, and Hauze. *Env. Sci. & Tech.*, 25(3), 481–488. Copyright 1991 American Chemical Society.

Because the metal-ion affinity of XFS 4195 is unusually high at very acidic pH, acid regeneration would understandably be inefficient, and this may be viewed as a major shortcoming against widespread use of these specialty chelating exchangers (with nitrogen donor atoms). This potential problem can be overcome easily by using ammonia or other non-ionized ligands (organic amines) as a regenerant. Figure 3.21 shows two regeneration profiles of copper-loaded XFS 4195 under identical hydrodynamic conditions. Note that regeneration with 2% HCl was essentially ineffective, while desorption with 2% NH_3 was highly favorable. The unusual properties of chelating exchangers with multiple nitrogen donor atoms, like XFS 4195, can be summarized as follows:

(1) They offer very high metal-ion affinity under highly acidic conditions.

(2) Metal-ion affinity is enhanced with an increase in aqueous-phase electrolyte concentration.

(3) The metal-ion-loaded exchangers can be regenerated very efficiently with ammonia or other non-ionized ligands.

The above-mentioned properties offer opportunities for new areas of application of chelating exchangers; areas where other specialty chelating exchangers are not suitable. The details of unique metal-binding mechanisms of these exchangers are available in Reference [39].

HEAVY METAL REMOVAL FROM A SOLID PHASE

Until now, our discussion was confined to using ion exchangers for removal of heavy metal cations from a pure aqueous phase, i.e., there were no suspended solids in the wastestream. Also, the ion exchange resins involved were granular or spherical. Lately, however, challenging environmental separation problems are being encountered, where the medium is essentially a slurry or sludge with high-suspended solids content. A vast new area of application would open if the morphology of ion-exchanging polymers can be modified to remove heavy metals from a solid-phase background.

A widespread environmental problem is the disposal of sludges or treatment of soil contaminated with a minor fraction (often less than 5%) of heavy metals in the solid phase in an otherwise innocuous background of materials that are not important from a regulatory viewpoint. This problem stems from the fact that the heavy metals present can cause the sludge/soil to be designated as "hazardous" waste, thus greatly increasing the cost of disposal. Selective and targeted removal of the heavy metals from the background solid phase would constitute an efficient treatment process, as it would render the sludge nonhazardous, and may make it possible for the heavy metals to be concentrated and recycled/reused. The physical configuration of conventional ion exchangers (spherical or granular) makes their use inappropriate for such a case. A material identified in References [65–71] shows the potential of being appropriate under such conditions. It is a new class of sorptive/desorptive composite ion-exchanging material (CIM) that is available commercially as thin sheets (approximately 0.5 mm thick) and that is suitable for heavy metal decontamination from sludges/slurries. The morphology of the material—along with its physical texture and tensile strength—makes it compatible for use with sludges/slurries. This property of the material, when combined with the adaptation of the chemistry of the reactor and/or the creation of an electric gradient-induced ion-transport process, makes it possible to selectively remove heavy metals from a solid-phase background composed primarily of nonregulated elements.

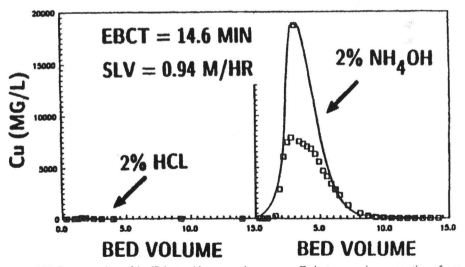

Figure 3.21 Demonstration of inefficient acid regeneration versus effecient ammonia regeneration of copper-loaded XFS 4195. Reprinted with permission from Reference [39], SenGupta, Zhu, and Hauze. *Env. Sci. & Tech.*, 25(3), 481–488. Copyright 1991 the American Chemical Society.

Characteristics of the Composite Ion-Exchanging Material (CIM)

The CIM is a thin sheet prepared by comminuting a cross-linked polymeric ion exchanger to a fine powder and fabricating it mechanically into a microporous composite sheet consisting of ion exchange particulate matter enmeshed in polytetrafluoroethylene (PTFE). During this mechanical process, the PTFE microspheres are converted into microfibers that separate and enmesh the particles [65]. When dry, these composite sheets consist of >80% particles (polymeric ion exchanger) and % PTFE by weight. They are porous (usually >40% voids, with pore size distributions that are uniformly below 0.5 μm). The ion exchange microspheres are usually <100 μm in diameter with total thickness ≈ 0.5 mm. As such, they are effective filters that remove suspensoids >0.5 μm from permeating fluids. Because of such sheet-like configuration, this material can be easily introduced into or withdrawn from a reactor with a high concentration of suspended solids, with the target solutes being adsorbed onto or desorbed from the microadsorbents. Figure 3.22 (top) shows the electron microphotograph of the composite IDA membrane, and the bottom figure provides a schematic depicting how the microbeads are trapped within the fibrous network of PTFE. Table 3.5 provides the salient properties of the composite IDA membrane used in the

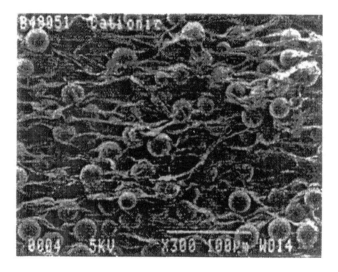

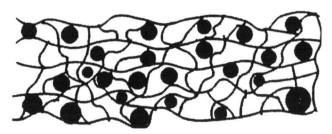

Figure 3.22 Scanning electron microphotograph (300×) of the composite ion exchange material (top) and schematic representation of the ion-exchanger beads present in a network of interlaced PTFE (bottom). Reprinted with permission from Reference [66], Sengupta and SenGupta. *Env. Sci. & Tech.*, 27(10), 2133–2140. Copyright 1993 American Chemical Society.

TABLE 3.5. Properties of the CIM.

Composition	90% Chelex Chelating Resin, 10% Teflon
Pore size (nominal)	0.4 µm
Nominal capacity	3.2 meq/gm dry membrane
Membrane thickness	0.4–0.6 mm
Ionic form (as supplied)	Sodium
Resin matrix	Styrene-divinylbenzene
Functional group	Imino-diacetate
pH stability	1–14
Temperature operating range	0–75°C
Chemical stability	Methanol, 1 N NaOH, 1 N H$_2$SO$_4$
Commercial availability	3M Corp., MN and Bio-Rad, CA

study. Note that the chelating microbeads constitute 90% of the composite membrane by mass. This feature allows the same level of performance to be achieved by the membrane as that of the parent chelating beads used in a fixed-bed operation. More details about characterization of the membrane are available in Reference [70]. It may be noted that this material differs fundamentally from traditional ion exchange membranes used in industrial processes like Donnan dialysis (DD) and electrodialysis (ED) because of its high porosity. DD and ED membranes have very low porosity and are strongly influenced by the "Donnan Coion Exclusion principle" [35], which means that anions are not permitted to pass through the cation exchange membrane and vice versa. However, in the case of a composite membrane, the large gaps between ion exchangers allow anions to pass through freely, even though it is a cation exchange membrane. The suspended solids that are >0.5 µm are not able to penetrate the skin of the membrane because of the pore size of the material, as explained above. However, water molecules and ions can easily move in and out of the thickness of the sheet, thus allowing for unimpeded ion exchange reactions between target ions in solutions (heavy metals in this case) and the counter-ions of the membrane, as shown schematically in Figure 3.23 and explained in detail by Sengupta [65–71]. After a design time interval, the membrane can be taken out and chemically regenerated with strong (3–5%) mineral acid solution.

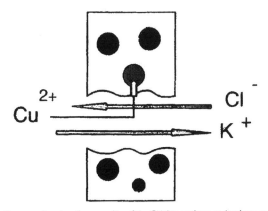

Figure 3.23 Schematic diagram showing the porosity of the CIM to cations and anions and selective entrapment of heavy metal cations. Reprinted from *Reactive Polymers*, 35, 111–134, Sengupta and SenGupta. Heavy-metal separation from sludge using chelating ion exchangers with nontraditional morphology, 1997, with permission from Elsevier Science.

Process Configuration

Figure 3.24 shows a conceptualized process schematic where a composite membrane strip is continuously run through the contaminated sludge (sorption step) and an acid bath (desorption step). Such a cyclic process configuration is relatively simple and can be implemented by using the composite membrane as a slow-moving belt. For a sludge containing heavy metal hydroxide, say $Me(OH)_2$, the process works in two steps:

(1) Sorption: the CIM, when in contact with the sludge, selectively removes dissolved heavy metals from the aqueous phase in preference to other nontoxic alkali and alkaline earth metal cations. Consequently, fresh heavy metal hydroxide dissolves to maintain equilibrium, and the following reactions occur in series:

$$\text{Dissolution: } Me(OH)_2(s) \Leftrightarrow Me^{2+} + 2OH^- \tag{20}$$

$$\text{CIM Uptake: } \overline{RN(CH_2COOH)_2} + Me^{2+} \Leftrightarrow \overline{RN(CH_2COO^-)_2 Me^{2+}} + 2H^+ \tag{21}$$

(2) Desorption: when the CIM is immersed in the acid chamber, the exchanger microbeads are efficiently regenerated according to the following reaction:

$$\overline{RN(CH_2COO^-)_2} + Me^{2+} + 2H^+ \Leftrightarrow \overline{RN(CH_2COOH)_2} + Me^{2+} \tag{22}$$

The regenerated CIM is then ready for sorption again, and the cycle is repeated. For such an arrangement to be practically feasible, the CIM sheet needs to be physically tough enough to withstand conveyor belt tension and resilient to the chemical forces that are created during sorption and regeneration. Previous studies conducted [68] in this area have confirmed that the CIM sheet is durable enough to withstand cyclical forces (physical and chemical) for more than 200 cycles.

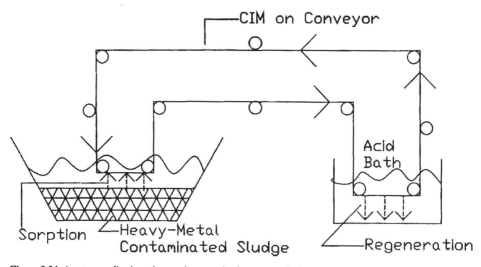

Figure 3.24 A conceptualized continuous decontamination process for heavy metal removal from a sludge reactor using CIM. Reprinted with permission from Reference [66], Sengupta and SenGupta. *Env. Sci. & Tech.*, 27(10), 2133–2140. Copyright 1993 American Chemical Society.

Cyclic Extraction of Heavy Metals Present in a
Background of High Buffer Capacity

In many situations regarding heavy metal contaminated soil/sludge, the background, nontoxic, nonregulated solid-phase compounds can interact chemically with the heavy metal dissolution. For example, if a heavy metal precipitate, $Me(OH)_2$, is present as a minor contaminant in a background of calcite soil, the mass of $CaCO_3$ (s) will be orders of magnitude higher than $Me(OH)_2$. Because of the high buffer capacity of $CaCO_3$, the pH of the system will be alkaline. This will result in the concentration of heavy metal in the aqueous phase being orders of magnitude lower than that of Ca^{2+}. Any physicochemical treatment process for such a scenario that primarily depends on the aqueous phase Me^{2+} concentration for separation from Ca^{2+} would be inefficient in this case. The CIM can be used in a novel scheme to achieve efficient heavy metal separation in such cases. The scheme entails introducing a selective, aqueous-phase ligand to the soil/slurry mixture. This ligand selectively complexes with the heavy metal cation and brings more of it from the solid phase to the aqueous phase, albeit as a heavy metal-ligand complex. After this, the chelating functional group of the CIM, which is more selective to the target heavy metal cation than the aqueous-phase ligand, breaks the aqueous-phase ligand-metal complex, and the heavy metal is transported to the ion exchange sites of the CIM.

Figure 3.25 shows the results of an experiment performed to simulate the above-discussed methodology. Cu(II) was the chosen heavy metal cation, oxalate is the aqueous-phase ligand and the functional group on the CIM is imino-diacetate. Cu(II) uptake isotherm is shown for two different oxalate concentrations. In both cases, there was no Cu(II) in the solid phase. To maintain this condition, the following procedure was adopted: the desired oxalate solution (0.005 M and 0.1 M) was prepared by adding $Na_2C_2O_4$ to DI water. The pH was adjusted to 9.0, and the maximum $\{Cu_t\}$ that can be supported for each oxalate solution was experimentally determined. This was done by adding CuO(s) in mass much higher than that obtained by theoretical calculations based on Cu complexes with OH^- and $C_2O_4^{2-}$. The system was allowed to attain equilibrium, after which a sample was taken from the reactor, filtered and analyzed for Cu(II). After determining the maximum $\{Cu_t\}$ for each case (8 mg/L for 0.005 M and 100 mg/L for 0.1 M), different amounts of CuO were added for each run after making sure that all the Cu added would remain in the aqueous phase. It can be observed that the uptake of Cu(II) by the CIM depends only on the total aqueous phase Cu(II) concentration; for $\{Cu_t\} < 8$ mg/L, concentration of Na^+ or $C_2O_4^{2-}$ in the system makes practically no difference in Cu uptake even though they are 20 times higher in one case than in the other. This is due to the extremely high affinity of the IDA functional group toward Cu as compared to that for Na^+, as has been discussed earlier.

For another experiment where CuO is always present in the solid phase, a mixture was prepared with 5% (w/v) sludge, fine sand (70%), $Na_2C_2O_4$ (21%), calcite (7.8%) and CuO (0.6%). The sludge pH was maintained at 9.0. CaC_2O_4 and $Cu(OH)_2$ were the controlling solid phases under these conditions. The added oxalate reacted with calcium to form calcium oxalate, a compound with low solubility. Precipitation of calcium oxalate reduced the aqueous-phase oxalate concentration to 0.05 M. After the formation of CaC_2O_4 (s), the cyclic process was started.

Figure 3.26 shows the recovery of Cu, Ca and C_2O_4 in the regenerant solution for the oxalate system. Although Cu is present primarily as $Cu-C_2O_4$ complex in the sludge phase, Cu recovery was significant and increased steadily with every cycle. Ca recovery was much lower and tended to approach an asymptotic concentration in the regenerant with an in-

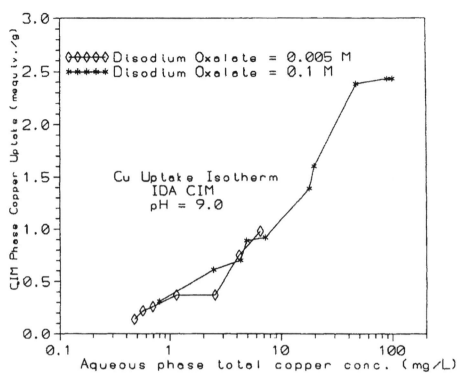

Figure 3.25 Copper(II) uptake isotherms for varying oxalate concentrations. Reprinted with permission from Reference [66], Sengupta and SenGupta. *Env. Sci. & Tech.*, 27(10), 2133–2140. Copyright 1993 American Chemical Society.

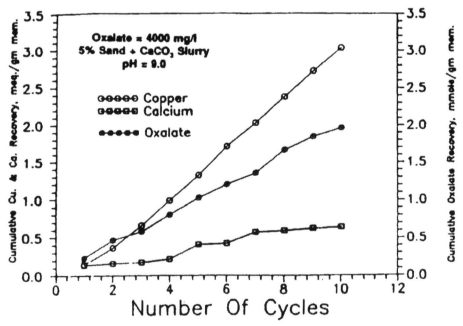

Figure 3.26 Cumulative copper, calcium and oxalate recovery with increase in number of cycles at alkaline pH. Reprinted with permission from Reference [66], Sengupta and SenGupta. *Env. Sci. & Tech.*, 27(10), 2133–2140. Copyright 1993 American Chemical Society.

crease in the number of cycles. Dissolved sludge phase concentrations of Cu and Ca remained fairly constant with the number of cycles (Figure 3.27), suggesting that they are controlled by solubility products of the solid phases. Heavy metal uptake by the CIM is substantial even when free Cu ion, Cu^{2+}, is practically absent, and most of the dissolved Cu exists primarily as anionic or neutral $Cu-C_2O_4$ complexes. Table 3.6 presents data about the computed percentage distribution of important Cu(II) species at three different oxalate concentrations based on stability constants data in the open literature [41]. Significant Cu uptake by the CIM under such conditions seems counterintuitive.

Also, Figure 3.26 shows a consistently increasing oxalate recovery in the regenerant. Oxalate exists primarily as the divalent anion $C_2O_4^{2-}$ at a pH of 9.0 and should be rejected by the chelating cation exchanger according to the Donnan exclusion principle. One possible explanation for this counterintuitive behavior could be that the suspended particles of the insoluble calcium oxalate and copper oxide were probably trapped in the large pores of the CIM and subsequently carried over to the regenerant phase, thus exhibiting significant copper and oxalate recovery. In order to eliminate such a possibility, another sorption isotherm was carried out in the absence of suspended solids at pH = 9.0 and total aqueous phase oxalate concentration of 4,000 mg/L but at varying dissolved copper concentrations. Figure 3.28 shows Cu and C_2O_4 uptakes under these experimental conditions free of suspended solids. Note that the C_2O_4 uptake by the CIM is significant, and Cu uptake increases with increased Cu concentration. These observations strongly suggest that Cu and C_2O_4 are removed from the solution (or sludge) phase primarily through sorption processes. The following binding mechanisms are identified as plausible for C_2O_4 and Cu(II) transport to the composite IDA membrane.

As already indicated, the neutral copper-oxalate complex $[Cu(C_2O_4)]^0$ was significantly

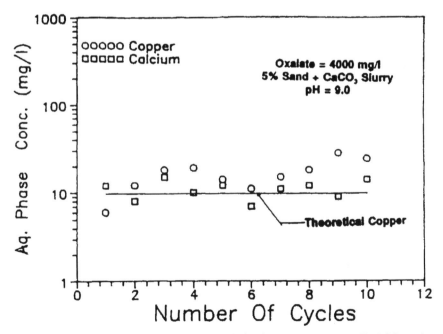

Figure 3.27 Dissolved copper and calcium concentration during the recovery process at pH = 9.0 for oxalate concentration of 4,000 mg/L. Reprinted with permission from Reference [66], Sengupta and SenGupta. *Env. Sci. & Tech.*, 27(10), 2133–2140. Copyright 1993 American Chemical Society.

TABLE 3.6. Percentage Distribution of Various Copper(II) Species.

Species	Oxalate = 0.005 M	Oxalate = 0.045 M	Oxalate = 0.1 M
Cu^{2+}	2.25 E – 04	4.37 E – 06	1.26 E – 06
$Cu(OH)^+$	2.25 E – 02	4.37 E – 04	1.26 E – 04
$Cu(OH)_2^0$	1.08 E + 00	0.02 E + 00	6.04 E – 03
$Cu(OH)_3^-$	2.25 E – 02	4.37 E – 04	1.26 E – 04
$Cu(C_2O_4)^0$	1.84 E + 00	0.26 E + 00	0.14 E + 00
$Cu(C_2O_4)_2^{2-}$	97.04 E + 00	99.72 E + 00	99.85 E + 00

present in the aqueous phase under experimental conditions, and only two of the four primary coordination numbers of Cu(II) are satisfied in this complex. Because these complexes are electrically neutral, they can permeate readily to the sorption sites containing nitrogen donor atoms, which can favorably satisfy the remaining coordination requirements of Cu(II). Figure 3.29 shows how electrically neutral 1:1 copper-oxalate complexes can be bound to the neighboring nitrogen donor atoms of the imino-diacetate moieties through formation of ternary complexes. This mode of sorption (i.e., formation of ternary complex through metal ligand interaction) is believed to be the primary pathway for sorption of oxalate onto the CIM and subsequent carryover into the regenerant as exhibited in Figure 3.26.

Although free Cu ions (Cu^{2+}) were practically absent under the experimental conditions of Figures 3.25 and 3.26, Cu uptake was quite significant. It is very likely that ligand substitution was a major mechanism by which Cu(II) was sorbed onto the chelating microbeads of the CIM for oxalate as an aqueous-phase ligand. Table 3.7 shows 1:1 and 1:2 metal-ligand stability constants of Cu(II) with oxalate and N-benzyl imino-diacetate (N-B-IDA), and the much higher ligand strength of IDA can be easily noted. Also, Cu(II) complexes are known

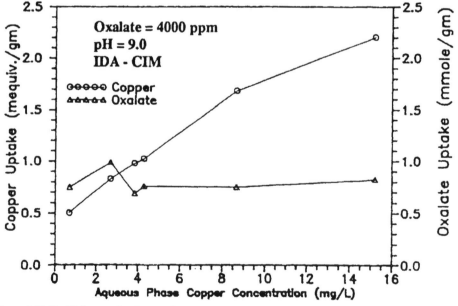

Figure 3.28 Equilibrium uptake of copper(II) and oxalate by the CIM at pH = 9.0. Reprinted with permission from Reference [66], Sengupta and SenGupta. *Env. Sci. & Tech.*, 27(10), 2133–2140. Copyright 1993 American Chemical Society.

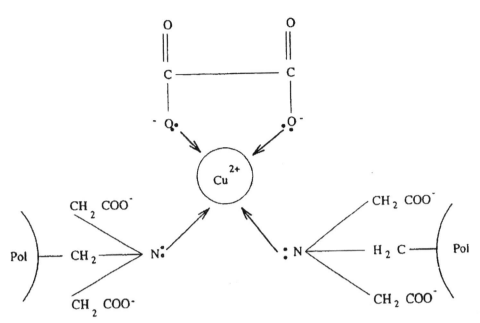

Figure 3.29 Sorption of neutral copper-oxalate complex onto chelating microbeads through formation of ternary complex. Reprinted with permission from Reference [66], Sengupta and SenGupta. *Env. Sci. & Tech.*, 27(10), 2133–2140. Copyright 1993 American Chemical Society.

to be labile. Therefore, the ligand substitution reaction, where oxalate in the aqueous phase is replaced by IDA in the exchanger phase, is favorable and results in an increased Cu(II) uptake as shown by the following reaction:

$$[Cu(C_2O_4)_2]^{2-} + \overline{R-N-(CH_2COO^-Na^+)_2}$$

$$\Leftrightarrow \overline{R-N-(CH_2COO^-)_2Cu^{2+}} + 2Na^+ + 2C_2O_4^{2-} \qquad (23)$$

Thus, ligand substitution and ternary-complex formation are the major mechanisms for heavy metal uptake in the case of oxalate as the aqueous-phase ligand.

From an application viewpoint, this concept is important because it proves scientifically that heavy-metal-laden sludges with high buffer capacity can be decontaminated even at alkaline pH in the presence of organic ligands without addition of acid to reduce pH; in fact, acid addition will not lower the pH because of high buffer capacity. The main concept employed here is to increase the aqueous phase concentration of the heavy metal with minimum increase of the concentration of the competing cation in the solid phase. The stability

TABLE 3.7. Stability Constants for Cu^{2+}-Ligand and Ca^{2+}-Ligand Complexes.*

Ligand	Metal: Ligand	Log K for Cu-Ligand	Log K for Ca-Ligand
N-B-IDA	1:1	10.62	3.17
N-B-IDA	1:2	15.64	—

*All values taken from Reference [41].

constant values of the two ligands (aqueous ligand added to the solid-phase ligand of the CIM functionality) with the metals of interest play an important role in the suitability of the process.

Another methodology that can be employed is the use of an ion exchanger that has very low affinity for Ca(II) compared to that for heavy metal, e.g., Cu(II). If one can identify a chelating functionality with affinity for heavy metal orders of magnitude higher than that for calcium, the need to add an aqueous-phase ligand would be obviated. In this regard, the earlier discussion regarding chelating polymers containing multiple nitrogen donor atoms is relevant. Chelating polymers such as XFS 4195 have very high affinity for strong Lewis acids and very low affinity for "hard" acids such as Ca(II). If such a functionality can be impregnated in the CIM, the process of cyclic extraction would be efficient.

SORPTION KINETICS OF THE CIM

Figure 3.30 shows almost identical uptake rates for Cu(II), Ni(II) and Pb(II) by the CIM, although they are not equally labile from a chemical reaction viewpoint. It is, therefore, likely that chemical reaction kinetics would not be the rate-limiting step. As discussed earlier, intra-particle diffusion is the most probable rate-limiting step with chelating ion exchangers. Also, because the PTFE fibers do not sorb any metal ions, solute transport by surface diffusion is practically absent. However, a significantly different physical configuration of the CIM may introduce additional diffusional resistance within it. As may be observed from the schematic diagram in Figure 3.31, fairly stagnant pore liquid is present in the channels of the CIM between individual microbeads, and the solutes need to be transported through this pore liquid for sorption or desorption. This additional resistance to sorption/desorption is likely to retard kinetic rate.

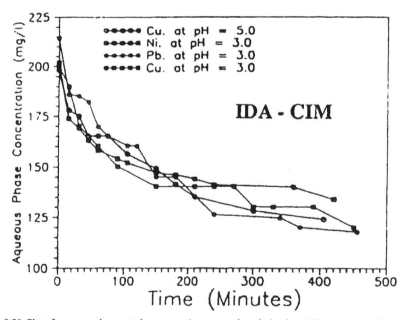

Figure 3.30 Plot of aqueous-phase metal concentration versus time during batch kinetic study of the CIM. Reprinted with permission from Reference [66], Sengupta and SenGupta. *Env. Sci. & Tech.*, 27(10), 2133–2140. Copyright 1993 American Chemical Society.

CIM

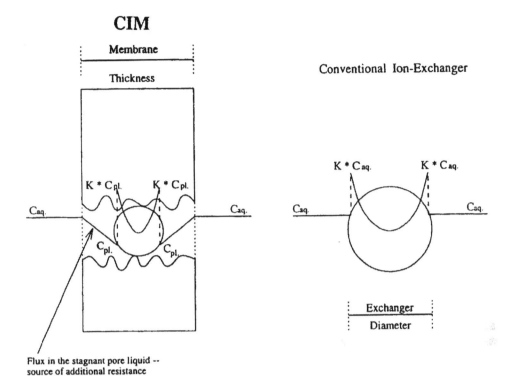

Flux in the stagnant pore liquid --
source of additional resistance

Figure 3.31 Schematic diagram showing the difference in flux curves between a conventional (spherical bead) ion-exchanger and the CIM. Reprinted with permission from Reference [66], Sengupta and SenGupta. *Env. Sci. & Tech.*, 27(10), 2133–2140. Copyright 1993 American Chemical Society.

Figure 3.32 shows the results of a batch kinetic study (fractional metal uptake versus time) comparing the parent chelating exchanger with the CIM under otherwise identical conditions. Fractional metal uptake, $F(t)$, is dimensionless and is defined as the ratio of the metal uptake $q(t)$ after time t and the metal uptake at equilibrium, q^0, i.e., $F(t) = q(t)/q^0$. Although the parent chelating microbeads were bigger in average diameter (50–100 mesh) than the microbeads within the CIM, the copper uptake rate, as speculated, was slower with the CIM. In order to overcome the complexity arising due to the heterogeneity of the CIM (chelating microbeads randomly distributed in nonadsorbing teflon fibers), a model was proposed [66,67] in which the thin-sheet-like CIM may be viewed as a flat plate containing a pseudo-homogeneous sorbent phase as shown in Figure 3.33. Under the experimental conditions, the following may be assumed:

(1) Surface of the CIM is in equilibrium with the bulk of the liquid phase.
(2) Total amount of solute in the solution and in the CIM sheet remains constant as the sorption process is carried out.
(3) Solute (heavy metal) has high affinity toward the thin-sheet sorbent material.

This is a case of diffusion from a stirred solution with limited volume. The CIM is considered as a sheet of uniform material of thickness $2w$ placed in the solution containing the solute, which is allowed to diffuse into the sheet. The sheet occupies the space $-w \leq x \leq +w$, while the solution is of limited extent and occupies the space $-w - a \leq x \leq -w, w \leq x \leq +w + a$.

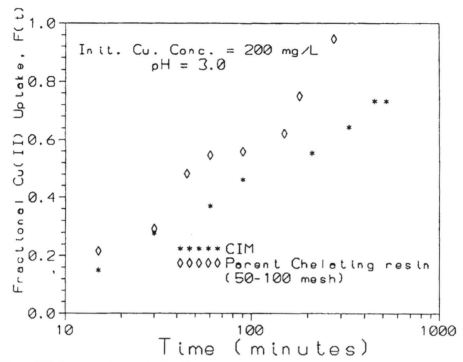

Figure 3.32 Copper uptake rates for the CIM versus the parent chelating exchanger under identical solution condition. Reprinted with permission from Reference [66], Sengupta and SenGupta. *Env. Sci. & Tech.*, 27(10), 2133–2140. Copyright 1993 American Chemical Society.

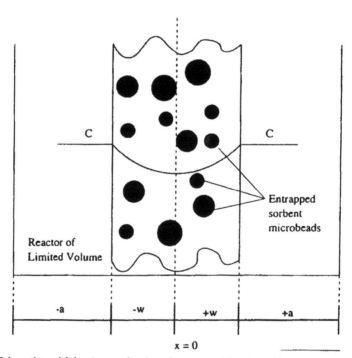

Figure 3.33 Schematic model showing sorption through an assumed flat plate with constant diffusivity from a reactor of limited volume. Reprinted with permission from Reference [66], Sengupta and SenGupta. *Env. Sci. & Tech.*, 27(10), 2133–2140. Copyright 1993 American Chemical Society.

The concentration of the solute in the solution is always uniform and is initially C_0, while initially the sheet is free from solute. Considering an apparent metal-ion diffusivity of within the CIM phase, the PDE that must be solved in this case of metal uptake through a plane sheet (the thickness of the CIM) from a solution of limited volume is given as follows:

$$\frac{\partial C}{\partial t} = \bar{D}\frac{\partial^2 C}{\partial x^2} \tag{24}$$

where

x = axial space coordinate in the direction of CIM thickness
C = concentration of solute in the solution

The IC of the above PDE is

$$C = 0, \quad -w < x < +w, \quad t = 0 \tag{25}$$

and the BC expresses the fact that the rate at which the solute leaves the solution is always equal to that at which it enters the sheet over the surfaces, $x = \pm w$. This condition is mathematically expressed as:

$$(\alpha / K)\frac{\partial C}{\partial t} = \pm\bar{D}\frac{\partial C}{\partial x}, \quad x = \pm w, \quad t > 0 \tag{26}$$

where K is the partition factor between the CIM and the solution, i.e., the concentration just within the sheet is K times that in the solution.

An analytical solution of this problem is given as [72]

$$F(t) = \frac{q_t}{q^0} = 1 - \sum_{n=1}^{\infty} \frac{2\alpha(1+\alpha)}{1+\alpha+\alpha^2 q_n^2} \exp\{-\bar{D}q_n^2 t / w^2\} \tag{27}$$

where the q_n values are the non-zero positive roots of

$$\tan q_n = -\alpha q_n \tag{28}$$

and $\alpha = a/(K \times w)$.

Figure 3.34 shows the results of three independent kinetic studies in which aqueous-phase copper concentrations are plotted against time. Two sets had different initial Cu concentrations (200 mg/L vs. 100 mg/L), while the third was a 10% fine-sand slurry. Solid lines indicate model predictions for an apparent copper ion diffusivity of 8.0×10^{-9} cm²/s. Note that although the exact morphology, i.e., distribution of voids and particles within the PTFE fiber, is not known, the suggested pseudo-homogeneous plane sheet model showed good agreement for the three sets of independent kinetic data produced with different initial concentration and level of suspended solids.

Cyclic Desorption of Heavy Metals from Ion Exchange Sites of Soil

Figure 3.35 shows the plot of percentage recovery of Cu(II) and the aqueous-phase

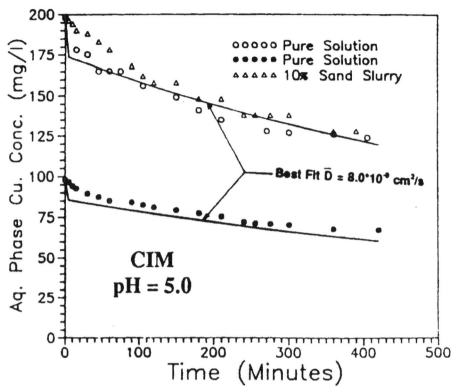

Figure 3.34 Comparison of kinetic model predictions (solid lines) with independent experimental data sets. Reprinted with permission from Reference [66], Sengupta and SenGupta. *Env. Sci. & Tech.*, 27(10), 2133–2140. Copyright 1993 American Chemical Society.

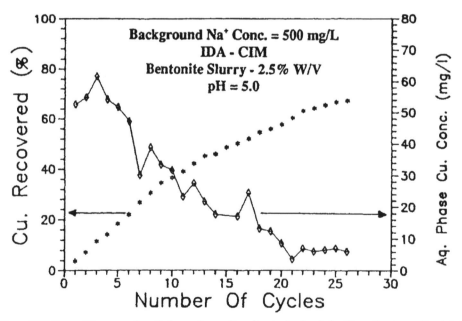

Figure 3.35 Copper(II) recovery from the ion exchange sites of bentonite clay during the cyclic process. Reprinted from *Reactive Polymers*, 35, 111–134, Sengupta and SenGupta. Heavy-metal separation from sludge using chelating ion exchangers with nontraditional morphology, 1997, with permission from Elsevier Science.

Cu(II) concentration versus number of cycles for the case of Cu(II) loaded bentonite. Note that 60% recovery of Cu(II) was achieved in less than 30 cycles.

Bentonite clays (aluminosilicates) derive their cation exchange capacity primarily from the isomorphous substitution within the lattice of aluminum and possibly phosphorus for silicon in tetrahedral coordination and/or magnesium, zinc, lithium, etc., for aluminum in the octahedral sheet [73]. Divalent metal ions are strongly held onto these ion exchange sites primarily thorough electrostatic interactions. Removal of heavy metals from such contaminated soils during the sorption step essentially involves the following consecutive steps:

(1) Desorption from the ion exchange sites of the soil into the aqueous phase by a counter-ion
(2) Selective sorption from the liquid phase onto the composite membrane

As in any sequential process with varying kinetic rates, the slower step will govern the rate of the overall process. Figure 3.36 compares the results of fractional desorption of Cu(II) from the ion exchange sites of bentonite at a pH of 5.0 using a 400 mg/L Ca(II) solution with that of fractional uptake of Cu(II) from a pure solution with an initial Cu(II) concentration of 200 mg/L carried out at the same pH value. From this figure, it can safely be concluded that the desorption rate of heavy metal from the ion exchange sites of clay is much faster than the sorption rate of the same by the composite membrane. For example, for the experimental conditions of Figure 3.36, time taken for 50% desorption from bentonite is

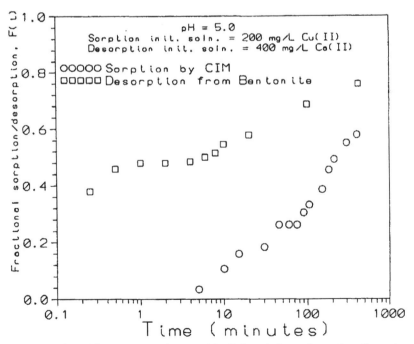

Figure 3.36 Comparison of fractional desorption rate of Cu(II) from the ion-exchange sites of bentonite with the fractional sorption rate of dissolved Cu(II) from pure solution by the CIM. Reprinted from *Reactive Polymers*, 35, 111–134, Sengupta and SenGupta. Heavy-metal separation from sludge using chelating ion exchangers with non-traditional morphology, 1997, with permission from Elsevier Sciences.

six minutes, while that for 50% sorption by the membrane (both expressed as a ratio of the equilibrium capacity) is $\simeq$ 240 minutes. Thus,

$$\frac{t_{1/2}\text{sorption}}{t_{1/2}\text{desorption}} \approx \frac{240}{6} = 40 \tag{29}$$

Therefore, in any system where two solid phases coexist, the sorption of heavy metal from solution is most likely to be the rate-limiting step. Also, any practical application would involve cyclic use of the CIM. Because the heavy metal is removed from the CIM into an acid solution with each cycle, the CIM is totally free of heavy metal at the beginning of each cycle. This creates a maximum possible concentration gradient between the aqueous solution phase and the CIM phase, thus facilitating faster kinetics. Therefore, for the case of heavy metal decontamination from ion exchange sites of clay, the kinetic limitations encountered would be due to the composition of the CIM and not from the desorption of heavy metal from the functional groups of the clay material.

Ion-Selectivity of Bentonite

Figure 3.37 shows the results of Cu/Ca selectivity of bentonite at a pH of 4.5 and Table 3.8 details coefficient and separation factor values for different composition of the aqueous and CIM phases. The aqueous-phase total cation concentration, $\{Ca^{2+}\} + \{Cu^{2+}\}$, remained almost the same. This proves that cation exchange was the major physicochemical process occurring.

For the Cu^{2+}/Ca^{2+} system, separation factor of copper over calcium is given by

$$\alpha_{Cu/Ca} = \frac{y_{Cu}x_{Cu}}{x_{Ca}y_{Ca}} \tag{30}$$

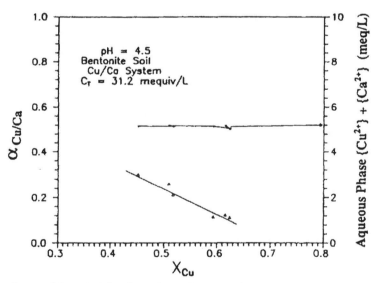

Figure 3.37 Copper-calcium selectivity of bentonite at pH = 4.5. Reprinted from *Reactive Polymers*, 35, 111–134, Sengupta and SenGupta. Heavy-metal separation from sludge using chelating ion exchangers with nontraditional morphology, 1997, with permission from Elsevier Science.

TABLE 3.8. Cu/Na Binary System.

y_{Cu} Dimensionless	x_{Cu} Dimensionless	$K_{Cu/Na}$ (gm/L)	$\alpha_{Cu/Na}$ Dimensionless
0.550	0.028	6389	42.43
0.600	0.031	7309	46.89
0.647	0.039	7283	45.16
0.739	0.044	11722	61.52
0.780	0.064	10009	51.85
0.794	0.087	7812	40.45
0.864	0.123	10532	45.30

Average $K_{Cu/Na}$ = 8722 (gm/L); average $\alpha_{Cu/Na}$ ≅ 47.66 (dimensionless).

where "y" represents the fraction of the element in the solid phase, and "x" represents the fraction in the aqueous phase.

From Figure 3.37 the following may be noted:

(1) $\alpha_{Cu/Ca} < 1.0$, meaning thereby that the bentonite has more affinity for Ca^{2+}. This matches the observation reported in literature [15,74] for strong acid-cation exchange resin. The primary reason attributed for calcium selectivity over copper is that Ca^{2+} ion causes less swelling of the resin than Cu^{2+}, or water uptake by the resin/clay is smaller for the Ca^{2+} form than for the Cu^{2+} form. This value of separation factor is in sharp contrast with $\alpha_{Cu/Ca} \approx 1,000$ for imino-diacetate functionality chelating resin [15].

(2) $\alpha_{Cu/Ca}$ decreases with increasing x_{Cu}. This phenomenon has been observed in almost all strong acid and strong base ion exchangers and is attributed to the nonuniformity of ion exchange sites [75].

(3) The aqueous-phase total cation concentration, $\{Ca^{2+}\} + \{Cu^{2+}\}$, remained almost the same in each case, proving that ion exchange between Cu^{2+} and Ca^{2+} is the major physicochemical process occurring. A similar procedure was carried out to determine selectivity of bentonite for the Cu/Na system. Because this is a heterovalent ion exchange system, the "selectivity coefficient" is thermodynamically a more rigorous parameter, and this value is reported in Table 3.8. Selectivity coefficient $K_{Cu/Na}$ can be mathematically expressed as follows:

$$K_{Cu/Na} = \frac{y_{Cu} \cdot x_{Na}^2 \cdot C_T}{y_{Na}^2 \cdot x_{Cu} \cdot Q} \tag{31}$$

where C_T = total aqueous-phase cation concentration, $\{Cu^{2+}\} + \{Na^+\}$, which ranged from 19 mequiv/L to 36 mequiv/L; and Q = bentonite cation exchange capacity = 0.5 meq/gm. Average separation factor values calculated for Cu/Ca and Cu/Na systems are as follows:

$$\text{Avg. } \alpha_{Cu/Ca} = 0.325$$

$$\text{Avg. } \alpha_{Cu/Na} = 47.66$$

From the previous observations, it can be concluded that the cation exchange behavior of smectite clay is similar to that of strong acid-cation exchanger. Thus, any system containing bentonite loaded with heavy metals in its ion exchange sites and the CIM can be mechanistically represented by the interaction of two cation exchangers: one with selectivity preference the same as that of strong acid-cation exchanger ($\alpha_{Cu/Na} \approx 50$) and the other with chelating functionality ($\alpha_{Cu/Na} \approx 10,000$) [15,67]. For the case of Cu(II) desorption from ion exchange sites of bentonite by addition of Na^+ in the aqueous phase, the exchange reactions involved can be summarized as follows:

$$\overline{(Z^-)_2 Cu^{2+}} + 2Na^+(aq) \Leftrightarrow \overline{2Z^- Na^+} + Cu^{2+}(aq) \tag{32}$$

$$\overline{R-N-(CH_2COO^-Na^+)_2} + Cu^{2+}(aq) \Leftrightarrow \overline{R-N-(CH_2COO^-)_2Cu^{2+}} + 2Na^+(aq) \tag{33}$$

Overall:

$$\overline{(Z^-)_2 Cu^{2+}} + \overline{R-N-(CH_2COO^-Na^+)_2} \Leftrightarrow \overline{R-N-(CH_2COO^-)_2Cu^{2+}} + \overline{2Z^- - Na^+} \tag{34}$$

where Z and R represent the bentonite clay and the CIM, respectively.

In order for the proposed process to succeed, the overall reaction [Equation (16)] has to be thermodynamically favorable. Considering ideality, the equilibrium constant for such a reaction in terms of soil-phase and membrane-phase considerations is given as follows:

$$K_{overall} = \frac{q_{Na}^Z \, q_{Cu}^R}{q_{Cu}^Z \, q_{Na}^R} \tag{35}$$

Superscripts Z and R denote the soil phase and the CIM phase, respectively, while q_{Na} and q_{Cu} represent the sodium and copper concentrations in the corresponding solid phase. Multiplying both numerator and denominator of the equation by C_{Na}/C_{Cu} (C_i denotes the aqueous phase concentrations of species i), we obtain:

$$K_{overall} = \left[\frac{q_{Cu}^R / C_{Cu}}{q_{Na}^R / C_{Na}} \right] \left[\frac{q_{Na}^Z / C_{Na}}{q_{Cu}^Z / C_{Cu}} \right] = \frac{\alpha_{Cu/Na}^R}{\alpha_{Cu/Na}^Z} \tag{36}$$

Thus, $K_{overall}$ is the ratio of the Cu/Na separation factor between the CIM and the bentonite clay. As discussed earlier, due to the presence of the chelating functional group (imino-diacetate moiety), the dimensionless Cu/Na separation for the CIM is about two to three orders of magnitude greater compared to bentonite. As a result, the overall process is quite selective for decontamination of clays with high ion exchange capacity. Thus, advantage is being taken of the fact that the bentonite clay ion-exchanging functionality has much less selectivity for Cu(II) over Na (by introducing a high concentration of Na in the aqueous phase), whereas the chelating CIM has very high affinity for Cu(II) over Na (due to which only Cu(II) is taken up by the membrane and concentrated in the regenerant solution).

METAL-ION SEPARATION

Continuous Annular Chromatography

We have discussed how metal ions can be removed from a dilute wastewater stream or a contaminated solid phase (soil/sludge). Frequently, we come across a situation where the contaminated water or soil/sludge contains several heavy metals. The methodologies outlined earlier can be effectively employed to remove all the heavy metal cations, but the corollary is that the regenerant solution will have a mixture of all the heavy metal cations. In other words, the heavy metal cations are not separated in the regenerant stream. Such a mixture of salts of toxic metals has but only a limited commercial value, and there is a pressing need for separating the heavy metal cations. Commonly encountered toxic metal cations, such as Cu^{2+}, Pb^{2+}, Ni^{2+}, Cd^{2+}, Zn^{2+}, etc., exhibit similar physical and chemical characteristics and, therefore, separating one from another poses obvious difficulty. Some progress has been made [31,76,77] in characterizing and synthesizing chelating polymers in order to enhance the relative selectivity between two given metal ions, especially in the presence of ligands. However, to date there has not been any commercial application of this scheme. A separation technique known as Continuous Annular Chromatography (CAC) developed at Oak Ridge National Laboratory (ORNL) seems to show promise.

In CAC, separation is achieved by using a slowly rotating annular bed of ion exchange resin, where feed is continuously introduced at a stationary point, while the eluent flows over the remainder of the annulus. The rotation of the sorbent bed coupled with the passage of the eluent causes the separated solutes of the feed to appear as helical bands, each of which has a characteristic stationary exit point. Figure 3.38 shows a conceptual diagram illustrating the key features of CAC, which is capable of achieving multicomponent chromatographic separation for a relatively concentrated mixture of metal salts. Various versions of a CAC apparatus have been constructed and tested experimentally at ORNL [78–80]. Multicomponent systems tried include separation of Ni, Cu and Co in ammoniacal solution, the separation of iron and ammonium in ammonium sulfate-sulfuric acid solution and the separation of hafnium from zirconium in sulfuric acid solution. For the above-mentioned examples, Dowex 50W-X8, a polystyrene strong-acid (sulfonic group)

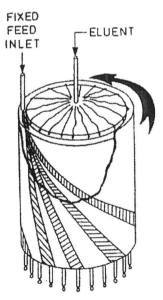

Figure 3.38 Conceptual diagram illustrating the operation of the continuous annular chromatographic unit. Reprinted from *Ion Exchange for Industry,* Streat (ed.). Byers, Sisson, and Decarli, The use of gradient elution in optimizing continuous annular ion exchange chromatography with applications to metal separations, p. 19, Ellis Horwood, Chichester, England, with permission from Society of Chemical Industries.

ORNL DWG 74-12638R3

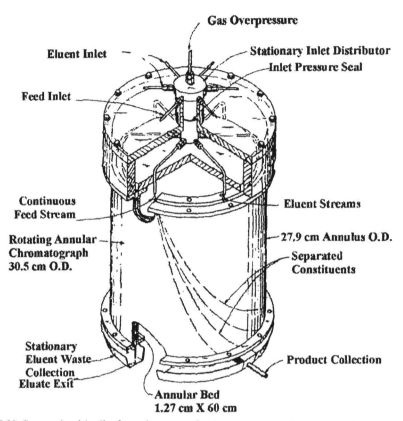

Figure 3.39 Constructional details of a continuous annular chromatographic unit used at Oak Ridge National Laboratory. Reprinted from *Ion Exchange for Industry*, Streat (ed.). Byers, Sisson, and Decarli, The use of gradient elution in optimizing continuous annular ion exchange chromatography with applications to metal separations, p. 19, Ellis Horwood, Chinchester, England, with permission from Society of Chemical Industries.

cation exchanger resin with 8% DVB cross-linking (Dow Chemical, MI, USA) was used as the stationary phase. Figure 3.39 provides some constructional details of a CAC apparatus built for bench-scale testing at ORNL.

Figure 3.40 shows separation of Cr(III) from Fe(III) with the CAC apparatus by Byers et al. [80] using ammonium as an isocratic eluent where the feed concentration was fairly high (5 gm/L). Theoretically, elution and resolution of various components being separated can be predicted quite precisely using steady-state material balance and rate-limiting equations. From Figure 3.40, one can note the excellent agreement of experimental results with the theoretical prediction (solid lines).

Using gradient elution, Byers et al. [80] have shown that the purity of Fe(III) and Cr(III) in the product can be increased by an order of magnitude. However, all of these studies were conducted with a strong acid-cation exchange resin, Dowex 50. Earlier discussion has indicated that chelating functionality may be used to improve the relative selectivity between two metal ions, especially in the presence of ligands. Use of appropriate chelating exchangers may, therefore, broaden the application potential of CAC in separating heavy metals. Figure 3.41 shows separation of Cu(II) and Ni(II) in a packed bed containing XFS

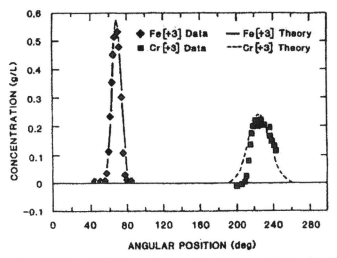

Figure 3.40 Separation of Fe(III) and Cr(III) in a continuous annular chromatograph using 0.4 M $(NH_4)_2SO_4$ as the eluent. Reprinted from *Ion Exchange for Industry*, Streat (ed.). Byers, Sisson, and Decarli, The use of gradient elution in optimizing continuous annular ion exchange chromatography with applications to metal separations, p. 19, Ellis Horwood, Chinchester, England, with permission from Society of Chemical Industries.

4195 (Dow Chemical, MI, USA); the bed was first loaded with a mixture of 200 mg/L Cu(II) and 200 mg/L Ni(II) solution and then regenerated with 2% HCl followed by 2% ammonia [64]. Use of such specific chelating polymers in CAC is likely to improve copper-nickel separation significantly. Thus, CAC—in conjunction with a proper choice of chelating polymers—may be a useful tool in accomplishing this desired goal of heavy-metal-cation separation. CAC apparatus is practically stationary and, hence, maintenance-free. Central treatment facilities, where wastewater streams and soils/sludges are collected and treated for heavy metal removal, are gaining momentum due to their overall economy [81] and are being encouraged by state and federal regulatory authorities. Using CAC in such facilities for separation and recovery of heavy metals may be a worthwhile venture in coming years.

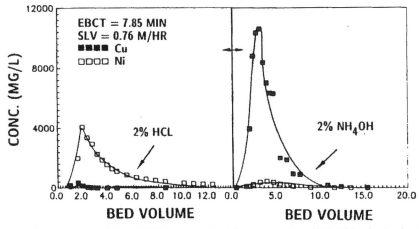

Figure 3.41 Separation of Cu(II) and Ni(II) in a packed-bed column containing XFS 4195 resin. Reprinted from Reference [64], Sengupta and Zhu, *AIChE Journal*, 38, 1, 153–157, 1992.

CLOSURE

Due to widespread use of heavy metals in industrial applications and because of their toxicity, removal of heavy metals from contaminated wastewaters and soils/sludges is a common environmental problem. The authors have attempted to distill the existing body of knowledge pertaining to chelating ion exchange and present technical issues of major importance. In comparison with other available technologies, chelating ion exchangers offer some unique advantages, namely,

- They can meet very strict heavy metal discharge standards that are often in the parts per billion range.
- They are very selective in heavy metal removal, resulting in low mass/volume of waste generated.
- They offer potential for recycle/reuse of the waste generated.

Conventional chelating ion exchangers suffer from one major drawback: they are very inefficient in highly acidic (low pH) solutions because the heavy metal ions suffer severe competition from H^+ ions. Recent work with specialty chelating exchangers containing nitrogen donor atoms has shown that it can be used in such situations with high separation efficiency. Although counterintuitive, higher competing ion concentration may even contribute to higher heavy metal ion selectivity in certain cases. Moreover, all this is achieved without any loss in regenerability.

Another limitation of conventional chelating exchangers is their morphology. These exchangers are synthesized as spherical beads with 0.3–1.2 mm diameter and are packed in a fixed-bed column through which the wastewater solution is passed. Thus, for many scenarios where the wastewater has high suspended solids or where the background phase is solid (soil/sludge), conventional ion exchange technology cannot be applied. The authors have presented detailed results of studies conducted with a novel material referred to as composite ion-exchange material (CIM) where the morphology of the exchangers has been changed (from a bead) to a thin cloth-like physical configuration. This arrangement allows the material to be used in soil/sludge decontamination cases, thus considerably extending the range of applicability of ion exchangers. This chapter provides experimental evidence regarding how one can tailor the aqueous-phase chemistry (by adding suitable combination of ligands) to remove heavy metal ions from the solid phase using CIM.

Finally, one aspect that has been gaining considerable interest is the reuse/recycle of the regenerant. In many situations, the wastewater or the soil/sludge contains multiple heavy metals. When a fixed-bed or CIM process is applied to this waste, all the heavy metals are removed and concentrated in the regenerant. However, in order to recycle/reuse this regenerant, the heavy metals are to be separated from one another. This chapter discusses continuous annular chromatography or CAC that has demonstrated potential in this respect, but additional studies are needed with a much wider variety of regenerant streams and different kinds of ion exchangers for field-scale trials.

REFERENCES

1 Adams, B. A. and Holmes, E. L. Brit. Patent, 450,309, 1936; U.S. Patent, 2,151,883, 1939.

2 D'Alelio, G. U.S. Patent, 2,340,110 and 2,340,111, 1944 to General Electric Co.

3 Reichenberg, D. and McCauley, D. J. Cation exchange equilibria on sulfonated polystyrene resins of varying degrees of crosslinking, *J. Chem. Soc.*, 2741, 1955.

4 Clifford, D. and Weber, W. The determination of divalent/monovalent selectivity in anion exchangers, *Reactive Polymers,* 1, 77, 1983.

5 Chu, B., Whitney, D. C., and Diamond, R. M. Toward anion exchange resin selectivities, *Inorg. Nucl. Chem.,* 24, 1405, 1962.

6 Kunin, R. Acrylic-based ion exchange resins and adsorbents, *Amber Hi-Lites,* No. 173, Rohm and Haas Co., Philadelphia, PA, 1984.

7 Sengupta, A. K., Roy, T., and Jessen, D. Modified anion exchange resins for improved chromate selectivity, *Reactive Polymers,* 9, 293, 1988.

8 Chromatography, electrophoresis, immunochemistry, molecular biology, HPLC. Biorad Chemical Division, CA, Price List M, April 1987.

9 Petruzelli, D., Passino, R., Santori, M., and Tiravanti, G. Chromium, aluminum, iron separation and recovery from tannery sludge by selective ion exchange, presented at the *Second International Symposium on Metal Speciation, Separation and Recovery,* Rome, Italy, May 14–19, 1989.

10 Carlson, L. Ion exchange treatment for removing toxic metals and cyanide from wastewater, U.S. Patent 4,321,145, 1982.

11 Greig, J. A. and Lindsay, D. Copper removal from zinc-bearing leach liquors, in *Proc. Int. Symp. Ion Exchange for Industry,* Streat, M., Ed., Ellis Horwood, England, 1988, 337.

12 Hessler, J. C. Recovery and treatment of metal finishing wastes by ion exchange process, *Ind. Water Waste,* 6, 75, 1961.

13 Cotton, F. A. and Wilkinson, G. *Advanced Inorganic Chemistry,* Interscience Publishers, New York, 1972.

14 Pepper, K. W. and Hale, D. K. *Ion Exchange and Its Applications,* Society of Chemical Industry, London, p. 13, 1955.

15 Roy, T. K. Chelating polymers: Their properties and applications in relation to removal, recovery and separation of toxic metal ions, M.S. Thesis, Civil Engineering Department, Lehigh University, Bethlehem, PA, 1989.

16 Waitz, W. H. Ion exchange in heavy metals removal and recovery, *Amber-Hi-Lites,* No. 162, Rohm and Haas Co., Philadelphia, PA, 1979.

17 Bolto, B. A. and Pawlowski, L. *Wastewater Treatment by Ion Exchange,* E & F. N. Spon, London, 1987.

18 Janauer, G. E., Gibbons, R. E., and Bernier, W. E. *Ion Exchange and Solvent Extraction,* Marinsky, J. A., and Marcus, Y., Eds., Vol. 9, Chap. 2, Marcel Dekker Inc., New York, 1985.

19 Woelders, J. A., Urlings, L. G. C. M., and Vanderpiji, P. P. In-situ remedial action of cadmium-polluted soil by ion exchange, in *Proc. Int. Symp. Ion Exchange for Industry,* Streat, M., Ed., Ellis Horwood, England, 169, 1988.

20 SenGupta, A. K., Millan, E., and Roy, T. Potential of ion exchange resins and reactive polymers in eliminating/reducing hazardous contaminants, *Proc. of the 2nd International Conference on Physicochemical and Biological Detoxification of Hazardous Wastes,* Atlantic City, NJ, May 3–5, 1988, 191.

21 Matejka, Z. and Zitkova, Z. The sorption of heavy-metal cations from EDTA complexes on acrylamide resins having oligo(ethyleneamine) moieties, *Reactive & Functional Polymers,* 35, 81–88, 1997.

22 Courduvelis, C. New developments for the treatment of wastewater containing metal complexes, *Proceedings AES 4th Conference on Advanced Pollution Control for the Metal Finishing Industry,* Florida, 1982.

23 Marton-Schmidtt, E. et. al. Separation of metal ions on ion exchange resin with ethylenediamine functional groups, *J. Chromatography,* 201, 73–77, 1980.

24 Loureiro, J. M. et. al. Recovery of copper, zinc and lead from liquid streams by chelating ion exchange resins, *Chem. Eng. Sci.,* 43, 1115–1124, 1988.

25 Warshawsky, A. Modern research in ion exchange, *Ion Exchange: Science and Technology,* NATO AISI Series, Rodrigues, A. E., Ed., Martinus Nijhoff Publishers, Boston, 1986.

26 Warshawsky, A. Extraction of platinum group metals by ion exchange resins, *Ion Exchange Pro-*

cesses in Hydrometallurgy, Streat, M. and Nadan, D., Eds., Soc. of Chemical Industry, John Wiley and Sons, New York, Vol. 19, 1987.

27 Hudson, M. Coordination chemistry of selective ion exchange resins, *Ion Exchange: Science and Technology,* NATO AISI Series, Rodrigues, A. E., Ed., Martinus Nijhoff Publishers, Boston, 1986.

28 Melling, J. and West, D. W. *Proceedings of the 4th International Conference on Ion Exchange Technology at the University of Cambridge,* England, July, 1984, 724–735.

29 Marinsky, J. Equations for the evaluation of formation constants of complexed ion species in crosslinked and linear polyelectrolyte systems, *Ion Exchange and Solvent Extraction,* Marinsky, J. and Marcus, Y., Eds., Vol. 4, Marcel Dekker, Inc., New York, 1973.

30 Helfferich, F. *Ion Exchange,* Chapter 4, Xerox University Microfilms, Ann Arbor, MI, 1961.

31 Zhu, Y., Millan, E., and SenGupta, A. K. Toward separation of toxic metal(II) cations by chelating polymers: some noteworthy observations, *Reactive Polymers,* 13, 241–253, 1990

32 Helfferich, F. *Ion Exchange,* Chapter 9, Xerox University Microfilms, Ann Arbor, MI, 1961.

33 Kokoszka, L. C. and Flood, J. W. *Environmental Management Handbook: Toxic Chemical Materials and Wastes,* Chapter 2, Marcel Dekker, Inc., New York, 1989.

34 Grinstead, R. R. and Paalman, H. H. Metal ion scavenging from water with fine-mesh ion exchangers and microporous membranes, *Environmental Progress,* 8,1, 35, 1989.

35 Burgess, J. *Ions in Solution: Basic Principles and Interactions,* Ellis Harwood Ltd., John Wiley & Sons, New York, 1989.

36 Calmon, C. Impact of improved ion exchange materials and new techniques on separation performance, *Adsorption and Ion Exchange,* AIChE Symp. Series No. 233, 80, 84, 1984.

37 Helfferich, F. *Ion Exchange,* Chapter 5, Xerox University Microfilms, Ann Arbor, MI, 1961.

38 Ouchi, A. Structure of rare earth carboxylates in dimeric and polymeric forms, *Coordination Chemistry Review,* 29, 1988.

39 SenGupta, A. K., Zhu, Y., and Hauze, D. Metal(II) ion binding onto chelating exchangers with nitrogen donor atoms, *Env. Sci. & Tech.,* 25(3), 481–488, 1991.

40 Morel, F. A. *Principles of Aquatic Chemistry,* John Wiley & Sons, New York, 1983.

41 Smith, R. M. and Martell, A. E. *Critical Stability Constants,* Vol. 2, Plenum Press, New York, 1974.

42 Nativ, M., Goldstein, S., and Schumuckler, G. Kinetics of ion exchange processes accompanied by chemical reactions, *J. Inorg. Nucl. Chem.,* 37, 1951, 1975.

43 Helfferich, F. and Plesset, M. S. Ion exchange kinetics: A non-linear diffusion problem, *J. Chem. Phys.,* 28, 411, 1958.

44 Hoell, W. Optical verification of ion exchange mechanisms in weak electrolyte resins, *Reactive Polymers,* 2, 93–101, 1984.

45 Boyd, G. and Soldanho, B. A. Self diffusion of cations in and through sulfonated polystyrene cation exchange polymers, *J. Am. Chem. Soc.,* 75, 6091, 1953.

46 Helfferich, F. Ion exchange kinetics, V: Ion exchange accompanied by reaction, *J. Phys. Chem.,* 69, 1178, 1965.

47 Schmuckler, G. Kinetics of moving-boundary ion exchange processes, *Reactive Polymers,* 2, 103, 1984.

48 Yoshida, H., Kataoka, T., and Ikeda, S. Intraparticle mass transfer in bidispersed porous ion exchanger, Part I, Isotopic ion exchange, *Can. J. Chem. Eng.,* 62, 422, 1985.

49 Yoshida, H., Kataoka, T., and Fujikawa, S. Kinetics in a chelate exchanger I., *Chem. Eng. Sci.,* 41, 2525, 1986.

50 Price, S. G., Helditch, D. J., and Streat, M. Diffusion or chemical kinetics control in a chelating ion exchange resin system, *Ion Exchange for Industry,* Streat, M., Ed., Ellis Horwood Ltd., Chichester, U.K., 1988.

51 Yoshida, H. and Kataoka, T. Intraparticle mass transfer in bidispersed porous ion exchanger, Part II, Mutual ion exchange, *Can. J. Chem. Eng.,* 62, 430, 1985.

52 Hoell, W. and Sontheimer, H. Ion exchange kinetics of the protonation of weak-acid ion exchange resins, *Chem. Engr. Sci.,* 32, 755, 1977.

53 Helfferich, F. Models of physical reality in ion exchange kinetics, *Reactive Polymers*, 13, 191–194, 1990.

54 Dorner, H. E. Chloride as a factor in mobilities of Ni(II), Cu(II), and Cd(II) in soil, *Soil Sci. Am. J.*, 42, 882, 1978.

55 SenGupta, A. K. A unified approach to interpret unusual observations in heterogeneous ion exchange, *Jour. Colloid Interface Sci.*, 123, 1, 201, 1988.

56 Zhu, Y., Millan, E., and SenGupta, A. K. Toward separation of toxic metal cations by chelating polymers: Some noteworthy observations, *Reactive Polymers*, 13, 241–253, 1990.

57 Bird, R. B., Stewart, W. E., and Lightfoot, E. N. *Transport Phenomena*, John Wiley & Sons, New York, 1960.

58 Wankat, P. C. Efficient fractionation by ion exchange, *Ion Exchange Science & Technology, NATO ASI Series*, Rodrigues, A. E., Ed., Martinus Nijhoff Publishers, Boston, 1986.

59 Helfferich, F. and Klein, G. *Multicomponent Chromatography: Theory of Interference*, Marcel Dekker, Inc., New York, 1970.

60 Kovach, J. L. *Handbook of Separation Techniques for Chemical Engineers*, Chapter 3, Schweitzer, P. A., Ed., McGraw-Hill Inc., New York, 1979.

61 Brown, C. J. U.S. Patent 4,673,507, 1987.

62 Brown, C. J. and Fletcher, C. J. The Recoflo short-bed ion exchange process, *Ion Exchange for Industry*, Streat, M., Ed., Ellis Horwood, Chichester, 1988.

63 Brown, C. J. Acid and metals recovery by Recoflo short-bed ion exchange, *Separation Processes in Hydrometallurgy*, Davies, G. A., Ed., Ellis Horwood, Chichester, 1987.

64 SenGupta, A. K. and Zhu, Y. Metals sorption onto chelating polymers: a unique role of ionic strength, *AIChE Jour.*, 38, 1, 153–157, 1992.

65 Errede, L. A. et. al. Swelling of particulate polymers enmeshed in poly(tetrafluoroethylene), *J. App. Poly. Sci.*, 31, 2721–2737, 1986.

66 Sengupta, S. and SenGupta, A. K. Characterizing a new class of sorptive/desorptive ion exchange membranes for decontamination of heavy-metal-laden sludges, *Env. Sci. & Tech.*, 27(10), 1993.

67 Sengupta, S. A new separation and decontamination technique for heavy-metal-laden sludges using sorptive/desorptive ion exchange membranes, Ph.D. Dissertation, Lehigh University, Bethlehem, PA, 1993.

68 SenGupta, A. K. and Shi. B. Selective alum recovery from clarifier sludge, *Jour. AWWA*, Vol. 84, 1992.

69 Sengupta, S. and SenGupta, A. K. Solid phase heavy metal separation using composite ion exchange membranes, *Hazardous Waste & Hazardous Materials*, 13, 2, 1996.

70 Sengupta, S. and SenGupta, A. K. Heavy-metal separation from sludge using chelating ion exchangers with nontraditional morphology, *Reactive & Functional Polymers*, 35, 111–134, 1997.

71 Sengupta, S. Electro-partitioning with composite ion exchange material: An innovative in-situ heavy-metal decontamination process, *Reactive & Functional Polymers*, 40, 263–273, 1999.

72 Crank, J. *The Mathematics of Diffusion*, Oxford University Press, London, 1975.

73 Grim, R. E. *Clay Mineralogy*, Second Edition, McGraw-Hill Book Co., New York, 1968.

74 Bonner, O. D., and Livingston, F. L. Cation exchange equilibria involving some divalent ions, *J. Phys. Chem.*, Vol. 60, 1956.

75 Reichenberg, D. and McCauley, D. J. Properties of ion exchange resins to their structure. Part VI: Cation exchange equilibria on sulfonated polystyrene resins of varying degrees of crosslinking, *J. Chem. Soc.*, Vol. III, 1955.

76 King, J. N. and Fritz, J. S. Separation of metal ions using aromatic *o*-hydroxyloxime resin, *J. Chromatogr.*, 153, 504, 1978.

77 Gjerde, D. T. and Fritz, J. S. Chromatographic separation of metal ions on macroreticular anion exchange resins of a low capacity, *J. Chromatogr.*, 188, 391, 1980.

78 Scott, C. D., Spence, R. D., and Sisson, W. G. Pressurized annular chromatograph for continuous separations, *J. Chromatogr.*, 126, 381, 400, 1976.

79 Begovich, J. M., Byers, C. H., and Sisson, W. G. A high-capacity pressurized continuous chromatograph, *Sep. Sci. Technol.*, 18, 1167, 1983.

80 Byers, C. H., Sisson, W. G., and Decarli II, J. P. The use of gradient elution in optimizing continuous annular ion exchange chromatography with applications to metal separations, *Ion Exchange for Industry*, Streat, M., Ed., Ellis Horwood, Chichester, England.

81 Chen, J. J. Metro recovery systems—A centralized metals recovery and treatment facility in twin cities, U.S.A., *Proc. Second Int. Symp. Metals Speciation, Separation, and Recovery*, Rome, Italy, May 1989.

CHAPTER 4

Elimination of Heavy Metals from Wastewaters by Magnetic Technologies

MATTHIAS FRANZREB[1]
JAMES H. P. WATSON[2]

W HEN looking for magnetic technology-based processes for the treatment of water and sewage in conventional scientific literature, efforts often meet with no success. In most cases, only the simple example of the removal of the finest metal particles from the wastewaters of steel and iron industries is found. This reflects the fact that magnetic technologies have not yet found wide application in spite of their advantages. The few established fields of use include the removal of corrosion products from the condensate circuit of power plants and, in particular, of nuclear power plants [1], the separation of the finest metal particles from the wastewaters of steel and iron industries [2] as well as the protection of products and production facilities against the accidental introduction of metal particles. In addition, a number of applications have been implemented on the technical scale, although they have not yet been generally accepted on the market. This holds true for the removal of phosphate from municipal sewage [3], the so-called Sirofloc process for the processing of surface waters [4], and the separation of precipitated heavy metal compounds by high-gradient magnetic separation [5].

In addition to this relatively small number of technically implemented applications, several methods have already been tested on the laboratory or pilot scale. Based on magnetic technology, innovative approaches have been conceived to solve water technology problems. Most of the basic concepts underlying these processes were developed in the 1970s and 1980s. Often, however, industrial exploitation was prevented by the high costs of the magnetic systems available at that time. But, costs of superconducting magnetic coils and, in particular, of permanent magnet systems based on rare earths (e.g., neodymium or samarium), dropped considerably during the past decade, such that it is now considered worthwhile to again investigate and evaluate these approaches in a modified form, considering modern magnetic technology. Following a short introduction describing state-of-the-art

[1]Forschungszentrum Karlsruhe, Institute for Technical Chemistry, P.O. Box 3640, 76021 Karlsruhe, Germany, matthias.franzreb@itc-wgt.fzk.de

[2]University of Southampton, Department of Physics and Astronomy, Southampton SO17 1BJ, UK, jhpw@sotonac.uk

technologies, the sections that follow give an overview of the studies performed by the authors during the last decade in the field of separator development and its use for heavy metal elimination from water. The sections dealing with separator technology, the ferrite process, and magnetic seeding have been compiled by M. Franzreb. The section on the process of biomagnetic sorption was kindly written by J. H. P. Watson.

SEPARATOR TECHNOLOGY

To separate heavy metal-containing particles from aqueous suspensions, either drum-type magnetic separators or high-gradient magnetic separators are applied depending on particle size and magnetizability. A third group of separators is made up of the so-called open-gradient magnetic separators which so far have only been tested on the laboratory scale. These types of separators can differ by the underlying principle of separation, by the source of the magnetic field used and by a cyclic or continuous mode of operation. The physical fundamentals describing the interaction between the magnetic fields and matter, however, are the same in all cases and shall be outlined below.

Physical Fundamentals

All types of magnetic separators are based on the property of magnetic fields exerting a force on matter. The general relationship for this magnetic force F_m is as follows:

$$F_m = \mu_0 V_p M_p \text{grad } H \tag{1}$$

where μ_0 denotes the permeability constant of the vacuum, V_p the particle volume, M_p particle magnetization and grad H the gradient of the magnetic field strength at the position of the particle. Particle magnetization may be expressed by the magnetic volume susceptibility χ and magnetic field strength H as follows:

$$M_p = \chi H \tag{2}$$

where the volume susceptibility is a constant for diamagnetic and paramagnetic substances and a function among others of the particle shape and size as well as of the field strength for ferrimagnetic and ferromagnetic substances. The different orders of magnitude of volume susceptibilities of these substances are obvious from Table 4.1.

As the type of particles to be separated is usually given, the magnetic force achievable in a separator can be influenced by the prevailing field strength and, in particular, by its gradient [see Equation (1)]. Simple drum-type separators reach comparatively moderate values only for these parameters. So-called high-gradient magnetic separators on the basis of solenoids attain field strengths of 1–2 tesla and gradients of up to about 10^5 T/m. The limits of magnetic particle separation are, therefore, determined by HGMS technology. Consequently, the short discussion of the underlying physical laws below shall be restricted to this type.

The forces that are of decisive importance to the separation of magnetizable fine particles from waters are the magnetic force and the hydrodynamic resistance force. Assuming field distribution in the surroundings of a magnetically saturated wire and with the Stokes equation being valid, theoretical particle velocity caused by the magnetic force in the direct vicinity of the wire [6–9] may be approximated as follows:

TABLE 4.1. Susceptibilities of Various Substances.

Diamagnetic Substances:

Substance	$CaCO_3$	$Ca(OH)_2$	Cu	$PbCO_3$	NaCl	H_2O	Bi
$\chi \cdot 10^6$	−12.9	−8.11	−9.61	−19.1	−14.1	−9.04	−165

Paramagnetic Substances:

Substance	Al	Pt	$Cu(OH)_2$	$FeCO_3$	$FePO_4$	$MnCO_3$	Dy
$\chi \cdot 10^6$	20.7	279	508	3,097	2,625	3,894	64,860

Ferromagnetic and Ferrimagnetic Substances:

Substance	Iron (99.91%)	Iron (99.95%)	Steel (98.5 Fe)	Permaloy	Supermalloy	Magnetite	Nickel
χ (1 T)	2.15	2.15	2.1	≈1	0.8	0.61	0.618

$$v_m = \frac{2}{9} \cdot \mu_0 \cdot \Delta\chi \cdot M_s \cdot H_0 \cdot \frac{b^2}{a \cdot \eta} \qquad (3)$$

where b is the particle radius, $\Delta\chi$ is the difference between the volume susceptibilities of the particle and fluid, M_s ($A \cdot m^{-1}$) is the saturation magnetization of the ferromagnetic matrix wires of radius a (m), H_0 is the applied magnetic field ($A \cdot m^{-1}$) and η (Pascal·s) is the viscosity of the fluid (for water near room temperature, η is approximately 10^{-3} Pascal·s).

In a number of theoretical and experimental investigations, the ratio between this magnetic velocity and the actual flow velocity in the separator v_m/v_0 proved to be an important orientation value for the probable magnetic separation efficiency (this will be discussed later). At values of $v_m/v_0 > 1$, the separation behavior of an HGMS is similar to that of a classical deep-bed filter. This means that a relatively sharp loading front is formed in the separation matrix. In the course of time, this front is shifted along the filter in the outlet direction. At values of $v_m/v_0 \ll 1$, rather extended loading fronts are obtained, and magnetic separation remains insufficient.

Drum-Type Magnetic Separators

Drum-type magnetic separators on the basis of permanent magnets may be referred to as the "workhorses" of wet magnetic separation. Judging from the number of units installed and the volume flows reached, this group of magnetic separators represents the most frequently used magnet technology [10]. The separators consist of a nonmagnetic rotating drum, in which permanent magnets are arranged over an angle of 100°–180°. The poles directed to the outside alternate in polarity and are located as close as possible to the inner drum surface. In wet drum-type magnetic separators, the particle suspension may be passed either in a concurrent or in a countercurrent flow, relative to the drum rotation (see Figure 4.1). Usually, the permissible flow rates are smaller in the counterflow mode. On the other hand, a higher degree of separation is reached, and magnetic particles of smaller size (<100 μm) may be separated. In principle, drum-type magnetic separators on the basis of perma-

CONCURRENT MODE OF OPERATION   COUNTERCURRENT MODE OF OPERATION

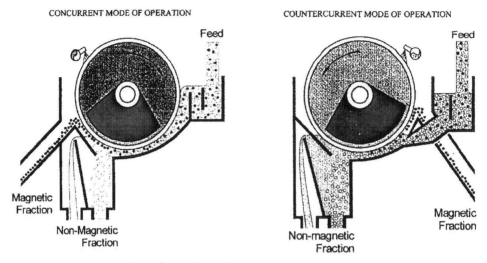

Figure 4.1 Setup and modes of operation of drum-type magnetic separators (courtesy of Steinert Elektromagnetbau).

nent magnets or solenoids are only suited for the separation of strongly magnetic particles, i.e., metal chips of low-alloy steels or magnetite, due to their comparatively small field strength and field strength gradient.

High-Gradient Magnetic Separators

The basic principle of high-gradient magnetic separation (HGMS) explained in Figure 4.2 is simple and similar to that of deep-bed filtration. A canister filled with a magnetizable separation matrix is introduced into the area of an external magnetic field. The matrix, symbolized by a single wire cross section in the figure, may consist of a loose package of rough steel wool or of wire meshes stacked on top of each other. The matrix wires bundle the external magnetic fields in their surroundings and, thus, generate areas on their surface that

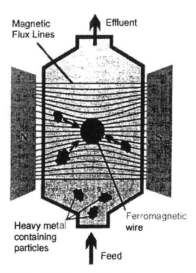

Figure 4.2 Basic principle of high-gradient magnetic separation.

strongly attract paramagnetic or ferromagnetic particles. The magnetic force acting on the particles may exceed the weight acting on the particles by a factor of more than 100. This results in filtration rates of more than 200 m/h.

With waters having a small proportion of low-magnetic, heavy metal containing particles, cyclically operated HGMS (see Figure 4.3) are best suited for reaching the results desired in terms of flow rate and separation efficiency. This is due to their compact design and high field strengths achieved. During the separation cycle, heavy metal containing particles deposit on the separation matrix. As a consequence, only a particle-free solution or a solution loaded with nonmagnetic solid particles only leaves the separator. The residence time of the separator may range from several weeks, e.g., when separating iron-containing corrosion products from condensate circuits, to a few minutes. If the capacity of the separation matrix is exhausted, solution supply is suspended. In case the particles separated represent a valuable product, a short rinsing cycle is often performed under the still operating magnetic field in order to remove impurities caused by the initial solution from the particles. However, if the treatment is aimed at separating heavy metal containing compounds, this step is not employed. The subsequent step takes place in a short and intensive backwashing phase in the counterflow mode with the magnetic field switched off. The separated particles are obtained in the form of a concentrate, with the solid content rarely higher than 5%. Finally, the magnetic field is switched on again and another separation phase may start.

Cyclically operating HGMS are produced by a number of companies, e.g., Eriez Magnetics Ltd. (Erie, PA, USA), Svedala Industri AB (Sala, Sweden) or Aquafine Corporation (Brunswick, GA, USA). Depending on their size and the maximum flux density desired, the separators are simply air cooled or equipped with a water or oil cooling system. The diameter of the filter matrix may be up to 3 m with the power consumption of the respective systems amounting to about 400 kW.

Continuously operating carrousel-type magnetic separators have been developed for high volume flows with a medium to high content of particles to be removed. In this type of magnetic separator, the separation matrix is located in a disk- or ring-shaped housing that slowly rotates around its axis in a continuous or stepwise manner (see Figure 4.4). One or several separation zones are located along the circumference. In these areas, the separation

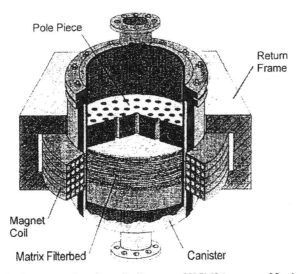

Figure 4.3 Sectional representation of a cyclically operated HGMS (courtesy of Svedala Industri AB).

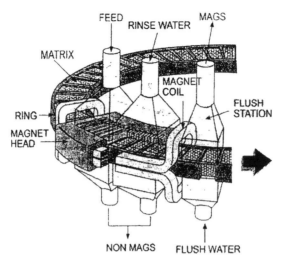

Figure 4.4 Continuously operating high-gradient magnetic separator (courtesy of Svedala Industri AB).

matrix is magnetized by an external magnetic field. The separation zone is followed by a rinsing zone that is located in the range of the magnetic field. It is followed by a flushing zone outside of the magnetic field. Due to the separation of the different zones, the processes of separation, rinsing and flushing may take place in parallel. This means that carrousel-type magnetic separators are operated under constant supply. The particles separated in the separation zone are transported to the rinsing and flushing zones by the rotation of the carrousel, which consists of a separation matrix and housing. A magnetic separator working in accordance with this principle was presented by Humboldt Wedag AG (Cologne, Germany) in the 1970s [11]. Other carrousel-type magnetic separators were developed by companies including Carpco, Eriez Magnetics Ltd, and Svedala Industri AB (formerly Sala Magnetics) [12]. All of these carrousel-type magnetic separators use solenoids as the source of the magnetic field. During the last two decades, however, performance of permanent magnets strongly improved such that their application now becomes increasingly interesting. Arvidson and Fritz developed a prototype of such a carrousel-type magnetic separator for use in ore processing [13]. The requirements to be complied with by a magnetic separator in wastewater treatment, however, strongly differ from those in ore processing. Separation efficiencies required usually exceed 90%, while the amount of rinsing water used may only be a fraction of the wastewater volume cleaned. In addition, shear forces inside the separator must be as small as possible in order to prevent the destruction of heavy metal hydroxide flocs for example. As a result, the suspension may not be supplied as a free jet, contrary to ore processing.

Taking into account the aspects mentioned above, a novel carrousel-type magnetic separator was developed within the framework of a cooperation of the Institute of Technical Chemistry and the Institute of Technical Physics of the Karlsruhe Research Center, Karlsruhe, Germany. This carrousel-type magnetic separator was to be especially suited for water treatment. While the first prototype of this separator still used a solenoid [14], the KMP2 prototype presented here is based on a rare-earth magnet. It is obvious from the schematic representation (Figure 4.5) that the core component of the separator is a rotating carrousel that is located inside a closed housing. The carrousel consists of a massive brass disk of 4 cm thickness. Along its circumference, 40 segments have been cut out by erosion. The individual segments are isolated from each other by 3-mm-wide walls. Inside the segments,

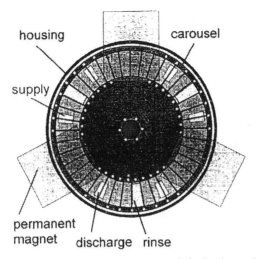

Figure 4.5 Schematic top view of the KMP2 prototype that is optimized with regard to separation efficiency.

matrix inserts made of a ferromagnetic wire mesh or steel wool are located. They are fixed by teflon seals that are screwed to the matrix frame at the top and bottom. In the area of the pole shoes of the permanent magnets, the housing is provided with openings for the supply and discharge of the solution to be cleaned as well as for rinsing. The opening for the supply of rinsing water is located opposite the opening for its discharge.

Figure 4.6 shows a side view of the separation zone in detail. As pointed out above, the splitting up of the filter wheel provides for a sealing of the individual matrix segments. Consequently, individual segments may be rinsed without influencing the separation process in the range of the magnetic field. It is evident from the figure, however, that the segments are not sealed from each other in the separation zone. By cuttings in the separator housing, which are located opposite each other in an offset manner, neighboring segments are connected. The cuttings are placed such that the suspension to be cleaned has to pass a number of matrix segments in a "zig-zag" manner while flowing from the supply to the discharge. Thus, the filter path available for the separation of the particles is increased by several factors as compared to a simple straight flow through the separation matrix.

If the filter wheel is rotated further by another segment, the filter insert moves beyond the supply range. It is sealed by the housing cover until it enters the range of a rinse opening after another rotation of the matrix frame. At the same time, an unloaded segment enters the range passed by the flow upstream of the discharge. The prototype developed represents the

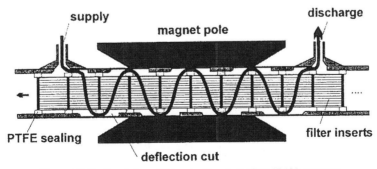

Figure 4.6 Sectional view of the separation zone of the KMP2 prototype.

Figure 4.7 Side view of the permanent magnet-based carrousel-type magnetic separator KMP2.

first continuous magnetic separator that approximates the principle of a counterflow of the separation matrix and the solution to be cleaned and allows complete control of the filtration rate inside the separator. Hence, continuous magnetic separation is achieved even with shear-sensitive precipitation products. Cleaning of the separation matrix outside the magnetic field allows for the use of permanent magnets. Thus, energy consumption of electric magnetic coils is avoided. Figure 4.7 shows a side view of the prototype. Table 4.2 gives a survey of the technical data of the prototype.

TABLE 4.2. Technical Data of the Carrousel-Type Magnetic Separator KMP2.

Maximum flux density of the rare-earth magnets	0.5 tesla
Energy product of the NdFeB material	270–320 kJ/m^3
distance between pole shoes (L × H × W)	150 × 102 × 62 mm
Carrousel diameter	500 mm
Energy consumption	40 W
Flow rate*	200–600 L/h
Separation efficiency*	80–96%
Filtration rate*	40–120 m/h

*The data refer to copper and phosphate elimination in combination with magnetic seeding. In the case of direct separation of ferrimagnetic microparticles, a slightly modified prototype (KMP3) reaches throughputs of up to 5–10 m^3/h.

Open-Gradient Magnetic Separators

In contrast to high-gradient magnetic separators, the so-called open-gradient magnetic separators (OGMS) do not possess a separation matrix. The magnetic force acting on the particles is exclusively based on the absolute strength and the gradients of the external magnetic field. It is a common feature of the OGMS used in ore processing that at least a partial component of the magnetic force resulting from the field gradients acts vertically to the flow paths of the fluid or the nonmagnetic particles. Hence, gradual demixing of the magnetic and nonmagnetic components takes place along the flow path. An open-gradient magnetic separator not based on this principle was first described by van Kleff [15]. In the simple setup he described, the magnetic force acts in an antiparallel direction to flow within a certain area (see Figure 4.8).

If the magnetic force acting on a particle exceeds the counteracting hydrodynamic drag forces, the particle is retained inside the area of the magnetic field. The area of high magnetic forces acts on these particles like a barrier that cannot be overcome. This is the reason this type of magnetic separator is referred to as a "magnetic barrier" by Franzreb [16]. During operation, the magnetizable particles increasingly accumulate in the magnetic separator. Due to the resulting, strongly increasing probability of floc collision, the magnetic barrier does not only cause a separation, but also improves the agglomeration and flocculation and, hence, leads to increased separation efficiency. To achieve stationary operation of the magnetic separator, it is necessary to provide for the accumulated particles being removed from the magnetic field area via a concentrate discharge. This gives rise to the problem that the particles have to overcome a "magnetic barrier" during concentrate discharge. The problem is solved by selecting a very small cross section of the concentrate discharge. This enables the strong hydrodynamic drag forces resulting from the high flow rate to overcome the magnetic forces. Figure 4.9 shows the first prototype of a magnetic barrier separator with improved sludge removal. The separation cell consists of a cylindrical PVC vessel that is tapered in the lower part. After lowering the magnetic separator into the room-tempera-

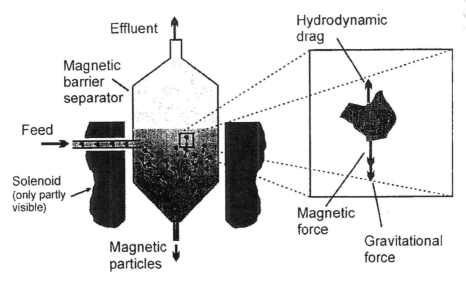

Figure 4.8 Principle of the magnetic barrier.

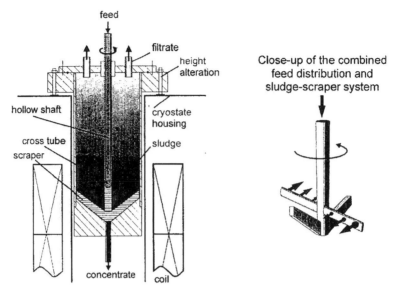

Figure 4.9 Prototype of a magnetic barrier with a sludge removal unit.

ture bore of a superconducting magnet system, the concentrate discharge of the separation cell is located on the same level as the center of the coil.

To promote sludge discharge, a removal unit is installed in the conical part of the separation cell. It is driven by an electric motor at a rotary speed of 1 min^{-1} via a hollow shaft. At the same time, the wastewater to be cleaned is supplied via this hollow shaft. For this purpose, a transverse tube with outlets on the sides is installed in the lower part of the hollow shaft, vertical to the axial axis. This and the slow rotation of the hollow shaft result in a homogeneous loading of the separator cross section with the inflowing suspension.

REMOVAL OF HEAVY METALS IN COMBINATION WITH CHEMICAL PRECIPITATION

Removal of heavy metals from wastewater appears to be an obvious field of use of magnetic technology, as the ferromagnetic properties of iron, cobalt and nickel are known to most engineers. However, ferromagnetism or ferrimagnetism is restricted to the metallic forms of the elements mentioned, sometimes to their alloys, and to a few iron oxides. Accordingly, magnetic separation of metal iron or steel particles represents probably the oldest and most common field of use of magnetic separators in water technology. In most cases, simple drum-type magnetic separators are applied, and high-gradient magnetic separators are used for very small particles. On the other hand, in wastewater treatment plants, the task of removing various dissolved heavy metal ions arises more frequently than that of separating metal particles. It is demonstrated by a balance of forces that, in contrast to the gas phase, a magnetic influence on ions dissolved in a liquid phase causes minute effects only. Due to the tendency of concentration balancing resulting from the Brownian movement, even a long influence of strong magnetic field gradients does not result in a marked local separation of the different heavy metal ions. In the case of dissolved heavy metal ions, application of magnetic separation technologies has to be preceded by a process step, in the course of which the heavy metal ions are converted into or bound to a solid compound.

The Ferrite Process

A method for the synthesis of magnetite, which has been known for a long time, is the precipitation of iron(II) ions by the addition of sodium hydroxide solution in the form of $Fe(OH)_2$, followed by oxidation of the precipitant with oxygen. The formed magnetite is an iron(II)/iron(III) mixed oxide having the solid structure of an inverse spinel. If bivalent heavy metal ions, e.g., Cu^{2+}, exist in the water during the oxidation reaction, they may be incorporated into the magnetite lattice under formation of the respective heavy metal ferrite and replace the iron(II) ions. This reaction is the basis of the so-called ferrite process. The heavy metal ions are incorporated into ferrites in accordance with the mechanism described above and are subsequently removed from the liquid phase by means of magnetic separators [17,18]. The heavy metals then form part of the crystal lattice and can hardly be eluted, even with a subsequent pH decrease. Due to the large surface area of the finely dispersed ferrite particles, ion exchange processes result in a slight sorption of heavy metal ions. To prevent later re-dissolution, the ferrite sludge remaining after magnetic separation is often subjected to a thermal treatment at 200–400°C. In the course of this thermal treatment, even those heavy metal ions that are initially bound by ion exchange only are incorporated into the crystal lattice.

The ferrite process may be described as follows:

- setting of a heavy metal/iron(II) mass ratio of 0.05–0.1
- increase in the solution temperature to about 65°C
- addition of sodium hydroxide solution up to pH 9–11
- oxidation with oxygen up to a redox potential of about 0 mV
- magnetic separation of the ferrite sludge generated by means of drum-type magnetic separators
- neutralization of the discharge

The ferrite sludge generated by this process usually has a specific magnetization of 70–80 A m^2 kg^{-1} and, hence, is only slightly less magnetic than pure magnetite. In the ideal case, the product may, therefore, be applied as a raw material for the production of low-cost ferrite applications. The advantages of the ferrite process include its large independence of the existing type of heavy metal mixture, the generation of a small amount of easily separable sludge with low water content and the incorporation of the heavy metals into the ferrite lattice. Moreover, the ferrite process allows low discharge concentrations to be reached even at high supply values. These advantages, however, are counteracted by a number of drawbacks such as the usually required addition of large amounts of Fe(II) salts, the resulting strong increase in salination, the slow kinetics and the required heating to about 65°C. Due to these drawbacks, the ferrite process has not yet been applied in Europe and the United States. In Japan, where it was developed, it is used in a number of fields, where treatment of comparatively small amounts of wastewater is required. Applications described in the literature include the disposal of chemical effluents from about 50 Japanese universities, the disposal of electroplating effluents as well as the experimental treatment of acid mine drainage with a high iron(II) content.

Studies on the Formation of Heavy Metal Ferrites

Within the framework of a thesis [19], it was attempted at the Institute of Technical Chemistry of the Karlsruhe Research Center to reduce the drawbacks of the ferrite processes by optimizing and varying the ferrite synthesis conditions. The large Fe^{2+}/Me^{2+} ratio required and the necessary reaction temperature of about 65°C were considered to be the major drawbacks. In the following sections, the influence of both parameters on ferrite formation will be illustrated using copper ferrite as an example. In addition, an alternative way of synthesizing ferrite shall be described. Here, good results are reached even at low temperatures. The experiments performed and the chemical relationships are presented in detail in References [20,21].

INFLUENCE OF THE TEMPERATURE AND THE HEAVY METAL/IRON RATIO

In the experiments described below, the ferrite process at 20°C was studied at variable Cu/Fe ratios. It was of particular interest to determine whether or not the partial formation of copper ferrite also takes place in the case of a copper surplus. In the case of a copper surplus, complete incorporation of the copper ions into a ferrite lattice is excluded stoichiometrically, but even partial ferrite formation would be sufficient to reach a mean magnetization of about 10 A m^2 kg^{-1}. Preliminary experiments demonstrated that an efficient separation of the reaction products by high-gradient magnetic separation was already achieved at magnetizations of this order. In this experiment, various amounts of copper sulfate were added to iron(II) sulfate solutions ($c_0 = 5$ mmol/L) in a thermostated stirred reactor ($V = 2.5$ L). By the subsequent addition of a stoichiometrically calculated amount of sodium hydroxide solution, iron(II) hydroxide and copper hydroxide were precipitated. The Cu/Fe mass ratios studied ranged between 0 and 4. Oxidation of the hydroxide suspension was performed by the injection of synthetic air until both the pH value and the redox potential reached a constant final value. It was demonstrated by an investigation of the reaction products that, under the conditions selected, copper ferrite or magnetite may even be formed at 20°C. This is confirmed by a comparison of the X-ray diffractograms of the products with the respective reflex positions and reflex intensities of ferrites (ASTM map 19-629). As shown in Figure 4.10, the reflexes caused by the ferrite lattice were found to be markedly reduced with an increasing Cu/Fe ratio. In addition, other reflexes appear. They may be assigned to the tenorite copper oxide (CuO) (ASTM map 45-937).

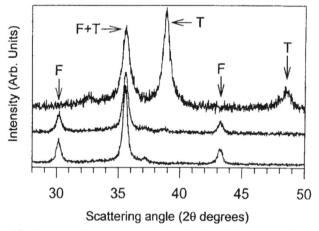

Figure 4.10 X-ray diffractograms of the reaction products at Cu/Fe = 1/10 (bottom), 1/7 (center) and 4/1 (top). Scattering angle for Cu/Kα radiation. The reflexes to be assigned to the ferrite and tenorite lattice are marked by "F" and "T," respectively.

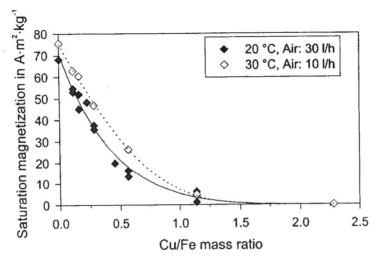

Figure 4.11 Specific saturation magnetizations of the reaction products as a function of the Cu/Fe mass ratio set at the beginning of the reaction at various temperatures and variable synthetic air volume flows.

Tenorite may have been generated from copper hydroxide during the reaction or during subsequent drying. Consequently, the presence of copper has a direct influence on the reaction in a way that amorphous iron(III) oxide hydrates, copper oxide and copper hydroxide are formed instead of the ferrite lattice. In the concentration range studied, the Cu/Fe ratio only is of importance, because the same products were obtained at various initial iron(II) concentrations. As expected, product composition directly affects the magnetic properties, because amorphous iron(III) oxide hydrate and copper hydroxide and copper oxide exhibit paramagnetic behavior. This results in a rapid decrease of the saturation magnetizations measured with increasing Cu/Fe mass ratio (Figure 4.11).

Having selected a limit value of 10 A m^2 kg^{-1}, it may be concluded that the desired medium saturation magnetization of the reaction products can be reached at a reaction temperature of 20°C and at Cu/Fe ratios of up to about 0.75. In comparison with the maximum Cu/Fe ratio of 0.1 of the initial ferrite process, this means a reduction of the required amount of iron(II) salt to about 13%. This high reduction leads to significantly worsened magnetic properties and the resulting necessity of separation by means of a high-gradient magnetic separator. In addition, the product generated no longer has the crystalline character of pure ferrites. A mixture of crystalline particles and amorphous precipitation products develops. Compared to pure precipitation of the heavy metals as hydroxides, however, the precipitant formed is more compact and easier to dehydrate. If another increase in the permissible Cu/Fe ratio is desired, this may be achieved to a certain extent by an increase in the reaction temperature and a reduction of the reaction rate by reduced air injection. As shown in Figure 4.11, satisfactory saturation magnetizations were reached at a reaction temperature of 30°C and an air volume flow of 10 L/h at Cu/Fe ratios of down to 1. These results show that partial ferrite formation may be achieved for a number of wastewaters without an addition of iron(II).

FERRITE FORMATION FROM FE(II)/FE(III) MIXTURES

As known from pure magnetite formation, synthesis from iron(II) and iron(III) hydroxide mixtures exhibits a much smaller temperature dependence than synthesis via oxidation of iron(II) hydroxide. This is why the so-called "mixing method" was also studied in connection with the ferrite process. For ferrite synthesis according to the mixing method, a molar ratio of $Fe^{3+}/(Fe^{2+} + Me^{2+})$ of 2:1 and a stoichiometric amount of sodium hydroxide solution for precipitation of all heavy metals as hydroxides were selected. The heavy metal/iron molar ratios given below refer to the total amount of iron used.

Figure 4.12 shows the saturation magnetizations of reaction products synthesized by the mixing method at variable Cu/Fe molar ratios and a reaction temperature of 5°C. Disturbance of ferrite formation by copper ions and the resulting effects on saturation magnetization qualitatively correspond to the effects of copper ions on the oxidation method results given for comparison. However, disturbance also occurs at smaller Cu/Fe molar ratios. Hence, insufficient saturation magnetizations are reached from a ratio of about 0.15. It must be pointed out that both synthesis methods can hardly be compared quantitatively due to the varying reaction temperatures. As known from experiments regarding pure magnetite synthesis, an exclusively paramagnetic product is obtained by the oxidation method under the experimental conditions prevailing at a reaction temperature of 5°C. Consequently, the mixing method only would be suited for a ferrite process with wastewaters of very low temperature.

The mixing method is found to be even more suitable in the case of ferrite formation from cadmium-containing wastewaters. As obvious from Figure 4.13, the oxidation method at 20°C down to a Cd/Fe molar ratio of 0.05 yields insufficient saturation magnetizations of the reaction products. In the case of ferrite synthesis according to the mixing method, however, satisfactory magnetic properties of the reaction products were achieved even at a reaction temperature of 5°C down to a Cd/Fe molar ratio of about 0.15.

Magnetic Seeding

In contrast to the ferrite process, heavy metal elimination by magnetic seeding largely corresponds to heavy metal elimination by conventional precipitation/flocculation with

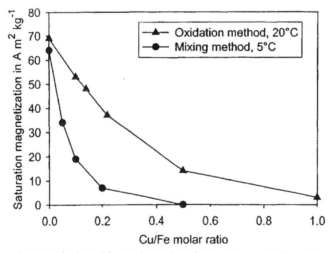

Figure 4.12 Saturation magnetizations of the reaction products from a copper-containing solution according to the "mixing method" in comparison with the values reached for the "oxidation method" (air volume flow: $30 \, l \cdot h^{-1}$).

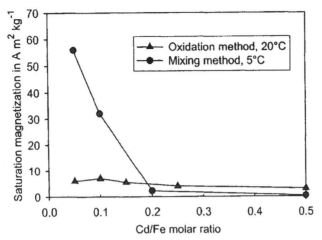

Figure 4.13 Saturation magnetizations of the reaction products from a cadmium-containing solution according to the "mixing method" compared to those obtained by the "oxidation method."

subsequent filtration. Figure 4.14 shows a schematic representation of the basic process steps in the separation of dissolved, particulate or already magnetic substances according to this principle. In addition, the figure indicates the possibility of returning the used seeding material. For heavy metal elimination, the path given for dissolved substances usually is of decisive importance. The heavy metal ions are first converted into a solid compound by precipitation. As precipitating agents, conventional substances such as sodium hydroxide solution, lime or sodium sulfide may be used. The microparticles formed are then combined to macroflocs in a flocculation stage under the addition of flocculants. In contrast to conventional precipitation, fine magnetic particles are added to the wastewater as seeding material. This is why the process is called "magnetic seeding." In most cases, the seeding material is natural magnetite with particles ranging in size from 5 to 50 µm. Its dosage usually amounts to 1–10 g per L wastewater. Due to the resulting high concentration of magnetite particles,

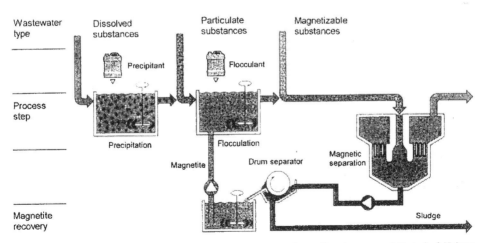

Figure 4.14 Basic setup for heavy metal elimination by magnetic seeding (courtesy of Oosterhof Holman Milieutechniek).

agglomeration of the heavy metal containing microparticles nearly always takes place on their surface. A composite consisting of a heavy metal compound and magnetite particles is generated. Due to their magnetic properties, the composite particles can be separated efficiently by magnetic separators even with high filtration rates. The cleaned filtrate leaves the separator. Depending on its pH value, it may have to be neutralized before it is discharged into the outlet channel. The concentrate that arises is subjected to another dehydration. As an option, the composite particles may also be separated again into magnetite particles and heavy metal compound by mechanical and chemical impacts. The magnetite may then be separated by means of drum-type magnetic separators, scrubbed, and eventually returned to the flocculation process.

SURVEY OF PROCESS VARIANTS DESCRIBED IN THE LITERATURE WITH REGARD TO HEAVY METAL ELIMINATION BY MAGNETIC SEEDING

One of the first literature sources where the possibility of heavy metal elimination by magnetic seeding was described was published by deLatour [22–26]. By the addition of 1,000 mg/L magnetite with a particle size of about 10 μm and a subsequent increase in the pH value to pH = 8 –10, a heavy metal elimination of up to 99% was achieved when using a solution with a copper concentration of 5 mg/L. The loaded magnetite particles were separated by high-gradient magnetic separation at filtration rates of up to 720 m/h. Experiments using similarly high amounts of magnetite were also performed by Anand et al. [27]. At sufficiently high pH values, initial separation rates of up to 99.9% were reached in simple breakthrough experiments. However, neither deLatour nor Anand provided any information on the separation efficiencies reached after several cycles. Terashima et al. used a somewhat higher heavy metal/magnetite ratio than deLatour or Anand [28]. In experiments with gas scrubber water from a waste incineration plant, for instance, 500 mg/L magnetite were applied. Precipitation was accomplished by the addition of calcium polysulfide, $FeCl_3$ (50 mg/L Fe^{3+}), and a non-ionic flocculation aid (1 mg/L). While most of the flocs formed were separated by sedimentation within a period of 5 min only, fine cleaning of the remaining portion was carried out using high-gradient magnetic separation at a filtration rate of 100 m/h and a field of 0.8 T. The obtained separation efficiencies for various water constituents are presented in Table 4.3.

Another process variant for heavy metal elimination using natural magnetite as a seeding material has been commercialized by the company Oosterhof Holman Milieutechniek (Arnhem, The Netherlands; formerly Envimag B.V.). In this process, 180 m^3/h wastewater with 1 mg/L Cu are cleaned to residual concentrations of 10 μg/L Cu. Precipitation/flocculation takes place under the addition of sodium sulfide (10 mg/L S^{2-}) and a polyelectrolyte (2 mg/L). The addition of 1 g/L magnetite during flocculation allows the particles generated to be separated by means of a specially developed high-gradient magnetic separator at a flux density of 0.2 tesla. The rinsing water fraction amounts to 0.2%. Energy consumption of the facility is 4.5 kW.

TABLE 4.3. Results of the Treatment of Heavy Metal-Containing Gas Scrubber Water by Precipitation Using Calcuim Polysulfide, Sedimentation and Fine Cleaning with HGMS [28].

	Cd	Pb	Zn	Cr	Cu	Mn	Fe	Hg	B	F	TOC	SS*
Supply [mg/L]	0.35	4.8	17.0	1.0	0.7	0.2	4.1	15.0	13.1	58	18	27
Discharge [mg/L]	0.01	0.05	0.05	0.11	0.30	0.24	12.1	0.003	10.7	45.1	8.5	4.7

*SS: suspended solids.

As an alternative to the use of natural magnetite, Hencl proposed a synthetically produced magnetite suspension for use as seeding material [29]. Experimental investigations of this type are reported by Choi, Calmano and Förstner [30,31]. In the experiments, a magnetite sludge generated from $FeSO_4$ with concentrations ranging from 150 to 700 mg/L Fe_3O_4, was added to the heavy metal containing model sewage or wastewater from electroplating. After setting the pH by means of sodium hydroxide solution and a subsequent flocculation time of about 20 min, the magnetite-containing flocs formed were separated by sedimentation that was accelerated by magnetic coils. Compared to conventional NaOH precipitation, use of the synthetically produced magnetite sludge yielded a better heavy metal elimination and smaller sludge volumes. Due to the high magnetite concentrations used, however, industrial exploitation of this variant of magnetic seeding appears to be promising only in combination with effective magnetite recovery. But, just for these synthetically produced magnetite particles, recovery is very difficult due to the small particle size of <1 μm and would require considerable expenditure. The approach chosen by the author and discussed below was, therefore, based on the premise of reducing the concentration of the magnetite sludge used as seeding material in the wastewater to such an extent that magnetite recovery was no longer necessary.

STUDIES WITH REGARD TO HEAVY METAL ELIMINATION BY MAGNETIC SEEDING AND SUBSEQUENT HIGH-GRADIENT MAGNETIC SEPARATION

The studies on magnetic seeding allow the conclusion to be drawn that natural magnetite has been the most frequently used seeding material. This may be attributed to the low costs and the excellent magnetic properties of magnetite particles having a size of about 5–50 μm. However, these advantages are counteracted by the small specific surface area of natural magnetite. To obtain a sufficient seeding effect, high dosages that often exceed 1 g/L wastewater have to be applied. The use of such high concentrations, however, makes it necessary to provide for an effective magnetite recovery. Only then can commercial exploitation be ensured. The approach preferred by the author is to use smaller amounts of synthetic magnetite without making any recovery attempt. Due to the very small size of magnetite particles produced by wet chemical processes, they can be suspended easily in the wastewater to be treated. During subsequent precipitation/flocculation, the resulting high particle concentration serves as an effective crystallization aid. The magnetite particles are enclosed in the developing flocs. Due to the small absolute mass of the dosed magnetite particles, however, the floc density is hardly increased. Hence, the sedimentation rate is only slightly accelerated. Moreover, mean floc susceptibility only reaches values that are characteristic of paramagnetic substances. The flocs generated largely keep the character of shear-sensitive, amorphous precipitation products. Therefore, a simple separation of the precipitation products by drum-type magnetic separators is not possible. High-gradient magnetic separators are required.

The results of preliminary experiments performed on laboratory scale with regard to this approach of a strongly reduced magnetite dosage can be found in Reference [32]. The experiments comprised the measurement of the breakthrough curves of simple, cylindrical high-gradient magnetic separators that were located in the magnetic field of an electro-magnet or superconductive magnet system. The separation efficiency achieved by a high-gradient magnetic separator in real applications, however, is additionally determined by the rinsing efficiency of the separation matrix. Due to the high field strengths prevailing during the separation process, the Weiss fields within the magnetite particles are oriented. As a result, the particles still possess a certain residual magnetization after the de-excitation

of the magnetic field. This residual magnetization and the existence of hardly accessible locations in the rigid separation matrix are responsible for the fact that filter recleaning by
rinsing never meets with 100% success. It is of decisive importance as to whether the remaining loads accumulate in the course of the separation cycles and, hence, lead to increasingly worse separation results, or whether a stationary state with a constant and very small
residual load develops after a few cycles. To answer this question, tests were performed under cyclic operating conditions at an automated test facility with a strongly reduced magnetite dosage. Figure 4.15 shows a simplified schematic representation of this facility. As the
source of the magnetic field, an electro-magnet was used. A short-term maximum field of 1
tesla and a maximum flux density of about 0.5 tesla were reached during permanent operation.

The results of an experiment with 24 separation and rinsing cycles are plotted in Figure
4.16. The test suspension contained magnetite-containing copper hydroxide [c(Cu) = 50
mg/L] with a copper/iron mass ratio of 1. The relatively high magnetite dosage was selected
among others to rapidly obtain reliable results in case of a continuous accumulation of magnetite in the filter.

It is evident from Figure 4.16 that stationary filter behavior develops as of the second
separation cycle, i.e., upon a single rinsing process. The term "stationary" here does not refer to the discharge concentrations within a separation cycle, of course, but to the filter behavior averaged over several cycles. For rinsing, about five bed volumes of filtrate were
spent. The total filtrate/concentrate ratio produced amounted to 4:1. The mean discharge
concentration of copper was about 0.7 mg/L during permanent operation. This means that
the process studied does not ensure that the concentration remains below the limit value of
0.5 mg/L specified by German law in spite of the high rinsing water fraction of 20%. The
reasons include the remaining residual load of the separation matrix even after the completion of the rinsing process as well as the rapid breakthrough of the copper hydroxide flocs
during the separation cycles. An alternative to the use of cyclically operating high-gradient
magnetic separators is the continuously operated carrousel-type magnetic separator. The
setup and functioning principle of the carrousel-type magnetic separator were presented in

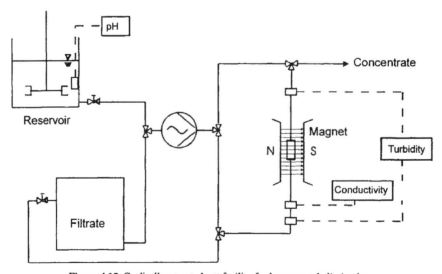

Figure 4.15 Cyclically operated test facility for heavy metal elimination.

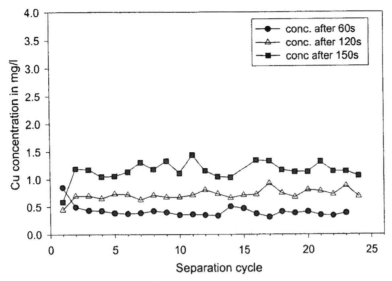

Figure 4.16 Results of the cyclic operation of a HGMS for the separation of magnetite-containing copper hydroxides. The values measured reflect the copper concentration at the discharge after 60, 120 and 150 s of each separation cycle. Cu:Fe = 1, $c_0(\mathrm{Cu})$ = 50 mg/L, B = 0.61 tesla, v_0 = 67 m/h.

the section on "High-Gradient Magnetic Separators." As obvious from the results given in Figure 4.17 for the example of magnetite-containing copper hydroxides, however, the filtrate concentration also exceeded the limit values specified by law. Furthermore, the filtrate/concentrate volume ratio of about 5:1 turned out to be not much better than in the case of a cyclic operation, in spite of the reduced supply concentration.

The experiments allow the conclusion to be drawn that using the process developed is only reasonable in cases where the heavy metal supply concentration is below about 5–10 mg/L. It is interesting to note that the competing seeding process, which is based on the use

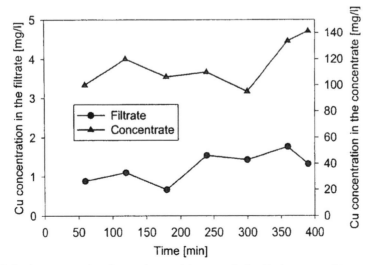

Figure 4.17 Continuous separation of magnetite-containing copper hydroxides by a carrousel-type magnetic separator. $c_0(\mathrm{Cu})$ = 25 mg/L; $c_0(\mathrm{Fe_3O_4})$ = 28 mg/L; B = 0.4 tesla; v_0 = 38 m/h.

of high dosages of natural magnetite, so far has only been accepted for one case in which the heavy metal concentration supplied amounted to about 1 mg/L. The restriction to wastewater of low concentration therefore seems to apply to both approaches used for heavy metal elimination by high-gradient magnetic separation.

INVESTIGATIONS WITH REGARD TO HEAVY METAL ELIMINATION BY MAGNETIC SEEDING AND SUBSEQUENT FLOC SEPARATION BY MAGNETIC BARRIERS

In view of the difficulties associated with the rinsing process of a high-gradient magnetic separator, which were discussed in the previous section, it was of interest to find out to what extent the development of a continuously operating magnetic barrier might represent an alternative for the separation of magnetite-containing heavy metal hydroxides. The model water studied contained 100 mg/L copper ions, which means that the concentration exceeded the investigated concentrations for HGMS. The maximum flux density gradient in the magnetic barrier amounted to 46 tesla/m at a set flux density of 5 tesla. When using the magnetic barrier, the filtrate/concentrate volume ratio may be controlled by adequately setting the capacities of the respective pumps. For a ratio of 20:1, a concentrate volume fraction of 5% is obtained, which corresponds to a rinsing water fraction of 5% in high-gradient magnetic separation. As obvious from Figure 4.18, a separation efficiency of 95% is achieved by the use of magnetic barriers at a concentrate volume fraction of 5%, even if the heavy metal concentration is comparatively high. When reducing the concentrate volume fraction to 2%, which corresponds to a volume flow ratio between the filtrate and the concentrate of 50:1, separation efficiencies exceed 90%. The copper contents in the concentrate reach about 5 g/L.

In the experiments, the magnetite dosage characterized by the Cu/Fe ratio supplied corresponded to the magnetite dosage in the case of HGMS. Due to the fact that magnetic barriers are particularly suited for the elimination of higher heavy metal concentrations, it was interesting to study whether or not the Cu/Fe ratio can be further increased without strongly reducing the separation efficiency. In the corresponding experiments (see Figure 4.19),

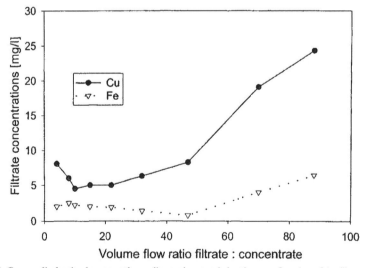

Figure 4.18 Copper elimination by magnetic seeding and magnetic barriers as a function of the filtrate:concentrate volume flow ratio. $c_0(Cu) = 100$ mg/L; $c_0(Fe_3O_4) = 125$ mg/L; Cu:Fe = 1.1; $B = 5$ tesla; $v_0 = 6$ m/h.

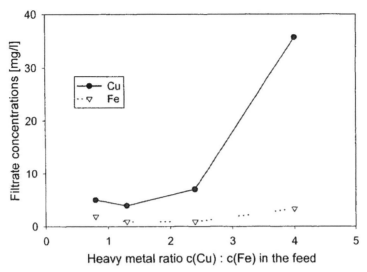

Figure 4.19 Copper elimination by magnetic seeding and magnetic barriers as a function of magnetite dosage, expressed as $c(Cu):c(Fe)$; $c_0(Cu) = 100$ mg/L; Q (filtrate):Q (concentrate) = 20; $B = 5$ tesla; $v_0 = 6$ m/h.

separation efficiencies exceeding 90% were reached up to a Cu/Fe ratio of 2.5, which corresponds to a magnetite dosage of 55 mg/L. When comparing the curves for Cu and Fe, it can be seen that the Cu/Fe ratio in the filtrate always exceeds that of the supply. This can be explained by the preferred separation of copper hydroxide flocs having an exceptionally high magnetite content.

Small filtration rates are a major drawback of magnetic barriers. The filtration rate of 6 m/h used in the examples above exceeded the value applied in the sedimentation basin for the separation of hydroxide flocs by about one order of magnitude, but it is unusually low in comparison with other magnetic separation processes. The main reasons include the fact that no separation matrix is used and that the magnetite content of the flocs to be separated is low. The aim of the test series represented in Figure 4.20 was to find out to what extent an increase in filtration rate is associated with a reduction of separation efficiency. For this purpose, the filtration rate was increased to 14 m/h, and the filtrate-to-concentrate ratio of 20:1 was maintained.

It was confirmed by measurements of the heavy metal concentrations in the filtrate that the copper and iron concentrations increased to more than 40 mg/L and 25 mg/L, respectively, at a filtration rate of about 11 m/h. In contrast to this, it was found out by an estimation of the forces acting on the flocs that the magnetic force acting in the direction of sludge discharge exceeds the hydrodynamic resistance force by several factors. A possible reason for the rapid increase in the heavy metal concentrations observed in the filtrate might be the slow kinetics of sludge compression in the magnetic barrier. The large amount of flocs entering the magnetic barrier in the case of high filtration rates and supply concentrations per time unit cannot be dehydrated sufficiently in the course of the mean residence time available. Consequently, the flocs are retained at the beginning, but the amount of sludge generated per time unit is larger than the volume flow of concentrate discharge. Hence, no stationary state develops in the magnetic barrier. The sludge bed in the lower part of the magnetic barrier grows with time and eventually enters the filtrate. With small supply concentrations, the permissible filtration rates are, therefore, expected to increase.

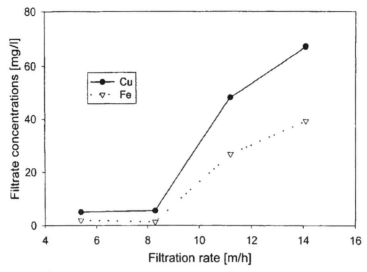

Figure 4.20 Copper elimination by magnetic seeding and magnetic barriers as a function of the filtration rate. $c_0(Cu) = 100$ mg/L; $c_0(Fe_3O_4) = 140$ mg/L; Q (filtrate):Q (concentrate) = 20; $B = 5$ tesla.

As far as the combination of magnetic seeding and magnetic barriers for heavy metal elimination from wastewaters is concerned, it must be noted that this process is well suited to produce concentrate volume flows of less than 5% of the supply volume, even at high supply concentrations. The resulting discharge concentrations, however, exceed the discharge limits specified by German law. Therefore, potential field of application of magnetic barriers is the preliminary cleaning of highly polluted wastewater, followed by an end filtration. It is interesting that the permissible concentration ranges of magnetic barriers and high-gradient magnetic separators nearly ideally complement each other in the case of heavy metal elimination. A high-gradient magnetic separator would reduce the residual concentrations of about 5–10 mg/L remaining in the filtrate of the magnetic barrier to below the limit value of 0.5 mg/L. Due to the low supply concentrations, working cycles of 10–20 minutes' duration would have to be expected. The concentrate arising during subsequent rinsing of the separation matrix could be returned to the supply of the magnetic barrier, as a result of which the total amount of rinsing water would be rather small. It might even be possible to combine both magnetic separation processes in a single facility. In this case, the magnetic barrier would use the range of high flux density gradients within a short, superconducting coil. The flux densities of the scattered fields outside of the coil would then be sufficient for subsequent high-gradient magnetic separation.

REMOVAL OF HEAVY METALS USING MICROORGANISMS AND MAGNETIC SEPARATION

Background

There is a large amount of literature focusing on the interaction between microorganisms and metal ions in solution and in the recovery of metals from solutions [33,34] but this section will consider only those methods that use microorganisms in conjunction with magnetic separation (HGMS) to remove heavy metals from solution. The metal ions may be taken up by microorganisms intracellularly, by adsorption of the heavy metal onto the cell wall of the microorganism and/or exopolysaccharides produced by the cell, by processes in which the heavy metal ions enter the microorganism and are precipitated and immobilized within the cell and by precipitation of a metal compound onto the surface of the microorganism. However, it has been been discovered that a number of the metal compounds that can be precipitated onto the surface of the cells are very powerful adsorbents for heavy metals. This section will consider only those precipitated compounds that have enough magnetism to be recovered by magnetic separation. The next section discusses the methods in which metal ions are adsorbed onto the cell walls. Intracellular methods, methods (aerobic and anaerobic) using biochemical precipitation onto the surface of the cells and methods that use the metal ion compounds precipitated by microorganisms to adsorb heavy metals will be discussed in the following sections. The last two sections will present the production and the properties of the strongly magnetic FeS adsorbent and will outline a number of applications where the technology may be used.

Adsorption of Metal Ions onto the Surface of Microorganisms

Adsorption processes in connection with magnetic separation are effective enough in many cases to be considered an attractive method of removing metal ions from solution. This can be illustrated by a simple calculation. Consider a spherical bacterium with a radius of 0.4 mm with surface area of 2.01×10^{-12} m^2. If each adsorbed ion occupies 0.1 nm^2 then the number of ions adsorbed is 2.01×10^7 ions. The volume of the bacterium is 2.68×10^{-19} m^3, so the number of ions per unit volume N is 7.5×10^{-25} m^{-3}. The magnetic susceptibility χ (SI) is given by the following:

$$\chi = \frac{Np^2\mu_B^2\mu_0}{3k_BT} \qquad (4)$$

where p is the number of Bohr magnetons , μ_B is the Bohr magneton 9.27×10^{-24} J $\times$ T^{-1}, k_B is the Boltzmann constant 1.38×10^{-23} JK^{-1} and T (K) is the temperature. The magnetic susceptibilities for various ions and the resulting magnetic velocities v_m [see Equation (3)] of a bacterium to which the different ions have been adsorbed are presented in Table 4.4. Also presented is a similar calculation for a spheroidal microorganism with a major semi-axis of 2 μm and a circular minor axis of 0.4 μm, but although the magnetic susceptibilities are very similar, the magnetic velocities of the spheroidal particles are much larger due to the larger particle size. If it is necessary to capture other metals that are weakly paramagnetic or diamagnetic, it would be necessary to share the adsorption centers between the two ions, which would further reduce the magnetic susceptibility and the processing velocity v_m. When $v_m \geq v_0$, the capture of the microorganism is very strong, and the separator behaves like a filter. The magnetic velocity is, therefore, an extremely important parameter in the design of magnetic separation or filtration systems.

There are no industrially viable processes in which the adsorption of metals to microorganisms is followed by high-gradient magnetic separation with the objective of the recovery of the metals. As shown in Table 4.4 this could be achieved in a number of cases but has not been done in practice. There are, however, some applications where the process is carried out with the objective of separating the microorganisms. Zabrowski et al. [35] used *Escherichia coli, Klebsiella pneumoniae, Proteus mirabilis, Psuedomonas aeruginosa, Staphylococcus epidermidis, Staphylococcus saprophyticus* and *Enterococcus faecalis* to

TABLE 4.4. Magnetic Velocity and Loading.

Spherical Microorganism: Radius, 0.4 μm				
Ion	p	$\chi \times 10^5$	Loading (%)*	v_m (mm·s^{-1})
Fe^{3+}	5.92	2.28	0.56	0.51
Fe^{2+}	5.42	1.92	0.56	0.14
Ru^{3+}	2.09	0.28	0.81	0.02
Nd^{3+}	3.8	0.94	1.45	0.07
Dy^{3+}	10.5	5.76	1.30	1.63
Spheroidal Microorganism: Major Semi-Axis 2 μm and Circular Minor Radius, 0.4 μm				
Ion	p	$\chi \times 10^5$	Loading (%)*	v_m (mm·s^{-1})
Fe^{3+}	5.92	1.82	0.45	84.3
Fe^{2+}	5.42	1.53	0.45	70.7
Ru^{3+}	2.09	0.23	0.81	10.5
Nd^{3+}	3.8	0.75	1.16	34.8
Dy^{3+}	10.5	5.75	1.30	265

*In the loading calculations, it was assumed that the microorganism had a density of 1,250 kg·m^{-3} and the loading value is the total weight of the metal ion uptake divided by the weight of the unloaded microorganism, expressed as a percentage.

adsorb Er^{3+} onto the cell walls from a solution of $ErCl_3$. The cells were then recovered from suspension by magnetic separation. No significant difference in magnetic deposition was observed between individual strains or between gram-positive and -negative bacteria. It was suggested that magnetic isolation of cells may find application in rapid total cell count determination, such as rapid urine screening [35]. In summary, in order to remove metals by adsorption followed by magnetic separation, the uptake must be high, effectively a monolayer, and if the target metal is not magnetic, then sufficient magnetic material must also be adsorbed. This chain of concurrent requirements means that the opportunities for commercial applications are limited.

There is another interesting approach that is to use magnetotactic bacteria [36]. These magnetotactic microorganisms possess flagella and contain structured particles of magnetite within intracytoplasmic membrane vesicles. Conceivably, these particles impart a magnetic moment to cells. This explains the observed migration of these organisms in fields as weak as 0.5 gauss. If magnetotactic bacteria can adsorb heavy metals, the extraction of the metal does not depend on the magnetic moment provided by the adsorbed metal but on the response of the internal magnetic moment to the applied field which can be low [37]. In subsequent work, Bahaj, Croudace and James [38] attempted to remove Co from solution. Starting from a 5 ppm solution of Co, after 96 h the uptake was 8%. Other metals may have a higher uptake.

Intracellular Absorption of Heavy Metals by Microorganisms

The limitations discussed above with respect to adsorption of material to the cell walls of microorganisms apply equally to intracellular adsorption with the added feature that poisoning of the cellular chemistry is more likely. Lloyd et al. [39] also found that anaerobic but not aerobic cultures of *Escherichia coli* accumulated Tc(VII) and reduced it to a black insoluble precipitate. Electron microscopy in combination with energy-dispersive X-ray analysis showed that the site of the Tc deposition was intracellular. They suggested that Tc precipitation was a result of enzymatically mediated reduction of Tc(VII) to an insoluble oxide. Formate was an effective electron donor for Tc(VII) reduction that could be replaced by pyruvate, glucose or glycerol but not by acetate, lactate, succinate or ethanol. Work on mutant types of *Escherichia coli* suggested a role for the formate hydrogenlyase complex in Tc(VII) reduction, and further observations confirmed that the hydrogenase III (Hyc) component of formate hydrogenlyase is essential and sufficient for Tc(VII) reduction. Further work reported by Lloyd et al. [40] showed that *Desulfovibrio desulfuricans* reduces Tc(VII) (TcO4-) with formate or hydrogen as electron donors, whereas for *Escherichia coli* the reaction is catalyzed by the hydrogenase component of the formate hydrogenlyase complex (FHL), in *D. desulfuricans* the reduction is associated with periplasmic hydrogenase activity. In order to consider this reduction of the Tc(VII) to an insoluble form as a potential process for the treatment of nuclear waste, the effect of the presence of concentrations of nitrate ions up to at least 2 M must be investigated. Tc(VII) reduction in *E. coli* by H_2 and formate was either inhibited or repressed by 10 mM nitrate. By contrast, Tc(VII) reduction catalyzed by *D. desulfuricans* was less sensitive to nitrate when formate was the electron donor, and was unaffected by 10 mM or 100 mM nitrate when H_2 was the electron donor. The optimum pH for Tc(VII) reduction by both organisms was 5.5, and the optimum temperature was 40°C and 20°C for *E. coli* and *D. desulfuricans,* respectively. Tc(VII) was removed from a solution of 300 nM TcO_4^- within 30 h by *D. desulfuricans* at the expense of H_2. *D. desulfuricans* showed greater bioprocess potential attributable to the more accessible, periplasmic localization of the enzyme. The relative rates of Tc(VII) reduction for *E. coli* and *D. desulfuricans* (with H_2) were 12.4 and 800 m·mole Tc(VII) reduced/g biomass/h. The more rapid reduction of Tc(VII) by *D. desulfuricans* compared with various *E coli* strains was also shown by Lloyd et al. [40] using cells immobilized in a hollow-fiber reactor. A 17 ml glass reactor that contained a fiber bundle composed of 12 Romicon XM50 acrylic hollow-fiber membranes of 1.1 mm internal diameter (Romicon, Woburn, MA, USA) was used throughout this study. The reactor was as described previously [39], and the inoculation with cells and its operation has been described by Lloyd and Bunch [41]. An aerobic starter culture of *E. coli* was used to inoculate the reactor using a method adapted from that of Lloyd and Bunch [41]. The starter culture (10 ml; prepared as described previously) was inoculated into the shell side of the reactor through an inoculation port, using a sterile 20 ml syringe fitted with a hypodermic needle. Cells were delivered into the outer matrix of the hollow fibers by pumping phosphate-buffered saline (PBS: 20 mM phosphate buffer, pH 7, in 8.5 mM NaCl) into the reactor from the shell side, through the walls of the hollow fibers and into the lumen of the latter at a flow rate of 40 ml/hour for one hour. The shell side of the reactor was then rinsed with PBS (flow rate of 60 ml/hour for 30 min) to remove residual cells in this compartment. The cells were grown in situ for 24 h, with growth medium [Oxoid Nutrient Broth Number 2 supplemented with glycerol, 0.5% (V/V) and fumarate, 25 mM] supplied at a flow rate of 7.5 ml/hour from the shell side of the reactor, which was operated in transverse mode. *D. desulfuricans* was grown in sealed serum bottles as described and was inoculated into the reactor by pumping 200 ml of washed cells in 20 mM MOPS buffer (pH 7.0) into the reactor at a flow rate of 50 ml/h. The final biomass concentration in the 17 ml reactor was equivalent to 5 g dry weight cells per liter reactor volume, which was directly comparable to that of the reactor containing *E. coli* as described previously [42]. During reactor operation, 20 mM MOPS-NaOH buffer (pH 7) containing 50 μM Tc(VII), with formate (25 mM) supplied as an electron donor, was pumped into the reactor from the shell-side port. The reactors were maintained at 30°C.

Precipitation of Metals and Metal Compounds onto the Surface of Microorganisms

INTRODUCTION

Macaskie and Dean [43] worked out a technique in which glycerol 3-phosphate is the carbon source. This substrate does not enter the cell and is cleaved by an exocellular phosphatase to give a high concentration of phosphate anions at the cell surface. These phosphate ions react with metal ions present in solution to form a precipitate on the cell surface of the microorganism. The scavenging of uranyl ions from solution was studied with two microorganisms grown on glycerol 3-phosphate to induce phosphatase activity in accordance with the procedure described by Macaskie and Dean [43] and others [44–46]. The cells were then incubated with glycerol 3-phosphate in the presence of uranyl ions. Uranyl ions are strongly paramagnetic, so the coated organisms can then be successfully concentrated by high-gradient magnetic separation (HGMS) [31,47,48]. The aerobic process and the anaerobic process consist of four main steps as follows:

(1) The production of a concentrate of the microorganisms and material they produce (from a culture of a selected strain of bacteria or yeast with the addition of suitable nutrients) in a batch or a continuous fermenter called the *breeder* takes place.

(2) The reaction between the microorganisms and materials produced by them, the metal ions, in the solution to be treated, and various added chemicals takes place in a tank called the *reactor.* The microorganisms become loaded with the metallic compounds with time and with a corresponding decrease of the concentration of metal ions in solution. Chemicals are added to pretreat (e.g., to adjust the pH) or condition the solution and then nutrients and microorganisms are added. Residence times of the order of one day or less are required for efficient metal removal.

(3) The microorganisms and their acquired metallic compounds are collected by high-gradient magnetic separation as the contents of the reactor are passed through the *separator.* Organic material, metal ions remaining in solution or other nonmagnetic particles pass through the separator unhindered.

(4) The material is collected in the separator on a ferromagnetic fine-wire or other matrix operating in a high magnetic field. This material can be recovered from the matrix in a concentrated sludge by flushing the matrix with the separator electromagnet switched off. This material is collected in the *sludge tank.* In some applications, the sludge contains the valuable material, and in others, it contains the waste in a highly concentrated form. It is assumed that this concentrate is suitable for treatment by the usual mineral processing methods.

AEROBIC PROCESS USING INSOLUBLE METAL PHOSPHATES

In the work reported here, the scavenging of uranyl ions from solution was studied with two microorganisms grown on glycerol 3-phosphate to induce phosphatase activity in accordance with the procedure described by Macaskie and Dean. The cells were then incubated with glycerol 3-phosphate in the presence of uranyl ions. Uranyl ions are strongly paramagnetic, so the coated organisms behave like a paramagnetic material in a magnetic field. The cells can then be successfully concentrated by high-gradient magnetic separation (HGMS).

Of the microorganisms studied, *Candida utilis* and *Bacillus subtilis* strains behaved in the manner observed by Macaskie and Dean in the *Citrobacter*; that is, they picked up significant amounts of material from solution and could be made to acquire magnetic moments of significant strength. A number of HGMS tests were performed using *Candida utilis* and *Bacillus subtilis* resuspended in saline solution at pH = 7. In the tests, a magnetic field of 5T was used to magnetize a stainless steel wool matrix with a wire diameter of 150 mm. A fluid flow velocity of 2 cm·s^{-1} through the matrix was used. The results are shown in Table 4.5. The results have been calculated assuming that $v_m = v_0$, which means that the value of c and, consequently, the weight of material accumulated are minimum estimates.

The release of phosphate ions at the cell surface as shown by Macaskie and Dean [43] and the authors gives rise to precipitates of metal phosphates on the cell surfaces of microorganisms. It is now of interest to estimate the minimum obtainable concentration of metal ions remaining in solution after the biomagnetic extraction process has been completed. Calculations have shown that very low ultimate levels are possible for materials of interest to the nuclear industry. For example, for Am^{3+} the ultimate level is 2.43×10^{-19} ppm, for UO_2^{2+} the level is 5.38×10^{-9} ppm and for PuO_2^{2+} the level is 4.37×10^{-9} ppm. A number of tests were performed on solutions of interest to the nuclear industry and treated with *Candida utilis* with an overall cell concentration of approximately 2.3 g/liter of solution. The cells were added together with glycerol 3-phosphate and remained in contact with the solution for 2 hours before filtration. A number of tests were done with high and low starting concentrations of mixtures of ions. From the low residence time, it is clear, in this case, that metabolism of the cell during the 2-hour period had little influence, and the ion removal achieved was due to phosphate ions that were produced by the cells earlier. For the best results, longer residence times are needed in the presence of nutrient and of glycerol 3-phosphate. From Table 4.6 it is clear that a considerable reduction of ion concentration can be achieved for a number of different ions simultaneously. This allows the deliberate introduction of a magnetic ion that can act as the carrier for the nonmagnetic ions.

ANAEROBIC PROCESS USING INSOLUBLE METAL SULFIDES

The application to a wider range of metal ions has been accomplished by including a process that precipitates metal sulfides onto the microorganism surface which again allows them to be recovered magnetically if a magnetic material is precipitated [48–51]. In this work it was found that heavy metal ions with insoluble sulfides are easily removed from solution to residual levels near 1 ppb. It was shown that organisms precipitate sulfides on their cell surface, and loaded cells can be recovered using high-gradient magnetic separation (HGMS). Further, each bacterium accumulates a considerable fraction of its own weight, and in many cases, three to four times its own wet weight of metal sulfides such as iron or uranyl. The sulfide preparation route has produced similar susceptibilities to the phosphate preparation route. The ultimate levels to which heavy metal ion concentrations can be re-

TABLE 4.5. Comparison between *Candida utilis* and *Bacillus subtilis* in the Accumulation of UO_2 Ions from Solution.

	Effective Particle Radius (μm)	Minimum Magnetic Suceptibility χ Consistent with the Results	Ratio of the Mass of UO_2 to the Mass of the Microorganism
Candida utilis	1.4 ± 0.1	$(5.1 \pm 1.2) \times 10^{-5}$	0.44 ± 0.2
Bacillus subtilis	1.7 ± 0.15	$(2.5 \pm 0.8) \times 10^{-5}$	0.31 ± 0.2

TABLE 4.6. Removal of Radioactive Elements by *Candida utilis* after a Residence Time of 2 Hours.

Low Level Initial Concentration (Bq·ml^{-1})				
Material	^{242}Pu	^{241}Am	^{134}Cs	^{85}Sr	^{60}Co
Feed	15.76	11.10	18.50	17.83	50.21
Sample 1	0.44	0.94	9.51	8.14	6.51
Sample 2	0.54	1.12	9.40	8.62	9.32

High Initial Concentration (ppm)						
Material	U	Ru	Sr	Co	Cs	Ce	Zr
Feed	9.5		11.	11.0	11.0	11.0	11.0
Treated sample	0.40	0.05	0.47	0.55	0.33	<1	<1

duced appears to be even lower than for the phosphates. For example, for Co^{2+} it is 1.77×10^{-16} ppm, and for Hg^{2+} it is 8.02×10^{-43} ppm. A series of experiments using *Desulfovibrio* cells to reduce the concentration of copper were carried out under the conditions outlined in Table 4.7. Various concentrations of cells, various initial concentrations of copper concentration and various incubation periods were used.

As shown in Table 4.7, a very large reduction of copper concentration is possible using this method. The smallest incubation period used, which was one day, still produced excellent results.

TREATMENT OF EFFLUENT STREAMS CONTAINING METALS

On the basis of these results on single insoluble metallic sulfides, a larger scale study was undertaken of some effluent streams from the precious metal industry. The experiments were conducted on a laboratory scale. Aliquots of effluent streams were added to samples of Postgate's medium C, the pH was readjusted to 7.5 and the samples were inoculated with an actively growing culture of the laboratory strain of *Desulfovibrio*. Aliquots were taken after 5 d for examination by ICP-MS. In general, the metal ions that formed sulfides at pH 7.5 were removed from solution, e.g., Ag, Hg, Pb, Cu, Zn, Sb, Mn, Fe, As, Ni, Sn and Al. However, other metals such as Rh, Au, Ru, Pd, Os, Pt and Cr were also removed. Si was reduced by about two-thirds. Elements such as Mg and Sr were not usually affected, as shown in Table 4.8.

TABLE 4.7. Results of Experiments to Extract Copper from Solution Using *Desulfovibrio* Cells. The Initial and Final Concentrations Are Shown in the Table Together with the Initial Cell Concentration and the Incubation Period.

Sample No.	Incubation Period (Days)	Density of Cells (mg·L^{-1})	Copper Sulfate Concentration	
			Initial (ppm)	Final (ppm)
1	3	14	1.00	<1
2	3	20	0.10	<1
3	3	16	0.01	<1
4	1	6	1.00	Not detected
5	2	10	1.00	Not detected

TABLE 4.8. Results of the Treatment of Precious Metal Effluents with *Desulfovibrio*. The Results Are Expressed in $mg \cdot L^{-1}$.

	pH 12.5			pH 8.75			pH 0.20			pH 1.43		
	Start	2 Days	5 Days	Start	2 Days	5 Days	Start	2 Days	5 Days	Start	2 Days	5 Days
Rh	0.6	0	0	23	2.0	1.5	0.2	0.2	0.1	2	0.4	0.1
Ag	15	2.2	0.5	0.9	0.1	0	0.6	0	0	0.6	0	0
Ir	0	0	0	1.9	1.4	1	0	0	0	0.5	0	0
Au	1	0.2	0.1	0.5	0	0	0	0	0	0.5	0	0
Ru	25	5	0.8	25	18	11	32	0.2	0.1	2.0	1.0	0.4
Pd	2.2	2	0.7	43	22	15	165	78	50	2.2	0.3	0.2
Os	0.1	0	0	0.5	0	0	0	0	0	0.0	0	0
Pt	5	4.3	1.5	61	42	30	32	0.1	0.2	23	0.3	0.2
Hg	2.5	–	0.2	1.6	–	0.8	12	–	0.3	1.2	–	0.2
Pb	30	–	0	1.0	–	0	1.2	–	0	3.5	–	0
Si	4,020	–	1,545	4,515	–	1,785	4,008	–	1,562	2,114	–	1,325
Cr	10	–	1	13	–	3	6	–	0.8	10	–	1
Fe	250	–	82	245	–	86	204	–	102	250	–	64
Ni	32	–	2	29	–	2	13	–	1	72	–	20
Cu	118	–	0	72	–	2	76	–	0	59	–	0
Zn	33	–	0.5	89	–	23	20	–	1.5	1,004	–	0
Sb	2	–	0.4	34	–	13	0.5	–	0.5	0.0	–	0
P	861	–	238	795	–	265	855	–	242	531	–	50
Mn	5	–	0.5	5	–	0.5	3	–	0.5	10	–	0.5
As	182	–	0.5	532	–	222	192	–	35	201	–	0.5
Sn	5	–	0.5	181	–	60	8	–	0.5	5	–	3
Al	14	–	2	17	–	9	5	–	8	15,208	–	40
Mg	7,053	–	7,429	7,203	–	7,400	7,116	–	7,100	14,148	–	1,350
Sr	19	–	19	20	–	20	20	–	18	15	–	14

The experiments were repeated several times with variation in absolute amounts but with constancy in the general trends. The results suggested that the precious metals, apart possibly from iridium, were being collected by this treatment process. A number of different effluents containing precious metals have been examined in this way. In all cases, the precious metals together with many others can be collected onto the surface coat of FeS produced by *Desulfovibrio*. The reason for this is that iron sulfide FeS is known to be a compound of variable composition behaving rather as a phase than as a stoichiometric compound, so this suggested that the FeS material precipitated by the SRB was acting as an adsorbent. There are two main advantages to using the microorganisms compared with the chemical precipitation of the materials with H_2S, namely:

(1) When other solid materials are present, they need not be incorporated into the precipitate as they are if precipitation takes place chemically.

(2) The residual metal ion concentrations can be many orders of magnitude lower with the microorganisms than with direct chemical precipitation with H_2S or an equivalent chemical process except at high values of pH.

The argument supporting this statement is as follows: the solubility of H_2S in water is such that at 20°C the solution formed is approximately 0.1 M. However, H_2S is an extremely weak acid, and $[S^{2-}]$ is strongly affected by the value of pH. Equation (5) shows the overall stoichiometry of the reaction.

$$2H_2O + H_2S \rightarrow 2H_3O^+ + S^{2-} \tag{5}$$

At a value of pH = 14.15, this reaction goes completely to the right, and consequently, at this pH, $[S^{2-}] = 0.1$, and the value of $[S^{2-}]$ can be estimated at other values of pH. In particular, at a pH = 10.14, $[S^{2-}] = 10^{-5}$ and at pH = 7, $[S^{2-}] = 7 \times 10^{-9}$ [52]. On the other hand, because of its metabolism, the surface of the sulfate-reducing microorganism (SRB) can be regarded as a pump for S^{2-} ions and, from the Michaelis-Menton constant for the process, we can estimate that the $[S^{2-}] = 10^{-5}$ in the neighborhood of the exterior of the microorganism over the range of pH at which the microorganism is active [47]. This means that for all values of pH < 10.14, the SRB can produce a lower residual level of metal ion concentrations than chemical precipitation. For a specific example, consider the precipitation of the metal ion Cd^{2+} at a pH = 7 with a solubility product of the sulfide of 5×10^{-29}, the residual concentration of the metal ion depends on the $[S^{2-}]$, and if precipitation occurs by H_2S, then because $[S^{2-}] = 7.1 \times 10^{-9}$ at pH = 7, then the lowest possible value of $[Cd^{2+}] = 7 \times 10^{-21}$. If precipitation occurs with SRB, then the lowest obtainable value of $[Cd^{2+}] = 5 \times 10^{-24}$ or more than 10^3 times lower. In normal practice, it is found that a residual ion concentration of approximately 1 ppm is obtainable for Cd^{2+} when precipitated with H_2S, whereas with SRB, we have obtained values less than 1 ppb. It is well known that chemically precipitated $Fe_{1-x}S$ can be used as an adsorbent for metal ions, however, in this chapter it has been shown that the effective surface area of the microorganism-produced adsorbent is between a factor of 10 to 100 greater. Further, it is clear that to obtain the $Fe_{1-x}S$ in the most magnetic form, the $[S^{2-}]$ must be kept high. This can only be achieved in the case of chemical precipitation if the pH is kept high, which is often inconvenient for other reasons connected with the specific process. Such problems arise for the production of the $Fe_{1-x}S$ by the microorganisms less frequently because the $[S^{2-}]$ in the precipitation region is high, but in the suspension at large, $[S^{2-}]$ can be controlled through the pH. For the long-term stability of the $Fe_{1-x}S$ adsorbent, it is better to store the material at high pH, which keeps the $[S^{2-}]$ high which keeps $[O_2]$ low and prevents loss of S^{2-} from the adsorbent. HGMS was used to collect the SRB and their attendant precipitates; it is selective, and a high processing rate is possible. Rates as high as $400 \, m^3h^{-1}$ are available with existing machines, and forces of $10^5 \, g$ to $10^6 \, g$ can be exerted when a reasonable amount, approximately 10%, of the magnetic phase of the FeS is present. At this processing rate, the cost per m^3 is approximately 7p [10¢ (US)], including capital depreciation over 10 years, maintenance and running costs.

The Adsorption of Heavy Metals and Halogenated Carbon Compounds from Solution by Iron Sulfide Materials Produced by Sulfate-Reducing Bacteria

INTRODUCTION

As described previously the metal ions that form sulfides at pH 7.5 were removed from solution, e.g., Ag, Hg, Pb, Cu, Zn, Sb, Mn, Fe, As, Ni, Sn and Al by *Desulfovibrio*. However, other metals such as Rh, Au, Ru, Pd, Os, Pt and Cr were also removed. Silicon was reduced by about two-thirds. A large number of the ions that were removed do not have insoluble sulfides, and the conclusion was that another powerful immobilization mechanism or adsorption process was present in addition to the precipitation of insoluble sulfides. In all cases, the precious metals together with many others were associated with the precipitated surface coat of FeS produced by *Desulfovibrio*. The reason for this is that iron sulfide FeS is known to be a compound of variable composition that behaves as a phase rather than as a stoichiometric compound. The structure of the FeS system is complex, and on the S-rich side, a wide range of compositions of the form $Fe_{1-x}S$ can exist, and the cationic defects can provide active adsorption sites. These structural factors in $Fe_{1-x}S$ are clearly reflected in the magnetic properties of the sulfides of iron which are quite variable depending on the exact nature of the Fe/S ratio [53]. The magnetic properties of the monoclinic pyrrhotite Fe_7S_8 were first examined over 75 years ago, and the crystallographic, thermodynamic and magnetic aspects continue to be studied today. Pyrite FeS_2 occurs at 33 at. % Fe. The Fe ions do not possess a localized moment. At 50 at. % Fe is FeS, troilite, in which all the Fe sites are individually magnetic. All of the available sites for Fe are occupied. Troilite is antiferromagnetic below the Neel temperature $T_N = 325°C$. In each unit cell there is zero magnetic moment, as the moments of the two Fe ions are well compensated. Between these two limits lies the region of the FeS, more accurately written $Fe_{1-x}S$. The generic pyrrhotite formula signifies iron cation vacancies, and x varies from 0 to 0.13. The region is quite complex. In addition to the ferrimagnetic monoclinic Fe_7S_8 ($x = 0.125$, 46.67 at. % Fe), there is a variety of closely related hexagonal and monoclinic crystal structures that are linked to the number and spatial order or disorder of the Fe cation vacancies (which become mobile in the 200–300°C range). Randomly frozen vacancies produce hexagonal pyrrhotite that will be approximately antiferromagnetic. Ordered vacancies such as at Fe_7S_8 produce a monoclinic structure that is ferrimagnetic. The material has an effective magnetic susceptibility of approximately 1.3×10^{-2}. Intermediate cases occur. The region of these intermediate pyrrhotites with Fe concentrations above about 47 at. % is structurally complex. For the reproducible production of the $Fe_{1-x}S$ or in order to produce the Fe_7S_8 phase or perhaps even Fe_3S_4 which is ferrimagnetic, the essential thing is to control the redox potential, which controls the oxidation state of Fe. This problem is also closely related to the second problem, that is, the variability of the magnetic properties of the microorganisms themselves and, in relation to the first problem, the oxidation state of the $Fe_{1-x}S$ which the microorganism produces originally at, near or within the cell surface in the periplasmic space. The redox potential within this space is controlled by the cytochromes, whose number distribution between the different cytochromes is characteristic of the particular species of SRB. For example, cytochrome c_3 is the principal cytochrome of *Desulfovibrio vulgaris* which has a redox potential of approximately −0.2 V at pH = 7, and *Desulfotomaculum acetoxidans* contains c_7 which again has very low redox potential.

The iron sulfide was examined by transmission and scanning electron microscopy that showed that the material was composed of extremely finely divided material. Measurements of the grain size of the material indicated a surface area of 500–600 m^2g^{-1}. When these electron micrographs were examined closely, the occasional organism was seen. The microorganisms appear to produce a material with a large specific area that compares well with the surface area of highly absorbent activated charcoal.

THE PRODUCTION AND PROPERTIES OF THE ADSORBENT

The adsorbent was produced using sulfate-reducing microorganisms in a 100 ml chemostat. The chemostat was first sterilized in an autoclave and then flushed with oxygen-free nitrogen and then isolated from atmospheric oxygen to establish and maintain anaerobic conditions. The temperature was maintained at 32°C by placing the chemostat in a water bath with temperature accurately controlled at 32°C. Postgate's medium C [54] with excess Fe was used to grow the sulfate-reducing bacteria. The chemostat was first partly filled with the medium and then incubated with 20 ml of a pure culture of sulfate-reducing bacteria. After 24 h, the medium was continuously added to the chemostat at a dilution rate of 0.1/h. Large quantities of the adsorbent were produced and could be collected from the bottom of the chemostat for the adsorbent studies. Measurements on the adsorbent showed that it contains 15% solids by weight of $Fe_{1-x}S$. The composition of the material produced in this way was approximately $Fe_{0.96}S$ with the Ni-As structure identified by EXAFS [55,56]. The Fe in this structure is octahedrally coordinated. The magnetic susceptibility is 8.2×10^{-5} which is large enough to be extracted by high-gradient magnetic separation [31]. There are two known polymorphs of $Fe_{1-x}S$ at this composition, namely, a NiAs type superstructure in which the S forms a hexagonal close-packed lattice with the Fe occupying the octahedral sites; the principal distances are Fe-S 2.27 Å; Fe-S 2.39 Å; Fe-Fe 2.94 Å and Fe-Fe 3.65 Å [55]. The second polymorph is based on a PbO-type structure, the principal distances being Fe-S 2.13 Å and Fe-Fe 2.50, 3.68 Å. Examination of the data for the Fe/Metal systems reveals that the EXAFS data is consistent with the X-ray distances determined by Keller-Besrest and Collin [55] for the Ni-As structure. Based on the cost of the nutrient and large-scale manufacturing processes of such single-cell protein production, the estimated cost of the adsorbent is approximately 0.1 p per gram. For example, this means that 1.3×10^7 liters (13,000 tonne) of water containing 10 ppm of Hg can be reduced to 2 ppb of Hg by 1 tonne of the adsorbent costing £1,000.

EXPERIMENTAL RESULTS FOR HEAVY METALS ADSORBED BY IRON SULFIDE

For each metal, a series of 100 ml samples was prepared, each containing 10 ppm solution of the metal ion to be studied, and the pH was adjusted to pH = 7. Various amounts of the adsorbent were added to each of the solutions so that the ratios of the weight of wet adsorbent to that of the metal ion lay between 100:1 and 1:1. At hourly intervals, approximately 2 ml was taken of each supernatant which was filtered and then analyzed for metal ion concentration using atomic absorption spectroscopy. Many of the adsorbed species were examined using EXAFS that indicated that the ions examined were chemisorbed, and, as discussed below, all the adsorption studies were consistent with ions forming bonds with the FeS that were too strong to be affected by thermal energies at temperatures around 300 K. Considering the metal X-ray absorption K-Edge data, the Cd/Fe species was the best resolved and it was concluded that the Cd sits in sites similar to those found in the Wurtzite form of CdS [57]. The Pd/Fe species was similarly well rationalized with the Pd basically siting in a PdS2 or PdS site (Pd/Fe Pd-S 2.32 Å; PdS2 Pd-S 2.29 Å; PdS Pd-S 2.34 Å) [57]. In the Ru/Fe system, however, the Ru K-Edge EXAFS unfortunately did not give distances that corresponded to those in its binary sulfide RuS_2 (RuS_2 Ru-S 2.35 Å; Ru/Fe 2.16 Å, etc.). The element Ru forms no other binary sulfides, so this suggested that a completely different structure exists for the Ru/Fe species or that the system is not a sulfide one when the ion is chemisorbed to FeS (i.e., sulfate, thiosulfate, etc.) [57]. The uptake of cadmium by the adsorbent is typical of metals that have been studied. When a small amount of adsorbent is added, the concentration of metal ions falls quickly, and then the concentration remains constant in time. Larger additions of adsorbent to the initial solution result in a faster drop with time and then levels and becomes independent of time at a lower concentration. However, there is a point when a critical amount of adsorbent is added and the concentration no longer levels but continues to decrease with time. All of these features can be seen in Figure 4.21. The value of adsorbent that appears to be the critical value for a starting concentration

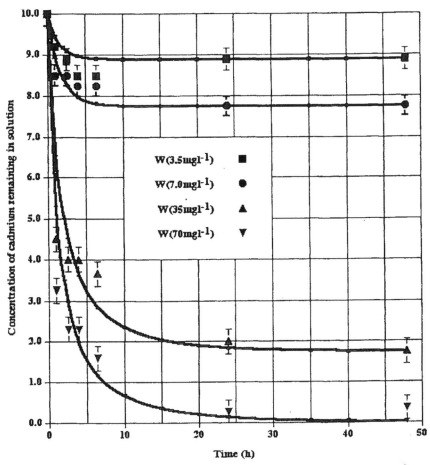

Figure 4.21 The concentration of Cd remaining in solution after various times in contact for various amounts of the wet adsorbent. Each concentration is shown in brackets in units of ppm dry weight. The starting concentration is 10 ppm of Cd.

of 10 mg·L^{-1} is approximately 35 mg·L^{-1} of adsorbent. This is consistent with the results from the extended X-ray absorption spectroscopy (EXAFS) that indicates that all the ions studied are chemisorbed and implies that binding energy is much greater than thermal energies, so after binding takes place, the ions do not diffuse back into the solution by thermal processes. Further, it implies that there is a certain capacity for the chemisorbed ions per unit area of adsorbent, so if in the amount of adsorbent added there are more sites than there are ions in the initial solutions, eventually all of the ions will be adsorbed. As shown in Table 4.9, the lowest level of Cd^{2+} obtained was 8 ppb corresponding to 2.22 mmole/liter, and at that point, the amount of Cd^{2+} acquired by the adsorbent corresponded to 109 mg/g (dry) or 0.97 millimole/g (dry). Although the overall behavior of the metal ions is the same, there is a big difference between the transition and the nontransition metals. The nontransition metals take up roughly 1–2 millimoles per gram of adsorbent, and using the estimated value for the adsorbent area [56], this means that the maximum area that the nontransition metal ions occupy is approximately 0.4 nm^2 per ion, whereas the transition metals can occupy 0.1 nm^2 per ion.

TABLE 4.9. Results for the Adsorption for a Number of Metals. The Metal Ion Uptake Is
Presented in the Second and Third Columns as Milligrams and Millimole per Gram
of Adsorbent Added. W_d Is the Ratio of the $mg \cdot L^{-1}$ of Adsorbent Added to the Initial
Concentration of the Metal Ions in $mg \cdot L^{-1}$. The Last Two Columns Show the
Minimum Concentration of the Metal Ions Remaining in Solution Expressed
as ppb ($mg \cdot m^{-3}$) and as $micromole \cdot L^{-1}$.

Metal Ion	Metal Ion on the Dry Adsorbent		W_d	Residual Concentration of Metal Ion	
	$mg \cdot g^{-1}$	$millimole \cdot g^{-1}$		(ppb)	$micromole \cdot L^{-1}$
Hg^{2+}	133.3	0.665	7.5	2	0.01
Cd^{2+}	143	1.27	7	8	0.071
Co^{2+}	414	7.02	2.4	60	1.02
Cr^{3+}	365	7.01	2.16	2,125	4.1
Cu^{2+}	138.6	2.18	7	300	4.72
Ni^{2+}	377	6.42	2.6	200	3.41
Pb^{2+}	194	0.94	5.1	60	0.29

Another interesting feature in Figure 4.21 is the area of occupancy on the adsorbent by
Cd^{2+} ions for the different adsorbent additions. For an addition of $3.5 \, mg \cdot L^{-1}$, the ions at saturation occupy $0.3 \, nm^2$; for an addition of $7 \, mg \cdot L^{-1}$, the average area occupied by each ion
is $0.29 \, nm^2$; for $35 \, mg \cdot L^{-1}$, the average area occupied falls to $0.4 \, nm^2$; and for $70 \, mg \cdot L^{-1}$, the
average area occupied falls to $0.67 \, nm^2$. There are at least two obvious explanations and a
third if the first and second are both present.

The first explanation is that there are a number of different sites that have different binding energies, some of which are the order of thermal energy so adsorption and desorption
occur, and at low ion concentration, these sites are largely empty. The second is the possibility that there are a total number of sites per unit area of adsorbent where chemisorption can
take place without subsequent desorption. For the case where there are less ions in solution,
chemisorption sites are eventually added, and the concentration of ions in solution should
become zero. The area occupied by each ion would be larger than for the case where there
are more ions in solution than there are chemisorption sites added. In this latter case, the
area occupied by each ion should reach the same minimum value. An examination of the results for cadmium suggest that the second explanation is more likely or more dominant, and
the adsorbent additions of $3.5 \, mg \cdot L^{-1}$ and $7 \, mg \cdot L^{-1}$ indicate that the number of sites added is
less than the number of ions in solution, whereas additions of $35 \, mg \cdot L^{-1}$ and $7 \, mg \cdot L^{-1}$ contain more chemisorption sites than the number of ions in solution.

Results on the other metals studied were similar qualitatively, differing in the rate of uptake and the relative amount of adsorbent required. The overall results for a number of metals studied are shown in Table 4.9. There is a distinct difference in the behavior of the transition elements compared with the nontransition elements. Assuming that the surface area of
the iron sulfide is approximately $500 \, m^2 g^{-1}$, the maximum coverage of Cd^{2+} is with one
atom occupying $0.4 \, nm^2$ compared with Cr^{3+} where the maximum coverage is with one ion
occupying $0.1 \, nm^2$.

The results presented here show that the microorganism-produced $Fe_{1-x}S$ is an excellent
adsorbent for a wide range of metal ions from solution. Although the magnetic susceptibility is not very high at 8.2×10^{-5}, the material was adequately removed from suspensions using high-gradient magnetic separation. The specific uptake of metal ions from solution is
high when compared with other adsorbents, and the residual levels to which the concentra-

tion of the ions is reduced is very low, in the case of mercury, it is as low as 2 ppb. There is evidence to suggest that the process is one of chemisorption, except in the case of chromium where the binding to the adsorbent appears to be considerably weaker.

The Production and Properties of Strongly Magnetic Iron Sulfide

INTRODUCTION

Although the process described above works well immobiliszing Ag, Hg, Pb, Cu, Zn, Sb, Mn, Fe, As, Ni, Sn, Al, Rh, Au, Ru, Pd, Os, Pt and Cr from solution, before this process was applied consistently to large-scale problems, it was found possible to improve the magnetic properties of the FeS material, dramatically reducing the cost of recovery of the FeS and the attendant adsorbed metal ions using HGMS. From the theory of capture of paramagnetic particles, the performance of the separator is controlled by v_m/v_0, the ratio of the magnetic velocity [Equation (3)] to the fluid velocity v_0. The magnetic properties of the sulfides of iron are quite variable depending on the exact nature of the Fe/S ratio as shown by Schwarz and Vaughan [53], and in particular, there are strongly magnetic compounds with compositions Fe_7S_8 and Fe_3S_4.

To calculate the total magnetic separation processing cost, the magnetic velocity has been calculated using Equation (3) and the magnetic susceptibility data obtained by Schwarz and Vaughan [53]. In the calculations, the particle radius b was 1 mm, the ferromagnetic matrix wire had a saturation magnetization of 1.35×10^6 A·m^{-1} (= 1.7 T) and a radius $a = 100$ mm. The applied magnetic field was 3.98×10^6 A·m^{-1} (= 5 T). Next, it was assumed that the process velocity was equal to the calculated values of magnetic velocity, and the suspension was applied to the specifications of CRYOFILTER Superconducting Magnetic Separator HGMS 5T/280 produced by Outokumpu, Carpco Division (Jacksonville, FL, USA). This superconducting magnetic separator conceived and experimentally verified at ECC International (Paris, France; now Imerys) by Watson, Clark and Windle [34,58] has come to dominate the clay processing industry since 1989 with machines throughout the world. The price of the machine was depreciated over ten years to give an hourly cost, and to this maintenance, running and staff costs were added. This was then divided by the process rate in tonne per hour assuming the process velocity was equal to v_m. These calculations are shown graphically in Figure 4.22 and illustrate how dramatically the processing cost is reduced as the magnetic susceptibility increases.

IMPROVING THE MAGNETIC PROPERTIES OF THE $Fe_{1-x}S$ ADSORBENT

It is clear from the previous section that there is much to be gained by improving the magnetic properties of the adsorbent, and in order to accomplish this, Watson and Ellwood used a novel method [59, 60]. The redox potential, which controls the oxidation state of Fe, for the production of the $Fe_{1-x}S$, the pH which controls the concentration of sulfide ions in solution [S^{2-}] and the other anions, cations and complex can also affect the magnetic product produced by the microorganisms. Small changes in the composition have a large affect on the magnetic susceptibility of the $Fe_{1-x}S$ material. In an interesting paper, Freke and Tate [61] reported the almost complete removal of iron as insoluble sulfide from ferrous and ferrous-ferric solutions, and they found under certain, poorly specified conditions, ie., type of microorganisms, redox potential E_h, although it appears E_h were very low, etc., the formation of substantial amounts of magnetic iron sulfide. The method proposed by Watson and

Ellwood was to grow a mixed culture of SRB in the solution to be treated in a chemostat at various dilution rates. The effluent passes to a magnetic separator where the most strongly magnetic materials are retained along with their attendant microorganisms. The weakly magnetic components are not retained. The magnetic adsorbent fraction together with the attendant microorganisms are then fed back to the chemostat. Thus, if a microorganism yielding an undesirable product has the smallest doubling time by using a high dilution rate, this organism's effective growth rate can be reduced with respect to a slower growing one producing a more magnetic product.

The influent to the reactor was based on the nutrients used by Freke and Tate [61] using 4–5 ml of 70% solution of sodium lactate per liter as the main carbon source. The solution was sterilized by passing it through a 0.2 μm filter, rendered free of O_2 and the redox potential E_h was reduced to at least –100 mV accomplished by the addition of Na_2S (a 5 mM solution of Na_2S has a value of $E_h = -220$ mV). The inoculum was obtained from the anaerobic region of a semi-saline estuarine sediment by collecting and mixing material taken from different levels within the region. As shown in Figure 4.23, the microorganisms washed out of the culture vessel go to a magnetic separator that is tuned to collect only the most strongly magnetic material, within which are the microorganisms responsible for the production of this highly magnetic material. This material and the attendant microorganisms are fed back to the chemostat. Those microorganisms producing the desired magnetic product will thereby be selected and preferentially multiplied, while with an increased dilution rate, those not producing magnetic material will be washed out. The principles used in this novel method can be applied to many other systems.

In the first application of the method, the $Fe_{1-x}S$ initially contained only a few percent of strongly magnetic material, however, after proceeding for three months, the $Fe_{1-x}S$ contained over 80% of strongly magnetic material. When after three months a second chemostat, without magnetic feedback, was inoculated with material from the feedback chemostat, magnetic material was produced by the second chemostat immediately. The structural and magnetic properties of the magnetic iron sulfide material have been exam-

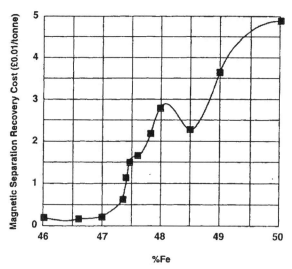

Figure 4.22 The magnetic separation processing cost for $Fe_{1-x}S$ in the region between 50% Fe and 46% Fe. Included in the cost is the price of the machine depreciated over 10 years plus the running and maintenance costs. The calculations used the data of Schwarz and Vaughan [53].

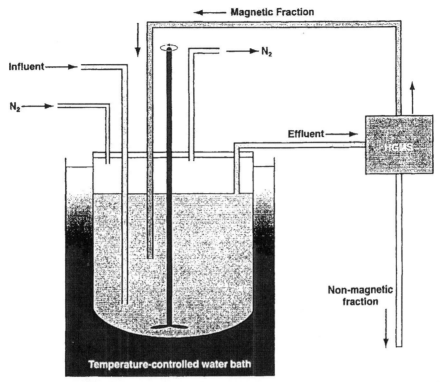

Figure 4.23 Chemostat for growing magnetic adsorbent using feedback of the most magnetic fraction of the microorganisms. The setup is shown for anaerobes.

ined in detail using transmission electron microscopy, surface area measurements, magnetic measurements, extended X-ray absorption fine-structure (EXAFS) spectroscopy and X-ray absorption near edge structure (XANES) spectroscopy at Daresbury and neutron scattering at the ISIS facility of the Rutherford-Appleton Laboratory [56,62].

Higher magnification electron microscopy reveals that strongly magnetic iron sulfide consists of small, irregular particles or clusters of particles, some of which were identified as greigite. The vast majority appear entirely featureless with no clear grain boundary or distinctive morphology. However, a few areas produce weak, diffuse, continuous or spotty powder-type ring SAD patterns, indicating a poor degree of crystallinity with small crystalline domains randomly oriented within the selected area. The lattice spacings of these rings are consistent with greigite [63,64], but the pattern is weak and incomplete, so unequivocal identification as greigite is difficult. The superimposed SAD patterns (hexagonal array of spots and diffuse rings) suggest that the sulfide probably consists of an intimate mixture of poorly formed greigite and mackinawite [56,62]. The material was largely paramagnetic at 290 K with a small hysteresis and an average magnetic susceptibility $\chi = 7 \times 10^{-4}$ (SI), approximately a factor of 10 greater than material produced without the magnetic feedback chemostat. After the large particles assumed to be greigite had saturated, the magnetization became linear with the applied field showing no saturation up to the maximum applied magnetic field available of 12.5 tesla. Trials with the strongly magnetic $Fe_{1-x}S$ revealed that the material was as good an adsorbent as the less strongly magnetic material whose adsorptive properties were previously discussed.

Some Applications for the Adsorptive Properties
of Bacterially Produced Iron Sulfide

HARBOR AND RIVER SLUDGES

Hamilton [65–69] found that radioactive hot spots are formed in anaerobic estuarine muds and marshes near the Sellafield nuclear reprocessing plant in the northwest of England. The presence of these hot spots is strong evidence, occurring as they do in conjunction with carbonaceous layers. The evidence suggests that one powerful mechanism by which this can occur is the immobilization of the radioactive heavy metals and other heavy metals such as Fe and Mn by colonies of sulfate-reducing bacteria (SRB) such as *Desulfovibrio* that occur in anaerobic regions of the mud. Most of the radioactive metals that find their way into the anaerobic muds are paramagnetic, so when they are immobilized in the SRB cluster, the cluster becomes paramagnetic and may be removed by magnetic separation. Based on this work, a method of decontaminating anaerobic muds and sludge from heavy metals and perhaps fluoro- and chlorocarbon compounds and PCBs can be suggested. As can be seen in Table 4.10, the adsorbent is effective in the removal of many halogenated hydrocarbons. If the adsorbent compound is introduced into the mud together with iron sulfate, over a period of time, the population of SRB colonies will increase and the iron and other heavy metals will be precipitated and immobilized in the colonies. After the mud is dredged, the application of high-gradient magnetic separation will concentrate a substantial fraction of the heavy metals and perhaps the chloro- and fluorocarbon compounds into a small magnetic fraction by weight at the very high throughputs of 250 tonnes/hour which can easily be achieved with high-gradient magnetic separation using the superconducting magnetic separator from Carpco SMS.

TABLE 4.10. A Solution Containing Standard Materials Used in the Agricultural Industry Incubated with Adsorbent 24 Hours at a Concentration of 0.7 g/Liter.

Substance (conc. in ng/liter)	Initial Conc.	Concentration after 24 Hours	% Removal
Hexachlorobenzene	1,000	60	94
Heptachlor	1,000	<10	>99
Heptachlorepoxide (*cis*)	1,000	20	98
Aldrin	1,000	<10	>99
Endosulphan A	1,000	<10	>99
Endosulphan B	1,000	<10	>99
op DDe	1,000	<10	>99
pp DDE	1,000	80	92
op TDE	1,000	<20	>98
pp TDE	1,000	260	74
op DDT	1,000	<20	>98
pp DDT	1,000	<20	>98
Endosulphan sulfate	1,000	<20	>98
Carbetamide	1,000	340	66
Chlorotoluron	1,000	370	63
Fluoranthene	100	<5	>95
Benzo(ghi)porylene	20	<2	>90
Benzo(u)fluoranthene	20	<2	>90
Indeno(123 cd)pyrene	20	<2	>90
Benzo(b)fluoranthene	20	<2	>90
Benzo(a)pyrene	20	<2	>90

It was proven to be possible that a substantial fraction (> 80%–90%) of these hot particles could be removed using high-gradient magnetic separation (HGMS) [70]. This indicated that the immobilization and localization of magnetic ions has taken place in the hot spots.

CLEANUP OF MERCURY IN THE NETHERLANDS

A large chemical company had an industrial effluent containing 2,000 ppb of the toxic heavy metal mercury which usually was treated using an ion exchange system that was being repaired. This provided an opportunity to use the FeS adsorbent to treat the solution. The batch contained 5,000 liters of mercury-contaminated industrial effluent. In a single pass through the system, the mercury content was reduced from just over 2 ppm to approximately 5 ppb with an adsorbent loading of 20 mg/liter.

THE RECOVERY OF PRECIOUS METALS FROM WASTE STREAMS

A number of companies are required to recover valuable metals from waste streams. The solutions usually contain platinum, palladium, rhodium, ruthenium, iridium, gold, silver, iron, nickel and copper. Typically, for example, where initial concentrations of each were approximately 10 ppm, after a residence time of six hours in contact with the adsorbent at a concentration of 2 g/liter, all metals except for iridium, as shown by atomic adsorption spectroscopy, are reduced to less than 0.3 ppm. Analysis using ICP-MS often shows that platinum was reduced to 5 ppb.

RECYCLING ELECTRICAL EQUIPMENT

A number of recycling companies that specialize in recycling end-of-life electronic equipment have been developing cathode ray tube (CRT) disposal plants. The driving forces behind this project include an increasing awareness that landfill costs are rising at a significant rate and a growing consensus within the industry of the need to exercise "Duty of Care" in the disposal of CRTs. Coopers Lybrand were commissioned by the UK Government to investigate the rising costs and concluded that there would be a 135% increase in costs for disposal to landfill by the year 2000. This in itself is a sufficient reason to develop a process. However, there is also the likelihood that CRTs will be treated as special waste. Apart from the very large increase in cost, this will mean that an environmentally sound process for processing end-of-life CRTs to recover the elements becomes very attractive to those faced with their disposal.

The processes being developed are often wet processes that result in heavy metals being transported, in the form of a solution, to a central collection vessel. The typical concentrations of heavy metals are shown in Table 4.11. The adsorbent produced by the microorganisms can be used to treat these solutions. These processes would take place in a closed-loop process that would lead to recycling of heavy metals and, in particular, rare earth metals, contained in the coatings of the CRTs. In essence, this closed loop would comprise a feed supply of solution passing into a container in residence with the adsorbent. As a result of this microbial activity, heavy metals will precipitate onto the microorganisms which, after a time, can be removed by circulation through a magnetic separator. Backwashing of the magnetic separator will concentrate the microorganisms enveloping heavy metals in a third tank from which they can be removed. This biologically produced adsorbent process has

TABLE 4.11. Typical Solution Concentrations of Heavy Metals
Produced in the Washing Process.

Cu	Zn	Cd	BA	Sr	Pb	W	Hg	Ti	Mn
0.2	0.8	0.05	1.6	1.5	0.01	2.6	0.05	0.27	0.18

Units mg/liter.

numerous advantages over other technologies such as precipitation or concentration; primarily, as a closed-loop process, there will be minimal contamination of other media such as land, air and water, and consent levels will be fully complied with.

Conclusions on the Removal of Heavy Metals Using Microorganisms and Magnetic Separation

Effluents from a number of industrial processes contain low but toxic concentrations of metal ions in solution. The adsorption process followed by magnetic separation described in this chapter allows for the removal of metal ions to very low concentration, rapidly and inexpensively; further, it allows for the collection of these materials in a highly concentrated material form. Therefore, this process will have applications in the mineral processing industry and in the treatment of effluents from nuclear and other industrial plants.

REFERENCES

1 Rodgers, D. N., Marble, W. J. and Elliott, H. H. 1981. "Testing of High Gradient Magnetic Filter at the Brunswick Steam Electric Plant," Nuclear Engineering Division, General Electric Company Report NEDO-25444 81NED326, San Jose, CA.

2 Harland, J. R., Nilson, L. and Wallin, M. 1976. "Pilot Scale High Gradient Magnetic Filtration of Steel Mill Wastewater," *IEEE Trans. Magn.*, 12(6): 904–906.

3 Krumm, E. 1992. "Abwasserreinigung mit Magnetseparation," *Galvanotechnik*, 83(1): 213–217.

4 Anderson, N. J., Bolto, B. A., Dixon, D. R., Kolarik, L. O., Pristley, A .J. and Raper, W. G. C. 1982. "Water and Wastewater Treatment with Reusable Magnetite Particles," *Wat. Sci. Tech.*, 14: 1545–1546.

5 Tamaura, Y., Katsura, T., Rojarayanont, S., Yoshida, T. and Abe, H. 1991. "Ferrite Process; Heavy Metal Ions Treatment System," *Water Sci. Technol.*, 23: 1893–1900.

6 Watson, J. H. P. 1973. "Magnetic Filtration," *J. Appl. Phys.*, 44: 4209-4213.

7 Watson, J. H. P. 1991. "Magnetic Separation," in *Concise Encyclopaedia of Magnetic and Superconducting Materials*, J. Evetts, ed. Cambridge: Cambridge University Press, pp. 242–245.

8 Watson, J. H. P. 1994. "Selectivity and Mechanical Retention in the Magnetic Separation of Polydisperse, Mixed Mineral Systems—Part 1," *Minerals Engineering*, 7: 769–791.

9 Watson, J. H. P. 1994. "Status of Superconducting Magnetic Separation in the Minerals Industry," *Minerals Engineering*, 7: 737–746.

10 Svoboda, J. 1987. *Magnetic Methods for the Treatment of Minerals. Developments in Mineral Processing 8.* Amsterdam: Elsevier Sc. Publishers.

11 Süsse, W. 1973. Patent DE 2222003.7.

12 Oberteuffer, J. A. 1974. "Magnetic Separation: A Review of Principles, Devices and Applications," *IEEE Trans. Magn.*, MAG-10:223-238

13 Arvidson, B. R. and Fritz, A. J. 1985. "New Inexpensive High-Gradient Magnetic Separator," *Proceedings of the 15th Intl. Mineral Processing Congress,* Cannes., p. 317.

14 Franzreb, M., Kampeis, P., Franz, M. and Eberle, S. H. 1998. "Use of Magnet Technology for Phosphate Elimination from Municipal Sewage," *Acta hydrochim. hydrobiol.*, 26(4): 213–217.

15 van Kleef, R. P., Myron, H. W., Wyder, P. and Parker, M. R. 1984. "Application of Magnetic Flocculation in a Continuous Flow Magnetic Separator," *IEEE Trans. Magn.*, MAG-20(5): 1168–1170.

16 Franzreb, M. 1998. "Open Gradient Magnetic Separator for Removal of Heavy Metals from Waste Water Based on a Cryogen Free Superconducting 5 T Magnet," *Proceedings of the 15th Intl. Conf. on Magnet Technology.* Vol. 1. Beijing, p. 744.

17 Ito, S., Kikuchi, A. and Yoneda, N. 1985. "Continuous Treatment on Magnetic Separation of Heavy Metal Ions in Water," *AIChE Symposium Series, Separation of Heavy Metals and Other Trace Contaminants*, 243(81): 133–138.

18 Okuda, T., Sugano, I. and Tsuji, T. 1975. "Removal of Heavy Metals from Wastewater by Ferrite Co-Precipitation," *Filtration Separation,* Sep/Oct: 472–478.

19 Kampeis, P. 1998. "Chemische und verfahrenstechnische Untersuchungen zur Erzeugung ferrithaltiger Suspensionen im Hinblick auf einen Einsatz als Zusatzstoff für die Magnetseparation," Ph.D. Thesis, University of Karlsruhe, Karlsruhe.

20 Kampeis, P., Franzreb, M., and Eberle, S.H. 1996. "Bildungsbedingungen von Zinkferriten im Hinblick auf die Entfernung von Schwermetallen aus Abwässern durch Magnetseparation," *Acta hydrochim. hydrobiol.*, 24: 61–67.

21 Kampeis, P., Franzreb, M., Nesovic, M. and Eberle S. H. 1997. "Einfluß der Temperatur auf die Bildung von Magnetit als Zusatzstoff für die Magnetseparation," *Acta hydrochim. hydrobiol.*, 25: 173–178.

22 deLatour, C. 1973. "Magnetic Separation in Water Pollution Control," *IEEE Trans. Mag.*, Mag-9(3): 314–316.

23 deLatour, C. 1976 "Seeding Priciples of High Gradient Magnetic Separation," *J. AWWA,* August: 443–446.

24 deLatour, C. and Kolm, H. H. 1976. "High-Gradient Magnetic Separation. A Water Treatment Alternative," *J. AWWA,* June: 325–327.

25 deLatour, C. and Kolm, H. H. 1975. "Magnetic Separation in Water Pollution Control—II," *IEEE Trans. Mag.*, Mag-11(5): 1570–1572.

26 deLatour, C. 1976. "HGMS; Economics, Applications, and Seed Reuse," *J. AWWA* September: 498–500.

27 Anand, P., Etzel, J. E. and Friedlaender, F. J. 1985. "Heavy Metals Removal by High Gradient Magnetic Separation," *IEEE Trans. Magn.*, MAG-21(5): 2062–2064.

28 Terashima, Y., Ozaki, H. and Sekine, M. 1986. "Removal of Dissolved Heavy Metals by Chemical Coagulation, Magnetic Seeding and High Gradient Magnetic Filtration," *Wat. Res.*, 20(5): 537–545.

29 Hencl, V., Mucha, P., Orlikova, A. and Leskova, D. 1995. "Utilization of Ferrites for Water Treatment," *Wat. Res.*, 29(1): 383–385.

30 Choi, S., Calmano, W. and Förstner, U. 1994. "Untersuchungen zur Abtrennung von Schwermetallen aus Abwasser mit frisch hergestelltem Magnetit," *Acta hydrochim. hydrobiol.*, 22(6): 254–260.

31 Choi, S. 1993. "Untersuchungen zur Schwermetallelimination aus Abwasser durch die Ausfällung mit künstlichem hergestelltem Magnetit," Ph.D. Thesis. University of Hamburg, Hamburg.

32 Franzreb, M., Kampeis, P., Franz, M. and Eberle, S. H. 1996. "Einsatz von Magnetfiltern zur Abtrennung magnetithaltiger Schwermetallhydroxide," *Vom Wasser,* 87: 235–250.

33 Beveridge, T. J. and Doyle, R. J. 1989. *Metal Ions and Bacteria,* 1st ed. New York, NY: John Wiley and Sons, Inc., p. 461.

34 Poole, R. K. and Gadd, G. M. 1989. "Metal-Microbe Interactions: Volume 26," in *Special Publications of the Society for General Microbiology,* 1st ed: Society for General Microbiology, p. 133.

35 Zborowski, M., Malchesky, P. S., Jan, T.-F. and Hall, G. S. 1992. "Quantitative Separation of Bacteria in Saline Solution Using Lanthanide Er(III) and a Magnetic Field," *Journal of General Microbiology,* 138: 63–68.

36 Blakemore, R. 1975. "Magnetotactic Bacteria," *Science,* 190: 377–379.

37 Bahaj, A. S., James, P. A. B., Ellwood, D. C. and Watson, J. H. P. 1993. "Characterization and Growth of Magnetotactic Bacteria: Implications of Clean Up of Environmental Pollution," *J. Appl. Phys.*, 73: 5394–5396.

38 Bahaj, A. S., Croudace, I. W. and James, P. A. B. 1993. "Heavy Metal Removal Using Magnetotactic Bacteria," Presented at *Second International Symposium on Subsurface Microbiology*, Bath, England.

39 Lloyd, J. R., Cole, J. A. and Macaskie, L. E. 1997. "Reduction and Removal of Heptavalent Technetium from Solution by *Escherichia coli*," *Journal of Bacteriology*, 179: 2014–2021.

40 Lloyd, J. R., Thomas, G. H., Finlay, J. A., Cole, J. A., and Macaskie, L. E. 1999. "Microbial Reduction of Technetium by *Escherichia coli* and *Desulfovibrio* desulfuricans: Enhancement via the Use of High-Activity Strains and Effect of Process Parameters," Biotechnology and Bioengineering, 66: 122–130.

41 Lloyd, J. R., and Bunch, A. W. 1996. "The Physiological State of an Ethylenogenic *Escherichia coli* Immobilised in Hollow-Fibre Bioreactors," *Enzyme Microbial Technology*, 18: 113–120.

42 Lloyd, J. R., and Harding, C. L. 1997. "Tc(VII) Reduction and Accumulation by Immobilized Cells of *Escherichia coli*," *Biotechnology and Bioengineering*, 55: 55–510.

43 Macaskie, L. E. and Dean, A. C. R. 1984. "Cadmium Accumulation by *Citerobacter* sp," *J. Gen. Microbiol.*, 130: 53–62.

44 Watson, J. H. P. and Ellwood, D. C. 1987. "Biomagnetic Separation and Extraction Process," *IEEE Trans. Magn,* MAG-23: 3751–3752.

45 Watson, J. H. P. and Ellwood, D. C. 1987. "Biomagnetic Separation Process for Heavy Metal ions (Ghent)," Presented at *Downstream Processing in Biotechnology,* Ghent, Belgium.

46 Watson, J. H. P. and Ellwood, D. C. 1988 "A Biomagnetic Separation Process for the Removal of Heavy Metal Ions from Solution," Presented at *International Conference on Control of Environmental Problems from Metal Mines,* Roros, Norway.

47 Watson, J. H. P. 1989 "High Gradient Magnetic Separation-Solid-Liquid Separation," in *Solid-Liquid Separation*, L. Svarovsky, Ed., 3rd ed. London: Butterworth & Co (Publishers) Ltd.

48 Watson, J. H. P. 1998. "Superconducting Magnetic Separation," in *Handbook of Applied Superconductivity,* vol. 2, B. Seeber, Ed., 1st ed. Bristol, UK: Institute of Physics Publishing, pp. 1371–1406.

49 Watson, J. H. P. and Bahaj, A. S. 1991. "Extraction of Heavy Metal Ions Using Microorganisms and High Gradient Magnetic Separation," Institute of Cryogenics, University of Southampton, Research GR/E 35374, 30 Dec 1991.

50 Watson, J. H. P., Ellwood, D. C., Hamilton, E. O. and Mills, J. 1991. "The Removal of Heavy Metals and Organic Compounds from Anaerobic Sludges," Presented at *Congress on Characterisation and Treatment of Sludge,* Gent, Belgium.

51 Ellwood, D. C., Hill, M. J. and Watson, J. H. P. 1992. "Pollution Control Using Microorganisms and Magnetic Separation," Presented at *Microbial Control of Pollution,* University of Cardiff, Cardiff, Wales.

52 *CRC Handbook of Chemistry and Physics,* 1978–79. 59 ed. West Palm Beach, Florida: CRC Press, Inc.

53 Schwarz, E. J. and Vaughan, D. J. 1972. "Magnetic Phase Relations of Pyrrhotite," *J. Geomag. Geoelectr.,* 24: 441–458.

54 Postgate, J. R. 1984. *The Sulphate-Reducing Bacteria,* Second ed. Cambridge, England: Cambridge University Press.

55 Keller-Besrest, F. and Collin, G. 1990. "Structural Aspects of the Transition in Stoichiometric FeS: Identification of the High-Temperature Phase," *Journal of Solid State Chemistry,* 84: 194–210.

56 Watson, J. H. P., Cressey, B. A., Roberts, A. P., Ellwood, D. C., Charnock, J. M. and Soper, A. K. 2000. "Structural and Magnetic Studies on Heavy-Metal-Adsorbing Iron Sulfide Nanoparticles Produced by Sulfate-Reducing Bacteria," *Journal of Magnetism and Magnetic Materials,* to be published.

57 Watson, J. H. P., Ellwood, D. C., Deng, Q., Mikhalovsky, S., Hayter, C. E. and Evans, J. 1995. "Heavy Metal Adsorption on Bacterially Produced FeS," *Minerals Engineering,* 8: 1097–1108.

58 Watson, J. H. P., Clark, N. O. and Windle, W. 1975. "A Superconducting Magnetic Separator and Its Application in Improving Ceramic Raw Materials." Presented at *Eleventh International Mineral Processing Congress,* Cagliari, Sardinia, Italy.

59 Watson, J. H. P. and Ellwood, D. C. 1995. "Feedback Chemostat." British Patent Application, vol. GB 9516753.2.

60 Watson, J. H. P., Ellwood, D. C. and Duggleby, C. J. 1996. "A Chemostat with Magnetic Feedback for the Growth of Sulfate Reducing Bacteria and Its Application to the Removal and Recovery of Heavy Metals from Solution," *Minerals Engineering,* 9: 973–983.

61 Freke, A. M. and Tate, D. 1961. "The Formation of Magnetic Iron Sulphide by Bacterial Reduction of Iron Solutions," *J. Biochem. Microbiol. Tech. Eng.,* 3: 29–39.

62 Watson, J. H. P., Ellwood, D. C., Soper, A. K. and Charnock, J. 1999. "Nanosized Strongly-Magnetic Bacterially-Produced Iron Sulfide," *Journal of Magnetism and Magnetic Materials,* 203: 69–72.

63 Skinner, B. J., Erd, R. C. and Grimaldi, F. S. 1964. "Greigite, the Thio-Spinel of Iron; a New Mineral," *The American Mineralogist,* 49: 543–555.

64 Roberts, A. P. 1995. "Magnetic Properties of Sedimentary Greigite (Fe_3S_4)," *Earth and Planetary Science Letters,* 134: 227–236.

65 Hamilton, E. I. and Clifton, R. J. 1981. "CR-39, a New Alpha-Particle Sensitive Polymeric Detector Applied to Investigations of Environmental Radioactivity," *Int. J. Appl. Rad. Isotopes,* 32: 313–324.

66 Hamilton, E. I. 1981. "Alpha-Particle Radioactivity of Hot Particles from the Esk Estuary," *Nature,* 290: 690–693.

67 Hamilton, E. I. and Clarke, K. R. 1984. "The Recent Sedimentation History of the Esk Estuary, Cumbria, UK. : The Application of Radiochronology," *The Science of the Total Environment,* 35: 325–386.

68 Hamilton, E. I. and Stevens, H. E. 1985. "Some Observations on the Geochemistry and Isotopic Composition of Uranium in Relation to the Reprocessing of Nuclear Fuels," *Journal of Environmetal Radioactivity,* 2: 23–40.

69 Hamilton, E. I., 1989. "Radionuclides and Large Particles in Estuarine Sediments," *Marine Pollution Bulletin,* 20: 603–607.

70 Watson, J. H. P. and Ellwood, D. C. 1992. "The Removal of Heavy Metals and Organic Compounds from Anaerobic Sludges," presented at *Conference on the Remediation of Sediments,* Institute of Marine and Coastal Sciences, Rutgers University, Brunswick, NJ.

CHAPTER 5

Polyelectrolyte Enhanced Removal of Metals from Soils· and Other Solids

BARBARA F. SMITH[1]
THOMAS W. ROBISON[2]
NANCY N. SAUER[3]

INTRODUCTION

SYNTHETIC, water-soluble polyelectrolytes can be used to recover and concentrate dilute metal ions and other inorganic and organic solutes from aqueous solutions for analytical and process applications. Using commercially available polymeric backbones and a wide variety of synthetic methods, new polyelectrolytes with selective receptors for metal ion and other inorganic species have been prepared [1,2]. When these polymeric polyelectrolytes are used in concert with ultrafiltration (UF) membranes, the polymer-metal complex, or guest-host polymer complex, can be readily retained, recovered, purified and concentrated [3]. Water and small, unbound solutes pass freely through the UF membrane, becoming permeate, while bound solutes remain as the retentate. The permeate stream, which is reduced in or "free" of the solutes bound to the polymer, can be used in further processing steps or discharged. The sequestered or bound solute is released from the receptor sites by adjusting the chemical and sometimes physical conditions of the retentate solution. A second UF step (diafiltration) recovers the sequestered solute in a concentrated form for recycle or disposal and retains the soluble polymer for additional process cycles.

The concept of using water-soluble polymers to retain small ionic solutes in this way was first discussed in the late 1960s by Michaels [4]. Most of the work to date has involved developing chelating polymers for use with metal ions, but polymers with receptors for other ionic and molecular species are also being developed. A review [5] of this hybrid technology details its concepts and development and provides comparisons with other separation technologies. Another [6] report includes a number of case studies of the application of water-soluble chelating polymers used in conjunction with ultrafiltration, termed Polymer Filtration™ (PF), to a variety of aqueous process streams.

[1]Los Alamos National Laboratory, Chemistry Division, MS J964, Los Alamos, NM 87545, U.S.A., bfsmith@lanl.gov

[2]Los Alamos National Laboratory, Chemistry Division, MS J569, Los Alamos, NM 87545, U.S.A., trobison@lanl.gov

[3]Los Alamos National Laboratory, Chemistry Division, MS J514, Los Alamos, NM 87545, U.S.A., nsauer@lanl.gov

Some attractive features of PF relative to other separation technologies include the following:

(1) The reaction between the target solute and the receptor sites occurs in a single liquid phase, which tends to give rapid attainment of equilibrium relative to separations requiring phase transfer (e.g., across liquid-solid phase boundary).

(2) On a weight basis, more of the polymeric structure is used for solute binding, giving the material a relatively high capacity. There are no requirements placed on the polymer to maintain mechanical stability (i.e., cross-linking) as is required for macroporous resins.

(3) Polymer mixtures can be used to target a suite of solutes and contaminates on solids.

(4) The PF process removes colloidal material, which is often useful. For example, waste streams containing radionuclides sorbed on colloidal material in addition to dissolved solutes that bind to the receptor sites are removed in a single operation.

(5) Operations involving PF are directly scaleable.

(6) The PF requires relatively low operational pressures, usually 10–50 psi, and operates at room temperature.

Two features of a PF process that must be addressed for successful application are the potential fouling of the UF membranes and the single stage of equilibrium binding for each UF operation. Improvements in commercial UF technology over the last 10–15 years have made the use of water-soluble polymers with UF a reliable and cost-effective approach to separations. Modern UF technology minimizes fouling in a variety of ways, and there are many commercial designs tailored to particular applications [7]. Generally, the binding constants of the polyelectrolytes have been so large that only one stage has been required. If multiple stages of PF are needed to accomplish a desired separation, then staging of the PF units can be designed much like the staging of contactors in solvent extraction systems [8].

The properties of the water-soluble polymer are key to successful implementation of PF. The receptor sites must have the required selectivity and affinity for the target species to yield permeate that meets the specifications of the process. The polymer, before and after loading, must remain soluble under the process conditions. The polymer should be completely retained by the UF membrane but should not interact with the membrane to foul it excessively. The polymer must remain stable under the process conditions, and the viscosity of the polymer solution must allow reasonable flux rates to be maintained during the concentration and regeneration phases of the process. Regeneration of the polymer by release of the guest molecules or ions can be accomplished by a variety of processes, including competition for receptor sites by protons at lower pH, competition for bound metal ions by ligands in the wash solution and oxidation/reduction reactions of metal ions to change their affinity for the receptor site [5,6].

Rejection Coefficients

The relative efficiency with which a UF membrane retains or rejects a solute can be determined experimentally with each solute being assigned a numerical value between 0 and 1 that is called the rejection coefficient (σ). A rejection coefficient of 0 means that the solute freely passes through a UF membrane (permeate), while solutes with a rejection coefficient of 1 are completely retained (retentate). Metal ions, and small organic and inorganic molecules will pass freely through the membrane ($\sigma = 0$) unless their effective size is temporarily increased by binding to the polymer ($\sigma = 1$). In the case of a polymer-metal ion complex in which the polymer (P) is physically too large to pass through the UF membrane, the rejection coefficient of the metal ions (M^{n+}) in the presence of a complexing polymer (P) is a reflection of the equilibrium or stability constant (K_s) of the complex, which is a measure of the affinity of the polymer for a metal ion.

$$P + M^{n+} \rightleftharpoons PM^{n+} \tag{1}$$

$$K_s \frac{[PM^{n+}]}{[P][M^{n+}]} \tag{2}$$

Modes of Operation

Generally, there are two modes of operation in PF, the concentration mode and diafiltration [9]. In the concentration mode, the volume of the retentate is reduced by simple UF. The final concentration of any solute can be determined by

$$C_f = C_0 \cdot \left(\frac{V_0}{V_f}\right)^{\sigma} \tag{3}$$

where C_f is the final concentration of the solute, C_0 is the initial concentration, V_0 is the initial volume of solution and V_f is the final volume. If the rejection coefficient of the solute is 1, as would be the case for the water-soluble solute-binding polymers, then Equation (3) simplifies to Equation (4).

$$C_f = C_0 \frac{V_0}{V_f} \tag{4}$$

For two species in solution, for example, a polymeric-metal-ion species (PM) with concentration C_{PM} and a molecular impurity (A) with concentration C_A where $\sigma_{PM} \gg A$, the UF of the solution should result in the concentration and enrichment of PM as expressed by Equation (5). UF results in a significant degree of purification during concentration of the polymer-metal-ion complexes.

$$\left(\frac{C_A}{C_{PM}}\right)_f = \left(\frac{C_A}{C_{PM}}\right)_0 \cdot \left(\frac{V_0}{V_f}\right)^{-(\sigma_A - \sigma_{PM})} \tag{5}$$

In the diafiltration mode, the lower molecular weight solutes are removed at a maximum rate when the rejection coefficient equals 0. Regeneration solution (V_w) is added to the retentate at the same rate that permeate is generated so as to maintain a constant volume. Thus, the retentate is washed free of smaller solute. Theoretically, the percent solute (any dissolved species) remaining in the retentate can be calculated by using Equation (6),

$$C_f = C_0 \cdot e^{-(V_w/V_0)(1-\sigma)} \tag{6}$$

where V_w is the volume of solute-free liquid (volume-equivalents) added, which also equals the amount of permeate volume produced (V_p).

Table 5.1 gives the percent solute recovered as a function of volume equivalents flushed through the system for a rejection coefficient of 0. Theoretically, after five volume equiva-

TABLE 5.1. Percent Solute Removed and Retained during Polymer Regeneration as a Function of Volume Equivalents of Solution Added for a Rejection Coefficient (σ) of 0.

Volume Equivalents	% Solute Removal	% Solute Remaining
1	63.2	36.8
2	86.5	13.5
3	95.0	5.0
4	98.5	1.5
5	99.3	0.7
10	99.9	<0.1

lents of processed solution, >99% of the lower molecular weight solute is removed. Experimentally, however, rejection coefficients of 0 are not ordinarily observed. Even weak interactions between the solute and the water-soluble polymer or the UF membrane can yield a small retention value. Curves generated for low retention coefficients follow an exponential decay with each additional volume equivalent giving diminishing returns in percent solute removed, while higher rejection coefficients approach a flat linear response with less to no solute removal [6].

General Process Conditions

The concentration of the water-soluble polymer in solution typically ranges from 0.001% w/v to 20% w/v of final concentrated solution. It is sufficient, and in some cases desirable, to have only enough polymer in solution such that the polymer's solute loading approaches 90–100%; use of higher concentrations of the water-soluble polymer results in lower flux rates through the UF membrane during the concentration stage. The use of a high initial polymer concentration can sometimes cause aggregation of the polymer and reduced binding capacity. In this case, operation at lower initial polymer concentrations can allow more complete solute binding, and the polymer can be concentrated to higher final concentrations with overall improved performance.

For a semi-continuous concentration mode process, it is necessary to work at low polymer concentrations to maintain high permeate flux across the membrane during the concentration stage of the process. During diafiltration, the polymer concentrations will always be higher, but at this juncture, the volumes being treated are relatively small. The flux is dependent on the transmembrane pressure, which is generally in the range of 25 to 50 psi. However, the increase in flux with increased transmembrane pressure is limited by concentration polarization, and increases in flux are minimal beyond 50 psi for typical tangential-flow, hollow-fiber UF units [7,9].

Ideally, no polymer permeates through the UF membrane during the PF process. If there is any polymer breakthrough (polymer $\sigma < 1.0$), polymer will ultimately be lost from the system [10]. Polymer loss is unacceptable from a number of process perspectives: (1) the polymer must remain in the system to maintain its working concentration, (2) polymer contamination in the permeate may create downstream contamination problems, (3) loss of solute-loaded polymer that would carry bound solutes into the permeate can result in failure to achieve target discharge limits, and (4) valuable solute-polymer complexes would be lost from the system. If only 1 ppm of polymer was lost from a system that contained 1,000 ppm polymer, a 50% loss of material in approximately a million volume equivalents would be lost. At $\approx$2% polymer breakthrough, as has been reported in some experimental systems

[11], 50% polymer loss would occur in approximately 35 volume equivalents. This amount of polymer loss is unacceptable for a viable process.

To calculate a solute concentration factor (*CF*) for a PF process cycle will require knowledge of the process system, which includes the initial feed metal-ion concentration and volume, the final metal-ion concentration and volume, the size of the reactor and the initial polymer concentration [10]. The polymer concentration seldom exceeds 20% in a continuous process because of the reduction in flux rates. For example, if the reactor size is 100 L and the initial polymer concentration is 1% w/v, we can concentrate the reactor volume from 100 L to 5 L at the end of the concentration phase. If the solute concentration in the feed going to the reactor is 100 ppm (e.g., Cu, AW 63.5) and a 1% w/v polymer solution has a capacity of 0.25 g Cu per g of polymer and a 100 L reactor has 1,000 g of polymer, we can bind 250 g of copper, which represents 2,500 L of feed that can be treated in this one batch. Thus, the concentration of 2,500 L to 5 L is a *CF* of 500 ($CF = V_0/V_f$). In actual single-stage practice, <100% of the polymer capacity is used to avoid metal-ion breakthrough that might exceed the discharge limit.

Polymeric Electrolytes for Solids Cleaning

Soils and other solid debris are determined to be hazardous if they contain certain levels of leachable RCRA (Resource Conservation and Recovery Act) materials as determined under the provision of the U.S. Environmental Protection Agency (EPA) Act 40-CFR using a TCLP (Toxic Characterization Leaching Procedure) test [12]. RCRA hazardous materials can exist singly, as mixtures with other RCRA elements, as a mixture with radioactive metals such as the actinides (defined as mixed waste) or as a mixture with a variety of nonhazardous metals and inert materials. TCLP is the EPA's method for determining if a material requires disposal as hazardous waste. TCLP is performed by leaching 100 g of ca. 8 mesh material in two liters of pH 4.9 acetate buffer for 18 hours (EPA Method SW 1650), which is supposed to simulate landfill leaching conditions to determine if RCRA material could leach from soils or solids in landfills over long periods. Radioactive wastes, which also contain RCRA metals (e.g., mercury, chromium and lead), or mixed wastes, represent a difficult waste management problem for the U.S. Department of Energy (DOE) [13]. If mercury can be removed to a level such that the low-level mixed waste (LLMW) can pass TCLP at ≤ 25 ppb [14], the material can be buried as low-level radioactive waste (LLRW). The regulatory limit for lead and chromium in leachate from the TCLP test is 5 ppm.

Soils and solids contaminated with radioactive and/or toxic metals are among the most common and difficult remediation problems at many DOE, Department of Defense (DoD) and manufacturing sites [15–17]. Volume-wise, soils are the single largest "waste" problem within the DOE complex, with an estimated 200 million cubic yards of contaminated soil requiring remediation [18–20].

Present alternatives for remediation of toxic or radioactive metal-contaminated soils include excavation of the site and transport of the contaminated soil to a secure repository, immobilization of the metal(s) in place, physical separation (usually by density or size) of the more highly contaminated soil fractions or chemical extraction of the contaminant from the soil [20]. The first three technologies suffer from a number of problems. Transport and storage of the entire volume of contaminated soil is exceedingly expensive. Immobilization approaches often suffer from high dilution of the waste, high ongoing monitoring costs and poor public acceptability. Physical separation methods are effective only for contaminated soils in which a large fraction of the contamination is concentrated in a small volume of soil that can be separated by size or density. In addition, physical separation methods often give very limited volume reductions. There are several emerging technologies for soil remediation that appear promising but are not yet at the implementation stage, such as phytoremediation and electrokinetic remediation [21,22].

Review of innovative soil remediation technologies shows that contaminant extraction from soil is reliable, permanent and a publicly acceptable remediation option [23–28]. This method relies on repeated washing of the soil with an extractant solution that binds the contaminant and washes it free of the soil. This technique is frequently used in the mining industry to remove residual valuable metal from low-grade or nearly depleted ores. The extractants are either acids, e.g., hydrochloric acid, or water-soluble organic compounds called chelators, which bind to the metal ions and solubilize them. Chelating extractants allow for the selective extraction of contaminating metal ions while eliminating soil disposal and landfill monitoring costs.

One of the principal barriers to chelator-based soil remediation is the identification of selective, readily regenerated and recovered extraction systems. Because polyelectrolytes form homogenous aqueous solutions, it is logical that they might have application as possible leaching agents to remove or recover metal ions from solids such as soils, debris and ores. There are many examples where simple water-soluble chelating agents have been used as leaching agents. For example, ethylenediaminetetraacetic acid (EDTA) has been used extensively to leach lead and mercury from soil [25–28]. EDTA is an excellent soil-leaching agent, but because of the copious amounts of solution required in the leaching process and because EDTA is difficult to regenerate, large amounts of secondary wastes are generated [29]. If soluble polymers can be used as the leaching agent, the leachate can be treated with UF to recover the polymer. The metals can be removed from the polymer-metal concentrate and recovered in concentrated form for proper waste management or reuse. The resulting soluble polymer in the retentate can be combined with the aqueous leach solution for further soil leaching. Surprisingly, until our own recent work, little work on the use of water-soluble metal-binding polymers for the removal of metal contaminants from soils had been done [30].

The following section describes a number of studies on the removal of RCRA metals from solids such as soils and hazardous debris. We have included a smaller section on the recovery of valuable metals from ores to demonstrate that solids-leaching with polyelectrolytes has applications other than waste treatment. Each case study gives some background as to the problem being addressed or the source of toxic metal pollution.

CASE STUDY 1: MERCURY REMOVAL FROM DEBRIS

Sources of Mercury in the Environment

Mercury, a RCRA metal, is found in numerous processes and waste streams at many DOE, DoD and commercial facilities. Mercury-containing wastes are generated in decontamination and decommissioning (D&D) and environmental restoration (ER) operations; it is found in old process and analytical wastes; and it is a contaminant in sludges [31], on soils [32,33], and on solid surfaces of debris [14]. In all cases, mercury can be found in a variety of oxidation states from elemental mercury(0) to ionic mercury(I) or (II).

DOE facilities at Oak Ridge, Tennessee, have many hundreds of tons of storm sewer sediments that were found to contain approximately 35,000 mg/kg of mercury(0) [34]. The old lithium processing facility, Y-12 Plant, Oak Ridge, is estimated to have several tons of mercury(0) in the piping system [35]. These systems represent a particular problem in that they have pipes and recesses that are not easily accessible to physical cleaning, and the plant is scheduled for decommissioning. Mercury(0) is also found in fluorescent light tubes, causing environmental problems at landfills, where an estimated 600 million fluorescent lamps are disposed of per year [36]. The three Oak Ridge, Tennessee, sites (Oak Ridge National Laboratory, Y-12 Plant and K-25 Site) alone have about 165,000 kg of crushed, mixed-waste fluorescence lamps in storage [36]. Though the amount of mercury(0) in new light tubes has been decreasing over the years, there is still a considerable amount contained in these tubes that could potentially be released into the environment. These mercury contamination issues at the DOE sites are unique in that they occur in radioactively contaminated facilities, and the wastes from these facilities are classified as mixed waste.

Industrial and municipal sources of mercury(0) and mercury(II) (e.g., HgO) include coal-fired plants [37], geothermal power plants [38], incinerator debris from medical research [39], mining operations [40,41], municipal incinerators [42,43] and sewage sludge plants [44,45]. The aqueous streams from scrubber waters in coal-fired utility plants, medical incinerators, commercial boilers and solid waste incinerators contain mercury(II). Other sources include dental facilities, chlor-alkali processes and measurement and control instruments [46,47]. All of these sources are responsible for releases of mercury(0) and mercury(II) into the environment [48,49].

Though mercury is dispersed into the environment from anthropogenic sources and has been for centuries even before the industrial revolution [50], there are many natural sources of mercury as well [51], including surface deposits of cinnabar, out-gassing of the earth via volcanoes and hot springs [52], and coal [53] and oil shale deposits [54]. The biogeochemical cycle of mercury causes it to continuously redistribute in the environment, and mercury is depositing in regions of the world that have no local industrial mercury sources [55,56]. If overall levels of mercury in the environment are to be reduced, it will first be necessary to reduce local and regional emissions by collecting mercury at anthropogenic emission-point sources and stabilizing them such that they cannot further contribute mercury to the biogeochemical cycle. Thus, spills, process and waste streams and emissions where mercury concentrations are already high are logical areas for mercury removal, recovery and stabilization.

Available Technologies to Treat Mercury-Contaminated Solids

Currently, two technology types are available for treating mercury-containing wastes: (1) those that remove mercury(II) and (2) those that remove mercury(0) [14]. The best demonstrated available technology (BDAT) for removing mercury(II) in aqueous waste streams is sulfide precipitation. Other technologies or materials that could be used and may, under certain circumstances, be preferred to precipitation are absorbers such as ion exchange resins, chelating ion exchange resins, reactive carbon, extraction chromatographic materials, or liquid-liquid extraction agents; concentration technologies such as reverse osmosis and evaporation; and electrical-based technologies, e.g., electrowinning and electrodialysis. The BDAT to treat pure mercury(0) and mercury(0) in some higher level waste streams (260 ppm) is amalgamation and retorting, respectively. Other technologies for mercury(0) removal include incineration/combustion with air stripping [43] or reaction with very strong oxidants, e.g., I_2, [34,57,58], or concentrated nitric acid. Though some of these processes function well with only certain mercury species, they may not remove mer-

cury to low enough levels, may not be compatible with certain debris types or may have serious waste by-product issues.

Technology Needs for Mercury Removal from Solids

There is a need to be able to mildly and selectively remove and concentrate mercury from
a variety of wastes or process streams, particularly low-level mixed waste streams that
could be incinerated or buried after treatment. Mercury needs to be removed/recovered to
meet discharge limits for process solutions, to meet soil and debris decontamination requirements and to pass TCLP ($\leq$25 ppb for mercury for burial). Though there is no regulatory level set, it has been stated that if mercury can be removed to the level of 1 ppm total
mercury in light debris [35], the debris could be incinerated in a low-level radioactive waste
(LLRW) incinerator instead of using scarce burial space.

Objectives have been to develop and optimize a mercury(0) oxidative-dissolution
method that is mild, rapid, selective and minimizes secondary waste; test a method for removing mercury(0), mercury(I) and mercury(II) from solid surfaces; optimize the concentration and recovery of mercury(II) from aqueous solutions including those produced by
dissolution of mercury(0); and to have a resultant waste form that will either meet incineration standards or can be buried as LLRW. The PF technology that we have been developing
specifically for low-level radioactively contaminated mixed waste has potential application
to mercury removal from industrial and commercial process streams, fuels such as coal,
soils and other materials that are not necessarily radioactively contaminated.

A number of water-soluble metal-binding polymers [5,59–61] have been shown to
readily bind with mercury(II) in dilute solutions, and some have been used to concentrate
mercury(II) using ultrafiltration membranes [62,63]. We have shown that polyelectrolytes
or water-soluble polymers can be used to leach mercury from solids [30]. The resulting
polymer-mercury complex can then be concentrated for secondary waste minimization.
Data indicate that mercury can be readily recovered from certain polymers, and the polymer
can be recycled for further use [62]. For our initial absorbency, recovery and HgO dissolution studies, we used polyethyleneimine, PEI (Polymin Water Free, BASF Corp., Charlotte,
NC, USA). PEI is a water-soluble polymer having amine ligands that bind mercury over a
wide pH range [59,64]. In general, even simple organic amines and diamines have demonstrated high mercury-binding affinity [65–67], but they are not effective for solids leaching
because they are not easily regenerated and recovered

Debris Leaching Studies

ADSORPTION OF POLYELECTROLYTE ON DEBRIS

To use soluble polymers to leach mercury from solid surfaces and debris, it is first necessary to determine if the leaching agent is extensively absorbed to the highly variable debris
surfaces. Polymer absorption and subsequent loss is important to consider, as it could potentially lead to poor mercury removal and recovery. It would serve little purpose if the
polymers were able to readily complex mercury in solution or on surfaces, but at the same
time, remain irreversibly sorbed to the debris.

We prepared surrogate debris and tested for absorption of the polymer onto the individual components of the debris. Table 5.2 gives the composition of the surrogate debris as recommended by the DOE Mixed Waste Focus Area (MWFA) [35]. The individual compo-

TABLE 5.2. Composition of Surrogate Debris.

Graphite
Aluminum Firebrick
Lead Brick
Wood
Stainless Steel
Nonferrous Metals (copper, aluminum)
Paper
PPE (50% PVC/50% non-PVC) personal protective equipment
Carbon Steel
Vermiculite
Cement
Sand
HDPE (high-density polyethylene)
Wood
PVC (polyvinylchloride)

nents were contacted with dilute PEI solutions. Measuring ultraviolet-visible absorbance at $\lambda = 278$ nm for polymer quantification, we looked for decreases in PEI concentration in the solution. As shown in Table 5.3, no significant polymer loss was indicated for any of the surrogate materials included in this study. Most of the materials were within approximately $\pm 2\%$ of the target polymer concentration. Two of the samples, the paper mixture and white paper, each exhibited slightly elevated polymer recoveries. These samples were examined in more detail, and it was determined that they contained an interfering constituent, which added slightly to the absorbance readings.

TABLE 5.3. Polymer Absorption Studies on a Variety of Debris Types.*

Material	UV/Vis Absorbance	Calculated wt. Percent
Polymer control	1.5699	0.975
Sand	1.6218	1.008
Vermiculite	1.6079	0.999
Paper	1.7177	1.068
Barley bag	1.6025	0.996
White paper	1.7405	1.070
Aluminum	1.6167	0.996
PVC	1.6192	0.996
Wood	1.6656	1.024
HDPE	1.6296	1.002
PPE	1.5491	1.021
Concrete	1.5123	1.004
Graphite	1.5123	0.981
Carbon steel	1.5596	1.011
Stainless steel	1.5451	1.000
Lead	1.5257	0.999

*Conditions: The individual surrogate debris components (15 g each) were placed in 250 ml plastic Nalgene bottles. The bottles were filled to 100 g with the 1% w/v PEI solution that had been adjusted to pH 5 with HNO_3. Each sample was shaken for 1 hour in a 37°C water bath. The polymer solution was filtered (0.45 μ, Millipore), and the sample was analyzed with a diode array UV/visible spectrometer (Hewlett Packard, Model 8452). The polymer analysis was conducted by mixing copper acetate (0.1 M), acetate buffer (0.1 M, pH 5.8) and PEI to volume, and the mixture was analyzed and compared to a standard curve. Each sample theoretically contained 10,000 ppm PEI at pH 5.00 and the absorbance was measured at 278 nm. Each sample was tested by making a 100-fold dilution (0.250 ml in a 25 ml volumetric flask) in order to be within the analytical range of the instrument.

A similar experiment was performed where composite synthetic debris components from Table 5.2 were tested for bulk leach solution and PEI recovery. Results are shown in Figure 5.1. Again, we observe excellent recovery with an overall 99.7% recovery of the polymer and 96.3% recovery of the water. It was expected that the vermiculite and other absorbing debris such as paper will retain some of the water (e.g., 15 g of vermiculite absorbed 53 g of water, and 11 g of Teri™ Towels absorbed 15 g of water in our hands). Though these two studies indicated that soluble polymer does not appear to be absorbed excessively on the debris or it can be readily washed from the debris, they do not necessarily prove that the polymer-metal ion complex is not absorbed to debris [68].

DISSOLUTION OF HgO IN THE PRESENCE OF DEBRIS

Previously, we demonstrated that PEI significantly enhanced the dissolution rate of such species as HgO [62,69]. These experiments were performed in the absence of debris, and the survey study used visual observation to determine dissolution time for HgO.

A similar experiment was performed, but in the presence of debris, to determine if the debris had an effect on the HgO dissolution rate. The concentration of mercury(II) in solution was determined as a function of time both in the presence and absence of debris. The results shown in Figure 5.2 indicate that the presence of the debris only slightly reduced the rate of dissolution, and the final, total dissolved mercury(II) reached similar plateaus. Mercuric oxide was rapidly solubilized in the presence of the polymer, with 62% dissolved in the time that it took to mix the materials and draw the first zero-time sample (about one minute). The PEI polymer-containing solution showed an approximately 10% negative bias during mercury analysis in the flow injection mercury system (FIMS), which is the reason for the apparent low mercury accountability for both experiments (observed approximately 900 ppm of the ca. 980 ppm mercury added). We did not digest the polymer solution before the solution was analyzed. The data were more erratic in the debris samples.

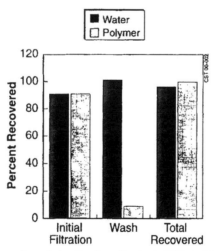

*Conditions: 15 g of each component was shaken together with a total of 750 ml of 1% w/v PEI solution at 50°C for 3 hours in a reciprocating water bath. The supernatant was vacuum filtered, and the debris was washed once with an equal volume of pH 5 water and was again vacuum filtered. The volume of water recovered was determined, and the polymer was quantified by the Copper Test in each recovered fraction. Polymer analysis was conducted by mixing copper acetate (0.1 M), acetate buffer (0.1 M, pH 5.8) and PEI to volume, and the mixture was analyzed and compared to a standard curve. Components included vermiculite, graphite, cement, HDPE, PVC, sand, wood and carbon steel.

Figure 5.1 Polymer and leachate solution recovery studies.*

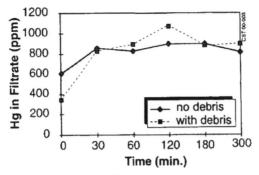

*Conditions: HgO (0.8055 g) was added to a 1 L Teflon™ bottle containing 760.5 ml of 1% PEI solution adjusted to pH 4 with HNO₃. A similar solution with 0.9038 g of HgO and ca. 200 g of debris from Table 5.1 (ca 15 g each) was used in 762 ml. The solutions were shaken at 50°C in a reciprocating shaker. Samples were removed, filtered through a 0.45 μ membrane, stabilized with 2% HNO₃/H₂SO₄ and analyzed using a mercury analyzer (FIMS, Perkin Elmer).

Figure 5.2 Determination of the effect of debris on HgO dissolution.*

Dissolution of Mercury(0)

MILD MERCURY(0) OXIDATION APPROACHES

Because PF is used to recover metal ions, not zero valent metal, it was necessary to bring mercury(0) into solution in a stable oxidation state (perform a mild oxidative dissolution). Optimization of the mercury(0) oxidation process was carried out to minimize reagent usage and secondary waste generation and to make the conditions as mild as possible. The rate of dissolution of mercury(0) was quite rapid under harsh conditions of 4 to 16 molar HNO_3, but it ceased as milder conditions (1 to 2 M HNO_3) were used [62,69]. Even though thermodynamic calculations, as shown in Equation (7) [70], indicate that mercury(0) is readily oxidized even at 1 to 2 M HNO_3, this oxidation was not observed experimentally, apparently because of slow kinetics [62,69].

$$NO_3^- + 4H^+ + 3e^- \rightarrow NO + 2H_2O \quad 0.957 \text{ volt (std. conditions, 25°C, 1 M acid)} \quad (7)$$

Mercury(0) oxidation is partially complicated by the possibility of two oxidation states, mercury(I) and mercury(II), both of which have similar oxidation potentials as shown in the potential energy diagram in Equation (8) [71]. Generally, conditions that oxidize mercury(0) to mercury(I) will oxidize mercury(0) to mercury(II), and mercury(I) readily disproportionates to mercury(0) and mercury(II). We have observed that, under some con-

$$
\begin{array}{ccc}
+2 & +1 & 0 \\
\hline
& 0.8535 & \\
\end{array}
$$

$$Hg^{2+} \xrightarrow{0.9110} Hg_2^{2+} \xrightarrow{0.7960} Hg$$

$$(8)$$

ditions, mercury(II) and mercury(I) coexist [62,69].

In our search for a mild oxidant that would not give a large secondary waste, hydrogen peroxide (H_2O_2) in acid was investigated as it has a reasonable oxidation potential as shown in Equation (9) [72].

$$H_2O_2 + 2H^+ + 2e^- \rightarrow 2H_2O \quad 1.763 \text{ volt } (25°C) \quad\quad (9)$$

Though the oxidation potential may be reasonable (thermodynamically favorable) with H_2O_2, the kinetics are unfavorable because H_2O_2 has been shown to be a poor oxidizer for mercury(0) in practice [73]. Under certain conditions of higher pH values, H_2O_2 is a reductant for mercury(II) (pH values larger than 6) [74]. Thus, to use H_2O_2 as a mild oxidant, it will be necessary to balance pH, H_2O_2 and chelator concentration, time, temperature and mixing. Conditions must be optimized such that once mercury(0) is oxidized and solubilized, it does not precipitate as the oxide or adsorb on surfaces of materials. Having a chelator in solution to capture mercury(II) may help eliminate absorption problems and will shift the equilibrium to favor oxidation of mercury(0) under milder, less acidic conditions [75,76]. It has been our premise that certain chelating ligands might also enhance the kinetics of dissolution by facilitating the electron transfer process [77].

Previously, we showed that mercury(0) can be solubilized under mild conditions in the presence of soluble polymer [62,69]. We showed the following for the H_2O_2 system:

(1) If the solution was not properly agitated, no oxidative-dissolution occurred.
(2) Mercury(0) in the presence of H_2O_2 but without polymer gave no dissolution under our mixing conditions (magnetic stirring, orbital shaker).
(3) Mercury(0) in the presence of soluble polymer but without H_2O_2 gave no dissolution (polymers do not act as oxidants).
(4) 2% H_2O_2, pH range of 1 to 4 with HNO_3, temperature 30–60°C, and 0.1 to 1% polymer gave mercury(0) dissolution times of 0.5 to 4 hour.
(5) Neither solid chelating resins nor molecular ligands with chelating functionality similar to that of the water-soluble polymers aided in the oxidative dissolution of mercury(0) in the presence of H_2O_2 under our reaction conditions [62,69].
(6) We showed that the presence of the polymer indeed enhanced the oxidative dissolution rate [62,69].

In previously reported work [62,69], we demonstrated that there are a number of water-soluble chelating polymers that can strongly bind mercury(II) such that when the polymer/mercury(II) complex is recovered by ultrafiltration, the permeate can meet the ≤25 ppb discharge limit set by the EPA. We showed that mercury(II) can be stripped from one of the polymers, WABOH-30, at pH 0 (see Figure 5.3), and that all of these polymers have a high capacity for mercury(II). Permeate concentrate from the diafiltration process can be readily precipitated with sulfide to form a stable HgS waste form. Though the complete oxidative stability of all the polymers to 2% H_2O_2 has not been proven, we showed that the breakthrough of mercury(II) was not substantially influenced by the presence of H_2O_2. We observed that PEI was more stable to H_2O_2 under acidic conditions, which is consistent with protonated amines being less reactive to oxidation [78]. If the polymer does not have long-term stability to H_2O_2, it may be possible to sacrifice some of the polymer during the oxidative-dissolution step of the process to enhance the rate of mercury(0) leaching.

STAINLESS STEEL LEACHING STUDIES AS DEBRIS SURROGATE

Tests were performed to determine if the conditions developed for solubilizing mercury(0) beads would aid in the dissolution of mercury(0) from the surfaces of stainless steel materials that represent contaminated pipes or other metal debris. Stainless steel coupons

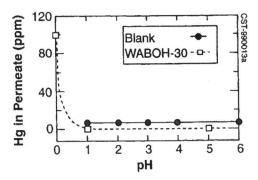

*Conditions: 1% w/v polymer, pH adjusted with HNO₃, starting with 100 ppm Hg(NO₃)₂ SPEX standards, filtered through Centricon-10 unit, permeate analyzed by FIMS. Blank has 10 ppm mercury(II).

Figure 5.3 Binding of mercury(II) as a function of pH with WABOH-30 polymer compared to a blank in the absence of polymer.*

contaminated with mercury(0) were mixed with several different leaching solutions, including several concentrations of HNO_3, for comparison to the polymer solution. A total of three washings was performed, and the results are summarized in Figure 5.4. Concentrated HNO_3 was very effective at removing mercury(0) (93–99%). In contrast, mild conditions, 0.1% WABOH-30, pH 2, 0.5% H_2O_2, solubilized 86% of the mercury(0).

ULTRASONIC IRRADIATION MIXING STUDIES

We could not use a magnetic stir bar for mechanical mixing for the coupon-leaching study, and we knew that the orbital shaker was ineffective as a mixing mode [62,69]. We had observed ultrasonic irradiation with a laboratory ultrasonic cleaning bath to be a reasonable mixing method (see Figure 5.4). Studies were performed with a more powerful ultrasonic bath to determine its effect on the rate of mercury(0) bead dissolution. The results were quite dramatic, and the mercury(0) oxidative dissolution rate was greatly enhanced.

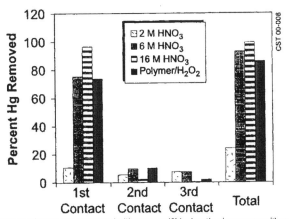

*Conditions: The stainless steel coupon was coated with mercury(0) by heating in an oven with excess liquid mercury on the SS metal. The excess mercury(0) was removed, and the coupons were immersed in the leachate solutions (0.1% WABOH-30, pH 2 with HNO₃, 0.5% H₂O₂), 2 M HNO₃, 6 M HNO₃, and 16 M HNO₃ in a glass bottle. The bottles were placed in an ultrasonic bath (45 KHz, 45 watts) for 2 hours at RT. The resulting solution was decanted and analyzed for Hg by FIMS. The leaching process was repeated two more times, analyzing the leachate for each contact. The coupons were dissolved in aqua regia to determine the remaining mercury for a mass balance.

Figure 5.4 Mercury(0) decontamination studies of stainless steel coupons.*

Table 5.4 shows that a 10–30 mg mercury(0) bead could be dissolved in four minutes, which was considerably faster than any of our previous experiments using magnetic stirring [62,69]. It was observed that the bead disappeared in one minute and a fine gray powder formed, then in the remaining three minutes, the powdery-gray material cleared. When the reaction was interrupted and chloride as HCl was added, the solution tested positive for mercury(I). At the end of the reaction time, the solution gave a negative chloride test for mercury(I). These experiments indicate that mercury(I) is formed either as an intermediate or as one of the reaction products, and that the polymer apparently rapidly promotes disproportionation, which could account for the formation of the fine gray material, which is probably finely divided mercury(0). A similar gray powder was observed when soluble polymer was added to a solution that was known to contain mercury(I). Mercury(0) dissolution using ultrasonic irradiation was not only rapid, but further tests indicated that it was rapid at room temperature. We could also lower the amount of H_2O_2 from 2% to 0.5% and still have rapid oxidative dissolution of the mercury(0). It is not clear whether we are observing enhanced surface area production on the mercury(0) bead as the ultrasonic energy rapidly disperses the bead, or if there is some sonichemistry occurring, such as the increased production of hydroxyl radicals from H_2O_2 and water sonolysis [79,80].

DEBRIS TREATMENT

An approach to debris leaching includes the processing steps shown in Figure 5.5. The debris is segregated, shredded or comminuted, pre-rinsed, leached, post-rinsed, drained, and analyzed by TCLP and 8 M HNO_3 leach to determine residual mercury levels. The leachate is concentrated using ultrafiltration, and the concentrated mercury(II) can be stripped from the polymer using a diafiltration process and stabilized as HgS in a final precipitation step. The regenerated polymer is recombined with permeate to reform the leaching solution, which is recycled for further debris leaching after it is adjusted for HNO_3 and H_2O_2 concentrations. Process controls included acid, nitrate and H_2O_2 concentration monitoring, buildup of mercury in the process and final TCLP on the debris.

PILOT-SCALE DEBRIS-LEACHING UNIT

A pilot-scale unit was built to perform mercury leaching from debris, and the schematic is shown in Figure 5.6. It consists of three subunit operations. The first subunit is the de-

TABLE 5.4. Mercury(0) Dissolution Survey with Ultrasonic Irradiation for Enhanced Mixing.*

pH	% Polymer	% H_2O_2	Dissolution Time
1	0	0	Incomplete (0.4%), >24 hours[1]
1	0.1	0	Incomplete (0.13%), >24 hours[1]
2	0.01	0.5	7 min
2	0.05	0.5	5 min
2	0.10	0.5	4 min[2]
2	0.10	0.1	38 min

*Conditions: 20–30°C, HNO_3, 30 mg Mercury(0), WABOH-30 polymer, 35 ml solution, NEARFIELD™ ultrasonic bath (Lewis Corp., ~16/20 kHz, 2,000 W).
[1]Used 50/60 KHz/1,010 W ultrasonic bath at 40°C.
[2]Gave a negative chloride test at end of time; the bead was gone in about 1 min, but the solution was gray to the time indicated.

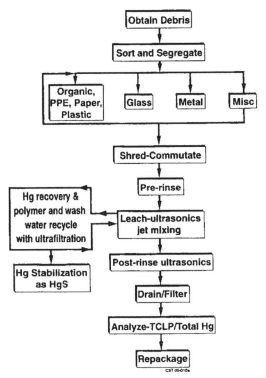

Figure 5.5 Diagram of debris leaching, mercury concentration and recovery cycle.

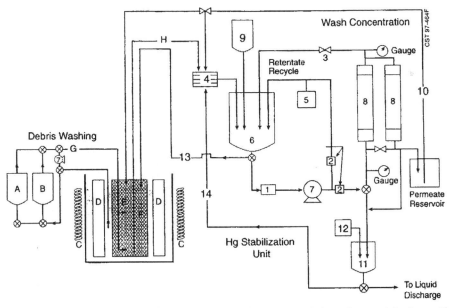

Figure 5.6 Schematic of a 25-gallon, pilot-scale debris-washing unit that includes an integrated flowing washing system, a PF concentration unit and a mercury stabilization unit. (A) Debris wetting and pre-rinse solution, (B) debris leaching solution, (C) cooling coils, (D) transducers for ultrasonic mixing, (E) jet mixers and circulation, (F) debris wash basket, (G) liquid transfer lines, (H) debris transfer lines. (1) Flow sensor, (2) solenoid valves, (3) back-pressure valves, (4) bag filter, (5) pH adjustment, (6) concentration vessel, (7) pumps, (8) UF-hollow-fiber membranes, (9) diafiltration solution, (10) wash solution recycle line, (11) precipitating tank, (12) sulfide reagent addition, (13) polymer recycle of wash line, (14) HgS to filter line.

bris-leaching tank, which can incorporate the pre-rinse operation and the leaching operation. The pre-rinse solution (water or dilute acid) and the leaching solution are stored in separate saddle tanks and pumped into the leach tank as needed. The purpose of the pre-rinse is to remove chloride salts that might trap mercury(I), as it is formed and other extraneous chemicals that may interfere with the mercury(0) oxidative dissolution process. The 25-gallon, stainless steel leach tank is equipped with cooling coils, two facing NEARFIELD™ ultrasonic transducers (16 and 20 kHz, 14 inches long by 19 inches high by 6 inches wide, Lewis Corporation, Oxford, CT, USA) placed 12 inches apart, and a multilevel basket to hold the debris. The unit is configured to allow for slurry pumping of soft debris to a filter bag for final collection if desired, or debris can be drained from the basket. The leach solution is circulated, and the return solution allows for jet mixing of the debris.

The second subunit is the ultrafiltration system that consists of a polymer concentration tank and two ultrafiltration membrane cartridges (Polysulfone-30,000 MWCO). It is equipped with a pH adjustment system, debris prefilter, flowmeters, diafiltration solution reservoir and its own pumping system. The third subunit of the system is a HgS precipitation tank and filter system to collect the HgS, which is equipped with its own recirculation pump. Photographs of the shredding and leaching/UF/precipitation units are shown in Figures 5.7 and 5.8.

Thus far, we have leached 7 kg of shredded glass, plastic, rubber and cloth under several conditions to determine optimum modes of operation of the washing unit for different types of debris. The material was shredded using a Taskmaster™ shredder (Franklin Miller, Inc., Livingston, NJ, USA) mounted on a cart. The shredder is vented into a hood and is equipped with a solids collection system and safety latches. Several processing issues are currently being addressed. Some debris types float and some sink, raising the issue of materials han-

Figure 5.7 Photograph of the debris shredder.

Figure 5.8 Photograph of debris leaching, ultrafiltration and sulfide precipitation unit.

dling. The correct mixing regime, how much jet and how much ultrasonic mixing to use is important, because jet mixing aerates the leach solution and decreases mixing efficiency of the ultrasonic irradiation [79]. The issue of leach-tank-loading is also important as some materials do not transmit ultrasonic energy, but absorb it. Thus, glass and metal do not appear to lose energy, but rubber, plastic and fabric absorb the ultrasonic energy.

The preliminary operation of the unit with 4 kg of glass indicated that we removed mercury from the glass material so that it readily met TCLP ($\leq$25 ppb) after about 1 to 3 hours of treatment. Our preliminary goal was to reach < 4 ppm remaining on the glass from an initial contamination of 28 ppm. Total mercury analysis before and after treatment indicated that 86%–99% of the mercury was removed.

Shredded plastic was placed directly into the temperature-controlled reactor, and samples were taken to determine mercury removal levels (see Table 5.5). We performed TCLP

TABLE 5.5. Results of Plastic Leaching with pH 2 Nitric Acid, 0.1% Polymer and 0.5% Hydrogen Peroxide.*

Material/Conditions	Hg Content µg Hg/g Debris	TCLP (ppb)
Preleached plastic	538.4	68.3
60-minute contact	52.2	—
120-minute contact	28.3	43.1

*Conditions: Plastic (112 g) was placed directly into the reactor, 15°C, ultrasonic irradiated at 60-minute intervals. Samples were removed, rinsed in water, dried and analyzed for the presence of residual mercury on the debris. Mercury content was determined by an 8 M nitric acid leach that was analyzed using Flow Injection Mercury System, FIMS (Perkin Elmer).

on the debris before and after leaching. Even though we had removed 93% of the mercury, 538.4 ppm down to 28.3 ppm, we could not pass the TCLP ($\leq$ 25 ppb for treated debris). The plastic that had the majority of the mercury removed did only slightly better on the TCLP test than the plastic that had not been treated at all. We hypothesized that because the process converts mercury(0) to mercury(II) and then removes the mercury(II) from the debris, perhaps we had left some mercury(II) or mercury(II)-polymer complex on the surface during the rinse step. Both mercury(II) and mercury(II) complex will be considerably more soluble in the acetic acid leach used in the TCLP than mercury(0). By treating the debris to convert all the mercury(0) to mercury(II), the rinse step becomes extremely important if we are to reduce mercury levels enough to pass the TCLP test. Further testing and optimization are in progress to get high mercury removal levels, to reduce processing times and to perform treatment cost estimates.

CASE STUDY 2: REMOVAL OF MERCURY FROM SOILS

An EPA surrogate soil (SSM, Synthetic Soil Matrix) used in a number of Oak Ridge National Laboratory tests to represent storm drain sediments was obtained [34,81]. It had little background mercury as determined by TCLP (2 ppb) and 8 M HNO_3 leach (1 ppb). We selected 100 mesh (150 micron) sieved soil-surrogate for spiking with various forms of mercury. Preliminary tests showed that the soil was alkaline with a large buffering capacity. The following experiments were performed to optimize leaching conditions and to segregate the mercury(0) oxidative-reduction reaction rate from simple mercury(II) desorption from soils rate.

Dependency of Mercury(II) Extraction on Leachate pH

Mercury(II) salts were used to separately test how well the polymer-containing leachate could remove mercury(II) sorbed on soil as a function of pH. In these tests, soil was spiked with mercury(II) as $Hg(NO_3)_2$ solution (4,000 ppm) to give 4 mg mercury(II) absorbed on the soil. The initial pH was varied, while the leachate-to-solid ratio was constant and high (250 ml leachate/g soil) and the mercury-to-polymer ratio was constant and low (see Table 5.6). For comparative purposes, these experiments were completed in a manner similar to previous work [82] where EDTA, KI, HCl and NaOH were used as leaching agents. The differences between these experiments were the starting soils [34] and the leachate.

As shown in Table 5.6, pH variation in the range of 1.5 to 6 (final pH 2.1 to 7.4) had little effect on the recovery of the mercury, which was extracted well in all cases when compared to the mercury(II) blank. There was slightly less extraction at the two lower pH values where polymer capacity is lower [62,69]. The mercury(II)-to-polymer ratio was 166 mg mercury/g polymer. This value is within the capacity range of the polymer (454 mg Hg to 114 mg Hg per g polymer in the pH range of 7 to 1) previously reported [62,69]. Thus, one of the criteria that must be met for appropriate soil leaching is that the capacity of the polymer must not be exceeded. The pH had little effect on the mercury leaching in a pH range between 2 and 7, though if H_2O_2 were present, we would have to work below pH 5 to maintain oxidizing reaction conditions [74]. This study indicated that the leachate-to-soil ratio may be important in that we observed good mercury(II) recovery with these high ratios (250 ml solution/g soil). In comparison to Wasay et al.'s [82] study with KI leachate, KI only extracted well at low pH values (<pH 4), whereas the polymeric system functioned well in the full pH range of 2 to 7. It is interesting to note that the WABOH-30 polymer gives approximately 10% positive bias in the mercury analysis on the FIMS analysis system, while previ-

TABLE 5.6. Results of the pH Dependency of Polymer Leaching
of Storm Drain Sediment Surrogates.*

Initial pH	Final pH	Hg(II)[1] (ppm)	Avg. Hg(II) (ppm)
pH 1.5	2.08 (2.03)	41.83 (41.52)	41.45
pH 2.0	2.32 (2.38)	39.39 (41.43)	40.41
pH 3.0	6.91 (6.93)	42.93 (43.30)	43.12
pH 4.0	6.69 (6.75)	41.83 (47.44)	44.63
pH 5.0	7.069 (7.16)	43.51 (45.53)	44.52
pH 6.0	7.42 (7.38)	45.00 (43.27)	44.14
Blank spike, no polymer, 2% H_2SO_4/2% HNO_3		39.98 (41.38)	
Control spike with pH 6 polymer		44.31	
Blank, pH 6 polymer		0.382	

*Conditions: 0.1 g of soil was spiked with 4 mg of mercury(II) as a nitrate solution (1 ml of 4,000 ppm) and aged 20 hours in Teflon™ centrifuge tubes. The leaching solution (0.1% w/v polymer adjusted to pH with HNO_3) was added to the soil (24 ml) and the mixture was shaken on a wrist arm shaker for 24 hours. The samples were centrifuged and the supernatant was removed to a 100 ml volumetric. The soil was washed two times with 25 ml of similar pH (HNO_3) solution, centrifuged and the wash was combined with the leachate in the volumetric. The volumetric was brought to volume with DI water, pH values were determined and analyzed by FIMS after the appropriate dilution.
[1]Sample duplicates are in parentheses.

ously, PEI gave approximately a 10% negative bias for undigested samples. We have observed in previous work that some polymers bias high and some low in the FIMS analysis method used [62,69].

Dependency of Mercury(II) Extraction on Leachate-to-Soil and Mercury(II)-to-Polymer Ratios

A study was performed where the amount of soil and mercury(II) was held constant while the volume of leachate was varied, thus, the leachate-to-soil and mercury(II)-to-polymer ratios were varied. There was an attempt to keep the pH constant, but it was difficult because of the large buffering capacity of the alkaline soil. The previous study indicated that the pH was not a major factor when leaching with polymer in the range of pH 2 to 7, so the pH changes were ignored. As shown in Table 5.7, there was full recovery of mercury(II) when the mercury-to-polymer ratio was less than 167 mg/g. When the polymer capacity was exceeded at 500 mg mercury/g polymer, the mercury(II) recovery was low, because the polymer-unbound mercury(II) was retained by the soil.

The leachate-to-soil ratios were quite small in the above experiment (see Table 5.7) and appeared to have influenced the level of mercury(II) leached, because the mercury-to-polymer ratio indicated that complete extraction was 100 mg/g instead of 167 mg/g as shown earlier. Thus, another experiment was performed where the mercury(II)-to-polymer ratio was kept constant (167 mg/g), and the leachate-to-soil ratio was varied in a wider range. The results shown in Table 5.8 indicate that the leachate-to-soil ratio had a major effect on mercury(II) leachability. When the leachate-to-soil ratio was at 250 ml/g, all the mercury(II) was leached, as shown previously (Table 5.7). When the ratio was reduced to 50 ml/g, there was a small reduction in the amount of mercury(II) leached from the soil, which decreased as the ratio decreased.

Dependency of Mercury(II) Extraction on pH in the Absence of Polymer

Wasay et al. [82] indicated that acidic and basic aqueous solutions gave little extraction

TABLE 5.7. Results of the Variation of the Mercury(II)-to-Polymer Ratio for Soil Leaching.*

Final pH	Leachate (ml)	Hg/Poly. (mg/g)	Leachate/Soil (ml/g)	Hg in Leachate (ppm)
5.53	20	100	40	26.90
3.79	15	167	30	23.48
3.20	10	250	20	20.91
3.21	5	500	10	12.54
Blank (no polymer)				24.50

*Conditions: 0.5 g of soil was spiked with 2.5 mg mercury(II) as the nitrate (0.25 ml of 10,000 ppm solution) and aged about 5 hours in Teflon™ centrifuge tubes. The leaching solution (0.1% w/v polymer adjusted to pH 2 with HNO₃) was added to the soil, and the mixtures were shaken on a rotary shaker (200 rpm) for 21 hours. The samples were centrifuged, and the supernatant was removed to a 100 ml volumetric. The soil was washed once with 20 ml of pH 2 (HNO₃) solution, centrifuged, and washed again with 10 ml of pH 2 solution, and the washes were combined with the leachate in the volumetric. The volumetric was brought to volume with DI water, pH values read and analyzed by a FIMS mercury analyzer after the appropriate dilution.

of mercury from soil in the pH range studied. We evaluated the extraction of the surrogate soil with dilute HNO_3 in the absence of polymer to compare with the literature data. The results are shown in Table 5.9. Again, we had difficulty maintaining constant pH values, but the data indicate that in the absence of polymer, between 30–50% of the mercury(II) remained adsorbed on the soil over a wide pH range, even at the high leachate-to-soil ratios of 250 ml/g. The soil used in this study released more mercury(II) into the aqueous solutions than did the soil in the Wasay et al. study [82]. The difference could be the lower organic content in the surrogate soil and the fact that the spiked surrogate soil was not aged.

Time Dependency of Mercury(0) Oxidative Dissolution and Extraction

Tests with Hg(0) spiked soil were performed. We used a leachate-to-soil ratio (200 ml/g) and mercury-to-polymer ratio (96 mg/g) that was nearer to the lower-acid capacity of the polymer (114 mg mercury/g polymer) at pH 1. The mixture was ultrasonicated at 10°C for 2 hours then shaken on an orbital shaker at 24°C for an additional 14 hours. The results are shown in Table 5.10. We can see that after 2 hours of ultrasonic irradiation that all of the mercury(0) was oxidized and extracted from the soil. Even though excess acid was added at

TABLE 5.8. Study Where Leachate-to-Soil Ratio was Varied for Extraction of Mercury(II) from Soil.*

Final pH	Diluted pH	Soil (g)	Hg/Poly. (mg/g)	Leachate/Soil (ml/g)	Hg in Leachate (ppm)
1.33	1.47	0.1073	167	250	44.06
1.47	1.76	0.5025	167	50	34.56
5.27	2.30	1.0014	167	25	27.75
5.32	2.34	1.5051	167	17	18.31
Blank (no polymer)					40.90

*Conditions: Soil was spiked with 4 mg mercury(II) as nitrate (1 ml of 4,000 ppm soln) and aged about six days in Teflon™ centrifuge tubes. The leaching solution (24 ml) (0.1% w/v polymer adjusted to pH 1.5 with HNO₃) and 0.3 ml of 8 M HNO₃ (to compensate for alkaline soil) were added to the soil, and the mixtures were shaken on a wrist arm shaker for 21 hours. The samples were centrifuged, the pH values read, and the supernatant removed to a 100 ml volumetric. The soil was washed twice with 25 ml of pH 2.0 (HNO₃) solution, centrifuged and the washes combined with the leachate in the volumetric. The volumetric was brought to volume with DI water, pH values read and analyzed by a FIMS mercury analyzer after the appropriate dilution.

TABLE 5.9. Acid Leach of Surrogate Soil with no Polymer Present.*

Initial pH	Final pH	Soil (g)	Hg/Poly. (mg/g)	Leachate/Soil (ml/g)	Hg in Leachate (ppm)
2.0	5.76	0.1000	167	250	28.43
2.0	5.36	0.1006	167	250	22.21
4.0	7.06	0.0999	167	250	20.65
4.0	6.98	0.0999	167	250	29.33
6.0	7.19	0.0997	167	250	31.01
6.0	7.18	0.1003	167	250	34.48
Blank (no polymer)					41.23

*Conditions: Soil was spiked with 4 mg mercury(II) as nitrate (1 ml of 4,000 ppm soln) in Teflon™ centrifuge tubes and not aged. The HNO_3 leaching solution (24 ml) was added to the soil, and the mixtures were shaken on a wrist arm shaker for 21 hours. The samples were centrifuged and the supernatant removed to a 100 ml volumetric. The soil was washed twice with 25 ml of same pH (HNO_3) solution, centrifuged and the washes combined with the leachate in the volumetric. The volumetric was brought to volume with DI water, pH values read and analyzed by a FIMS mercury analyzer after the appropriate dilution.

the beginning to compensate for the soil alkalinity, the pH went from 2.0 to 3.3 at the end of the extraction.

A timed leaching study was performed using 4 mg of mercury(II) as $Hg(NO_3)_2$ spiked on 1 gm soil under the same conditions described in Table 5.10. We observed that all the mercury(II) was extracted at the first sampling time of 0.5 hours, indicating that if the mercury is present in its oxidized form, it can be readily extracted if the conditions of polymer capacity and solution volume are met.

A number of general observations were made during these experiments. First, the soil surrogate was basic, making it difficult to maintain pH, and the soil enhanced the consumption or decomposition of H_2O_2 compared to a blank, where pH was easily maintained and H_2O_2 concentrations were not reduced. Although oxidation of the mercury(0) to mercury(II) on spiked soils appeared to occur with substoichiometric amounts of polymer, at least stoichiometric amounts of polymer were required to complex the mercury(II) or it would sorb to the soil [83]. Rate comparisons of oxidative dissolution were made with mercury(0) blanks, which contained similar amounts of leachate and mercury(0) but no soil. Mercury(0) in the presence of the soil oxidized slower than mercury(0) blank. The reduced

TABLE 5.10. Oxidative Dissolution of Mercury(0) in the Presence of Soil Surrogate.*

Time (hr)	pH	Hg in Leachate (ppm)	% Leached
0	2.0	—	—
0.5	2.5	17.6	16.6
1.0	2.7	65.6	62.1
1.5	3.0	95.3	90.5
2.0	3.3	104.7	99.5
14	—	105.5	100
Total possible Hg (corrected for 10% positive FIMS bias)	—	105.1	—

*Conditions: Soil (1.0 g) was spiked with 19.2 mg mercury(0) in a glass bottle and aged for two days. The leaching solution (200 ml of 0.1% w/v polymer, pH 2 HNO_3, 0.5% H_2O_2) was added to the soil, and the solution was spiked with an additional 0.5 ml of conc. HNO_3 to keep acidic. The mixture was ultrasonicated for two hours at 10°C. The sample was then shaken on an orbital shaker for 14 hours at 24°C. Aliquots (0.1 ml) were moved and diluted in 2% HNO_3/H_2SO_4 (9.9 ml) and analyzed by a FIMS mercury analyzer after the appropriate dilution. A blank (35.8 mg mercury) in the absence of soil was treated similarly.

rate could have numerous causes, such as the increase in pH, which potentially makes H_2O_2 a weaker oxidant; the decrease in the H_2O_2 concentration, which was observed in the presence of soil; and/or ultrasonic energy was absorbed by the soil and not transmitted to the bead, which would create poorer bead mixing.

Full-Scale Soil Leaching Unit

From the successful oxidative dissolution and extraction results, information determined in the preceding experiments and our general experience of working with solids, we have come to several conclusions about how an industrial mercury soil-leaching unit might be designed. One of the main questions is how to scale ultrasonic mixing. Figure 5.9 gives a schematic of how we think this process could be accomplished on a large scale. We propose using ultrasound trough units designed for soil or coal washing (made by Lewis Corporation). These units could be fitted with downward angled baffles along the length of both sides of the trough to allow for the physical segregation of the dense mercury(0) beads. The beads move down a trough at a faster rate than the soil and can be collected in a mercury catch trap placed before the end of the trough. The trap could be closed after the mercury(0) is captured and before the soil arrives. A variation on the baffle concept could be an auger moving countercurrent to the flow of the soil and leach solution. Again, we have observed that the mercury(0) beads flowed counter to the auger, while the solid remained stationary. In this way, the majority of the free mercury(0) can be segregated from the soil, requiring less polymer to remove the remaining mercury, which is mostly adsorbed on the soil as either mercury(0) or mercury(II). The trough length, tilt and number of baffles will determine the residence time of the soil in the process. Hoppers, weirs and conveyers can be used for soil movement and solid-liquid separation. It will probably be necessary to have a protective cover on the trough to prevent splash, mercury(0) vaporization and excessive air absorption into the leaching solution. We have observed that when the leaching solution was

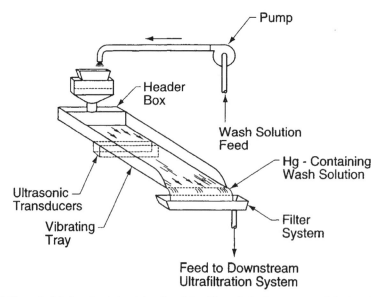

Figure 5.9 Conceptual design of an industrial-scale soil-leaching unit that incorporates mixing by ultrasonic irradiation and mercury(0) density separation.

aerated, the ultrasonic energy available for mixing was greatly reduced until the solution was degassed [79]. If the mixing is not efficient, the oxidative dissolution of mercury(0) does not occur.

CASE STUDY 3: REMEDIATION OF LEAD-CONTAMINATED SOILS

We undertook a systematic study to explore the application of Polymer Filtration to remediation of lead-contaminated soils. The soils used in these studies were from the 44-acre Cal-West (CW) Superfund site in southern New Mexico, where a former battery-cracking operation resulted in about 10,000 ppm average lead contamination in the soil. These typical southwestern soils contain only small amounts of clay and little organic material. The polymer system selected for study was polyethylenimine (PEI) functionalized with amino carboxylate groups (PEIC) initially prepared by Geckeler and coworkers by reaction of PEI with chloroacetic acid [84].

Baseline Extraction Studies

Simple aminocarboxylate-containing molecular compounds are commonly used for remediation of lead-contaminated soils [85–88]. Binding constants of EDTA and related aminocarboxylates for lead(II) are high ($>10^{16}$), and thus, these compounds are effective agents for removal of lead [89]. Polymeric sequesterants like PEIC should have similar high affinity for lead binding. For our studies, PEI was derivatized with two levels of chloroacetic acid to give fully functionalized polymer (PEIC-ff), where there is one equivalent of carboxylate for each PEI nitrogen, and partially functionalized polymer (PEIC-pf), where there is 0.25 equivalents of carboxylate groups for each PEI nitrogen. Table 5.11 shows the lead-loading capacities for PEI, PEIC-ff and PEIC-pf. All three systems have high capacities for lead(II).

Initial batch extraction studies using PEI or PEIC-ff and Cal-West soils were done to evaluate the efficiency of polymer extraction for our lead-contaminated soils. Twenty-four-hour batch experiments with an excess of each of the polymers were conducted. Lead removal determined by analysis of extractant solutions and digestion of treated soils showed that PEI had removed 36% of the lead from the Cal-West soil, and PEIC-ff had removed 88%. PEIC-ff was comparable to EDTA in its ability to bind and solubilize lead from soils. Given these results, subsequent experiments were designed to evaluate polymer functionalization, polymer molecular weight, extraction time and polymer-to-contaminant ratio for their effects on the efficiency of lead extraction.

Effect of Polymer, Contaminant Ratio and Extraction Time on Lead Extraction

The first series of studies compared the effectiveness of fully and partially

TABLE 5.11. Lead-Binding Capacities for PEI and PEIC of Varying Functionalization.

Polymer	Lead Bound (mmol/g Polymer)
PEI	1.4
PEIC-pf	1.9
PEIC-ff	2.3

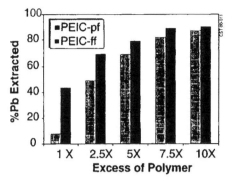

*Conditions: Soil (0.5 g) was extracted with 10 ml of polymer solution with the indicated polymer-to-contaminant ratio (0.06–0.6% w/v polymer) at pH 7.0 for 24 hours. Soils were centrifuged to separate polymer solutions. Ultrafiltration gave permeate and retentate, which were analyzed by ICP-AES. Analysis of extracted lead and residual lead remaining in the soils was determined by ICP-AES. A 0.5 g soil sample was placed in a Teflon PFA vessel with 10 ml concentrated HNO$_3$ (ultra-pure) and microwaved for 10 min at 80 psi and 574 watts (for six samples). The sample was allowed to cool, then mixed and the vessel vented under a hood before opening. A 1-ml aliquot was diluted to 10 ml and analyzed by ICP-AES.

Figure 5.10 Effect of polymer-to-lead ratio on the amount of lead extracted from Cal-West soils using PEIC-pf and PEIC-ff over a 24-hour period.*

functionalized PEIC for lead extraction. Cal-West soil was treated with each polymer at several polymer-to-contaminant ratios. Ratios were determined using the known lead capacities of each of the polymers and lead concentrations in the soils. Figure 5.10 shows lead removal after a single 24-hour extraction with PEIC-pf and PEIC-ff. At low polymer-to-contaminant ratios, PEIC-pf did not remove as much lead from the soils as did PEIC-ff. At higher polymer-to-contaminant ratios, there was little difference in the total amount of lead removed from the soil. It should be noted that in practice, high leachant-to-contaminant ratios are commonly used to assist the metal ion removal process. After treatment, the soils were evaluated using the TCLP test to determine if the soils were nonhazardous. The results of these tests are shown in Figure 5.11. A significant difference between the two polymers was observed. None of the soil samples extracted using a single four-hour extraction with PEIC-pf passed TCLP tests: leachable lead was above 5 ppm for each of the samples extracted, regardless of the amount of polymer used for the extraction. In contrast, soils extracted with PEIC-ff passed TCLP at polymer-to-contaminant ratios of 5

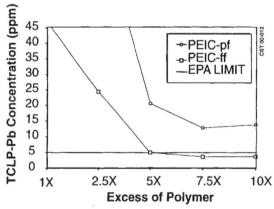

Figure 5.11 Results of TCLP tests on Cal-West soils extracted with PEIC-pf and PEIC-ff at varying polymer-to-lead ratios.

or higher. It is interesting to note that while lead removals for polymer-to-contaminant ratios of 5 were lower than those for 7.5 and 10 (79% vs. 89 and 90%, respectively), TCLP levels were not significantly different. A fivefold excess of polymer-to-contaminant ratio was used for all subsequent soil extraction studies.

The rate of lead extraction was also examined for both polymer systems. Figure 5.12 shows the time dependence of lead extraction for PEIC-ff and PEIC-pf for a single extraction with a tenfold excess of polymer. Clearly, longer extraction times resulted in higher overall lead removals for both PEIC-ff and PEIC-pf. While there was a significant difference between samples treated for 4 hours and those treated for 24 hours, little change was observed when going from a 24-hour to a 48-hour extraction period. TCLP results were consistent with this observation. For example, leachable lead levels were 4 ppm for the 24-hour PEIC-ff extraction and 2.4 ppm for the 48-hour extractions.

In all of our studies, soils treated with PEIC-ff consistently had higher lead removals than soils treated with PEIC-pf. However, also important in remediation of contaminated soils is the selective removal of a contaminant. In the case of the Cal-West soils, selectivity is an issue because of the high quantities of calcium in the soils. These alkaline soils contain significant concentrations of calcium (30,000 ppm), which could bind to the polymer and suppress lead extraction. Selectivity studies with both polymers showed that while PEIC-ff had higher total lead extraction, 100% compared to 96% for PEIC-pf, the fully functionalized polymer extracted 10 times more calcium than did PEIC-pf.

Investigations into the effect of polymeric chelators on lead removal included evaluation of lead desorption as a function of polymer molecular weight. For these initial studies, a low molecular weight PEIC (~2,000 Dalton) was prepared and compared to PEIC-pf and PEIC-ff (high molecular weight, >30,000 Dalton). Batch extraction studies of short duration (1 hour and 4 hours) were done to evaluate the access of the polymers to surface-bound metal ions. As shown in Figure 5.13, lead removal increased from 60–78% for the high molecular weight polymer and from 70–82%, respectively, for the low molecular weight polymer. In all studies, the low molecular weight polymer removed slightly higher amounts of lead from the soil than did the high molecular weight polymers over the same leaching period. However, the amount of lead removed between the two sizes of polymers is not large, suggesting that both systems had sufficient access to the surface-sorbed metals.

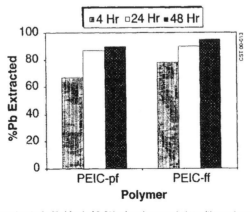

*Conditions: Soil (0.5 g) was extracted with 10 ml of 0.6% w/v polymer solution with a polymer-to-contaminant ratio of 10 at pH 7.0 for the indicated time. Soils were centrifuged to separate polymer solutions. Ultrafiltration gave permeate and retentate, which were analyzed by ICP-AES.

Figure 5.12 Time dependence of lead removal from Cal-West soils.*

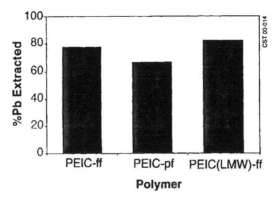

*Conditions: Soil (0.5 g) was extracted with 10 ml of polymer solution with a tenfold excess of polymer-to-lead concentration (determined by soil digestion prior to treatment) at pH 7.0 for 24 hours. Soils were centrifuged to separate polymer solutions. Ultrafiltration gave permeate and retentate, which were analyzed by ICP-AES. Analysis of extracted lead was done as described above.

Figure 5.13 Effects of polymer molecular weight on lead removal from Cal-West soils.*

Sequential Extraction Studies

A series of batch extractions, each with fresh polymer solution, were performed to determine conditions. These batch extractions consistently allowed treated soils to pass TCLP tests. Figure 5.14 compares the lead removals for sequential 4- and 24-hour extractions using PEIC-ff. Nearly quantitative removal of lead was observed after three 24-hour extractions. Treated soils were subjected to TCLP testing. The results of these tests are summarized in Figure 5.15. It is clear that equilibrium extraction is not reached during the 4-hour extraction period. Treatment of the soils for 24 hours had a dramatic effect on the amount of lead removed. After a single 24-hour extraction, the sample passed TCLP. After four sequential extractions of 24 hours each, digestion of the soils showed that residual lead concentrations were 500 ppb.

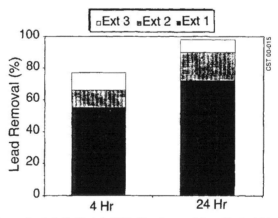

*Conditions: Soil (0.5 g) was extracted with 10 ml of a 0.6% wt/v polymer solution with a tenfold excess of polymer-to-lead concentration (determined by soil digestion prior to treatment) at pH 7.0 for either 4 or 24 hours. Four separate treatments on the same soil samples were done sequentially. Soils were centrifuged to separate polymer solutions. Ultrafiltration gave permeate and retentate, which were analyzed by ICP-AES. Analysis of extracted lead was done as described above.

Figure 5.14 Sequential extraction of Cal-West soils using PEIC-ff. A polymer-to-contaminant ratio of 5:1 was used for these experiments.*

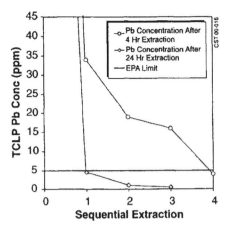

Figure 5.15 TCLP results for Cal-West soils exposed to sequential extractions of PEIC-ff.

Polymer Regeneration Studies

A critical characteristic of a soil remediation system is the regenerability of the extractant. For the soluble chelating polymers, the regeneration of the polymer system and the recovery of the lead were accomplished by pH adjustment. For both the PEIC-ff and PEIC-pf, all lead was released from the polymers at pH 1. Diafiltration of the solution allowed for the recovery of the polymer and isolation of a lead-containing solution for disposal. In order to demonstrate the utility of regenerated polymer extractants, a series of batch experiments was designed to test PEIC reuse. After the first extraction, the PEIC:lead solution was collected, the soil was washed to ensure that all of the polymer was removed, the lead was released using HCl and the polymer was diluted to the original extraction volume at pH 6.5–7.0. This regenerated polymer solution was then added to the next sample of Cal-West soil, and the extraction procedure was repeated. Three separate extractions reusing the PEIC were done, and as shown in Figure 5.16, the amount of lead bound was relatively consistent throughout the extractions. The observed difference in the lead removed from each sample resulted from variations in the soil samples. We were able to quantify the amount of polymer lost during the extraction and regeneration process spectroscopically: >98% of the polymer could be recovered.

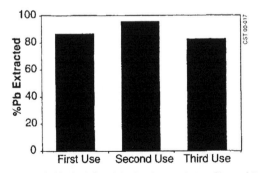

*Conditions: Soil (0.5 g) was extracted with 10 ml of a 0.6% wt/v polymer solution with a tenfold excess of polymer-to-lead concentration (determined by soil digestion prior to treatment) at pH 7.0 for 24 hours. After separation of the polymer solution from the soil, a fresh polymer solution was added to the soil, and the system was mixed for 24 hours.

Figure 5.16 Reuse of PEIC on three samples of Cal-West soil.*

CASE STUDY 4: REMOVAL OF TOXIC AND RADIOACTIVE
METALS FROM PAINT STRIPPER WASTE

In addition to soils, the DOE has significant quantities of debris or wastes classified as low-level mixed wastes, which will require treatment prior to disposal. Treatment of mixed wastes is daunting; few technologies exist for treating LLMW waste, and commonly, it is stored on-site. Because the toxic and radiological constituents are only a small fraction of the total waste, it is highly desirable to develop technologies that separate them and reduce the burden on storage and landfill facilities. An additional complicating factor is that often these wastes are heterogeneous with respect to physical form and chemical and radiological constituents.

A representative class of LLMW for general debris treatment studies is cellulosic materials (cloth, paper and other trash) contaminated with toxic metals, radioactive metals and organic constituents. An example would be paint stripper waste (PSW). This waste was historically generated at DOE facilities by painting over radiologically contaminated surfaces and subsequently removing the paint and radionuclides with organic solvent-based paint strippers and cheesecloth wipes. The resulting waste, a physically heterogeneous mixture of paint, paint stripper and cheesecloth, contains hazardous metals from the paint (often lead and chromium), radionuclides and organic solvents such as toluene and acetone. Treatment of this type of waste for proper disposal requires (1) destruction of the hazardous organic compounds to below regulatory levels, (2) separation and stabilization of hazardous metals to meet land disposal restrictions and (3) volume reduction. A multidisciplinary team at Los Alamos has demonstrated that a modular approach to treatment of PSW is highly advantageous. It was found that biodegradation in combination with chemical and physical techniques offers significant treatment advantages over other methods [90]. Three individual stages for the treatment of PSW were identified. Biological degradation of hazardous organics such as methylene chloride and enzymatic degradation of the cellulose materials with cellulase was examined. This later step significantly reduced the volume of waste. During the biological treatments, metals from the waste, including toxic metals such as lead and chromium as well as actinides, were solubilized in the sugar solution that resulted from enzymatic digestion of the cellulosic solids. These solutions were treated with soluble polymers and ultrafiltration (i.e., Polymer Filtration) to recover the metals in the third stage of the process [90].

Metals Solubilization

Initial testing on surrogate wastes indicated that the solubilization of paint-entrained metals occurred during cellulase digestion. Experiments on surrogate waste and PSW showed that, once released, lead and chromium could be collected with harvested sugars and removed from these solutions by PF to achieve RCRA discharge levels.

The action of cellulase on the heterogeneous waste caused mobilization of lead, chromium and radionuclides. During the course of 50 days of degradation, 4.5 mg of solubilized chromium, 8.2 mg of lead and 1,070 nCi of radioactivity (total alpha) were released from a 100-g sample of waste and collected during the harvest of sugars. Complete acid digestion of representative samples of the solid showed that 10.0 mg chromium and 16.3 mg lead remained in the treated residues. TCLP results showed that less than 1.0 ppm lead and 0.1 ppm chromium were leached from the treated materials. The TCLP limits are 5 mg/L for both lead and chromium, so the material was no longer RCRA-characteristic waste.

TABLE 5.12. Removal of Metals from Solutions from Enzymatic Degradation of Cellulose.*

Solution	Lead (mg/ml)	Chromium (mg/ml)	Total Alpha (pCi/ml)
Before Treatment	4.3	1.4	610
After PEIC-pf	<1.0	1.0	60
After PEI	<1.0	0.3	Not detectable

*Conditions: Sugar solutions were isolated by ultrafiltration of enzymatic digestion broths. Metal ion removal was done in two stages. In the first stage, PEIC (80 mg) was added to the sugar solution (50 ml), and ultrafiltration was used to isolate the polymer and bound metals. In the second step, PEI (37 mg) was added, the solution equilibrated, and the bound metals isolated by ultrafiltration. Final volume of the metal-containing solution was 2 ml.

Metals Recovery

Analysis of sugar harvests from PSW digestion showed that lead and chromium levels for the samples were at or below RCRA discharge limits. However, that may not always be the case, because metal concentrations would vary with the samples from different wastes. Because these solutions represent a significant secondary waste volume from this process, PF was used to polish these sugar solutions from enzymatic digestion. The sugar solution harvested on day 13 had the highest metal ion concentrations and, consequently, was selected to demonstrate polymer recovery of the metals. Two polymers were evaluated in this study, PEI and PEIC. Work in our laboratories had shown that PEIC extracted lead and chromium(III) from both solid matrices and aqueous solutions and that PEI removed metal oxyanions like chromate (CrO_4^{2-}). For PF treatment, a 50-ml aliquot of the day 13 sugar solution was mixed with PEIC. After ultrafiltration, inductively coupled plasma-atomic emission spectroscopy (ICP-AES) analysis of retentate showed that 80% of the lead, 30% of the chromium and 92% of the alpha emitters were bound to the polymer, leaving <1.0 ppm of lead, <1.0 ppm of chromium and 60 pCi/ml of radionuclides in the polished solution. Data are shown in Table 5.12.

As was shown in soil treatment studies, PEIC-pf has a high affinity for binding lead. Thus, it is not surprising that lead was removed to below our detection limit with a single extraction. In addition, ca. 90% of the radionuclides were removed. The low removal levels for chromium with PEIC suggests that this metal was present as an oxyanion (CrO_4^{2-}) in the sugar solutions. Subsequent treatment of permeate from the first stage of PF with cationic PEI removed the residual chromium and alpha emitters. In addition, the metals were concentrated during ultrafiltration into less than 5% of the original solution volume, thus dramatically minimizing the amount of hazardous waste requiring disposal. The resulting sugar solution contained nonhazardous levels of RCRA materials, no detectable radioactivity, and was suitable for discharge to the Los Alamos National Laboratory low-level wastewater treatment facility.

CASE STUDY 5: APPLICATION TO THE MINING AND MINERAL PROCESSING INDUSTRY

The previous discussions have demonstrated the effectiveness of using water-soluble chelating polymers in the PF process for the decontamination of various solid matrices from heavy metals. The ability to selectively remove the targeted metals from a complex matrix without the loss of polymer makes it reasonable to consider extending this concept to the recovery of metals found in mining and mineral processing industries. Polymers that show the ability to selectively complex to metals such as gold, silver and copper in solution have already been developed for other PF applications [1,3]. The use of PF to address the treatment of acid mine drainage problems has also been investigated [6]. The development of water-soluble chelating polymers showing the same high selectivities for targeted metals in ore deposits may offer innovative approaches to the extraction, concentration and recovery of these metals.

The mining and mineral processing industry has undergone significant changes in the past two decades. Increasing public awareness of environmental issues associated with mining, in addition to demands for technological changes to current mining methods to mitigate environmental liabilities, has made it more difficult for U.S. companies to remain competitive [91,92]. For example, the use of cyanide solutions to leach gold is coming under increasing regulatory restrictions out of concern for the effects that these solutions have on the environment. In 1998, the voters of Montana passed initiative 137, which called for the end to any new cyanide leach mining operations in the state. Under this initiative, no new or expanded mines can use cyanide in their metals extraction process. In excess of 100 million pounds of cyanide is estimated to be used annually in the U.S. to extract more than 90% of all U.S. gold [93]. Currently, the industry has no economically acceptable alternatives to this process. The potential impact these restrictions would have on the gold mining industry, if they were to be widely implemented, is clear.

In addition to the environmental issues, the mining industry is dealing with the gradual depletion of high-grade ore deposits and the high cost associated with mining these deposits by conventional methods. Traditional mining methods, such as underground mining, are becoming less profitable because of the expense of extracting metals from lower grade ores, which involves the treatment of much larger volumes of materials to maintain production levels. These issues are forcing the mining companies to reevaluate and implement innovative, more economically viable technologies for metals extraction and processing [94–96].

With the depletion of high-grade ore deposits, mining operations are increasingly turning to solution mining for metal extraction [97]. Solution mining can be defined as those operations that involve the extraction of metals from soil and rock using leach solutions, which includes the heap leaching of excavated ore, dump leaching of mine waste and the use of various in situ mining methods. Solution mining is generally much more cost effective than conventional methods for the extraction of low-grade ores. Because large volumes of solutions are used, fluid recovery and metal separation operations follow the extraction process, with the leachate being recirculated through the extraction process. Extraction times will vary with the nature of the deposit but often take anywhere from days to weeks before the ore becomes barren of extractable metals. The factor all these processes have in common is the use of aqueous solutions containing acids or other chemicals to extract the metals by flood or percolation leaching.

The most common method of solution mining is the heap leaching process, which is a part of almost all recent gold and copper mine operations. Heap leaching now accounts for more than 30% of the gold production in the U.S. and 15% of the annual world copper production [98]. Heaps are generated by the dumping of low-grade ore deposits from excavation mining onto impermeable graded leach pads. The pregnant heap is leached of metals by the percolation (0.003–0.005 gpm /ft^2) of solutions through the ore. The pregnant solutions are recovered by gravity after they permeate the heap matrix. The leach solution used for gold extraction is typically sodium cyanide in concentrations of 50–200 ppm with lime or sodium hydroxide added to provide alkalinity and to maintain a pH between 10 and 11. With the concentration of gold in low-grade ore being around 1–3 ppm, the pregnant leach solutions often reach concentrations of 0.2 ppm. Copper ores are leached in a very similar manner using dilute sulfuric acid as the leachate. If the copper deposits are located in limestone or dolomite formations, a significant amount of additional sulfuric acid is consumed, increasing the costs of the process. The leach times are often longer than they are for gold due to the much higher concentrations of copper in the ore (as much as 4,000 ppm), with the pregnant solutions reaching concentrations as high as 1,000 ppm.

In situ mining involves the extraction of nonferrous metals from an undisturbed ore deposit by injection of leach solutions through wells. Generally, these are site deposits where the ores are either low grade or too deep for excavation or subterranean tunneling to be economically practical. The deposits are surrounded by impermeable rock, which enables the leach solution to be contained. The metals are leached from the host rock as the solution moves through the fractures and pores. Often, it is necessary to rubblize the deposit with the use of explosives or hydraulic pressure to increase the permeability for leaching. The leachate is recovered through recovery wells drilled adjacent to the ore deposit. First commercially established in the 1970s, in situ mining has been used primarily in extracting copper and uranium and to a lesser extent for magnesium, molybdenum, nickel, cobalt, vanadium and zinc [97,99]. In addition to the much lower operational costs, in situ mining has the advantage of having less of an environmental impact on the mining sites [100].

The pregnant solutions from both the heap leach and in situ mining processes are sent to recovery operations. Typically, soluble gold-cyanide complex is recovered from dilute leaching solutions by the reduction and precipitation of elemental gold by the addition of zinc metal or by the adsorption of the cyanide complex onto activated carbon followed by stripping in hot basic solutions and electrowinning [101]. Activated carbon has a low capacity for the gold-cyanide complex, which necessitates the use of large volumes of carbon beds. Stripping time ranges from half a day to several days depending on the conditions used. A number of anion exchange resins have been evaluated as potential replacements for activated carbon but with little success. The exception is a guanidinium-functionalized resin. The recovery process for copper-pregnant solutions involves solvent extraction into a kerosene organic phase using various liquid ion exchange reagents (generally hydroxyphenyl oximes) [102]. The organic phase is then back-stripped with concentrated sulfuric acid, and the highly concentrated copper is collected by electrowinning. Solid stabilized emulsion formation is a common problem in this process.

The uses of water-soluble chelating polymers for mining applications seem to be most congruous to those solution-mining processes discussed above. Such applications may extend beyond metal extraction to include the metal recovery process. Because conditions used in solution mining are very similar to those used in soil decontamination, it is expected that a minimum amount of modifications for its implementation would be necessary. For the heap leaching of gold or copper ores, Figures 5.17 and 5.18 illustrate how the PF technology might be incorporated into the process. Dilute solutions of the water-soluble polymer are applied to the pregnant heap by conventional methods. The polymers selectively bind the targeted metal. The pregnant leachate is collected for concentration using ultrafiltration, and metal is recovered by releasing it from the polymer. The regenerated polymer solution is returned to the leach process for additional metal extraction.

If a successful polymer leaching process could be developed, it would offer a number of advantages. The most important is the replacement of cyanide and sulfuric acid leach solutions with relatively benign biodegradable polymers. The use of polymers would provide a physical means by which the leach solutions could be concentrated, giving metal solutions of sufficient concentration suitable for electroplating. In the case of copper, this would eliminate the need for the solvent extraction process.

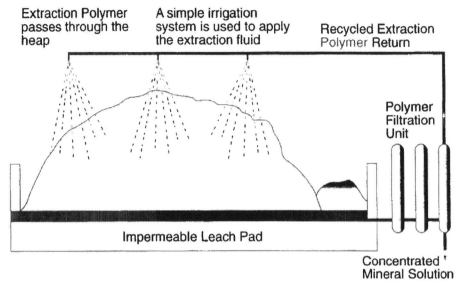

Figure 5.17 Cartoon of how PF technology might be used for aboveground heap-leaching of metals from ore

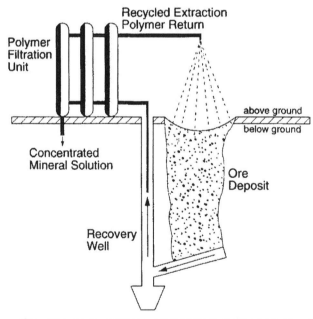

Figure 5.18 Cartoon of how PF technology might be used for belowground heap-leaching of metals from ore.

The in situ mining process (see Figure 5.19) would involve pumping the barren polymer solution down an injection well. As the solution passes through the ore deposit, the metals are extracted, with the pregnant polymer leach solution collected through the recovery well. This cycle may be continued until the polymer is loaded to capacity with the extracted metal. This solution is then treated in the same manner as in the heap leach process for the recovery of the metal.

In order to demonstrate the technical efficacy of the PF process for the leaching of a copper oxide ore deposit, a solution of a water-soluble polymer with high selectivity for copper was used in the recovery of metals from ore samples. A 100-ml solution of the polymer (4%) was contacted with the ore (20.7 g) containing copper oxide from Copper Queen Mine, Bisbee, Arizona, USA. After a 24-hour contact period with stirring, the polymeric solution was decanted, and fresh polymer (100 ml, 4%) added. After an additional 48 hours, the two solutions were combined and found to contain 203 ppm copper. The total amount of copper extracted from the rock was 71 mg, or about 0.4% by weight. As a measure of the efficiency of copper removal from the first two solutions, the treated rock was subsequently crushed and leached for additional copper using a polymeric (100 ml) solution. Analysis of this second solution after 24 hours of contact yielded only an additional 8.6 mg copper. Total copper removed was 78.6 mg. The polymer leach accounted for 90% of the total copper present in the ore. Analysis of the solutions for iron gave 0.9 mg of iron leached from the combined first two solutions and 0.6 mg in the second solution. We conclude that not only does the polymer show the ability to extract copper from ore in a selective manner, it was also able to effectively extract a significant percentage of the copper from the rock without the need for crushing.

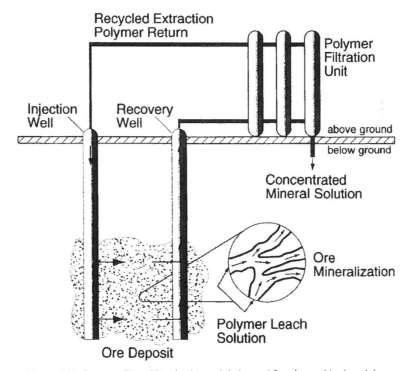

Figure 5.19 Cartoon of how PF technology might be used for advanced in situ mining.

SUMMARY

We have shown that PF can be used for the removal of metal ions from a wide variety of soils and debris for a range of different metal ions. With the judicious selection of a specific chelating polymer, metal-ion removal from solids can often be highly selective and efficient. Studies with both soils and complex debris streams have been done and have demonstrated that separation of the target metal ion(s) can be accomplished under relatively mild conditions with little or no loss of the polymeric extractant. In contrast to many molecular extractant systems, PF allows for the ready recovery, regeneration and reuse of the extractants through ultrafiltration. We have found that the advantages of PF for recovery of metals from solids (efficient extraction, high selectivity, ease of modular operation and extractant regenerability) make it an ideal technology for inclusion in multistep treatment or remediation processes.

LEGEND OF ACRONYMS AND SYMBOLS

σ = rejection coefficient
BDAT = best demonstrated available technology
CF = Concentration Factor
DOE = U.S. Department of Energy
D&D = decontamination and decommissioning
DoD = U.S. Department of Defense
EDTA = ethylenediaminetetraacetic acid
EPA = U.S. Environmental Protection Agency
ER = environmental restoration
FIMS = Flow Injection Mercury System
ICP-AES = inductively coupled plasma-atomic emission spectroscopy
kg = kilograms
LLMW = low-level mixed waste
LLRW = low-level radioactive waste
mg = milligrams
MWCO = molecular weight cut-off
MWFA = Mixed Waste Focus Area
PSW = paint stripper waste
PEI = polyethylenimine
PEIC = polyethylenimine functionalized with acetate groups
PEIC-ff = fully functionalized PEI polymer (1 carboxylate/PEI Nitrogen)
PEIC-pf = partially functionalized PEI polymer (0.25 carboxylate/PEI N)
PF = Polymer Filtration
ppb = parts per billion
ppm = parts per million
psig = pounds per square inch
RCRA = Resource Conservation and Recovery Act
SSM = synthetic soil matrix
TCLP = toxic characterization leaching procedure
K_s = stability constant
UF = ultrafiltration

ACKNOWLEDGEMENTS

The authors would like to acknowledge the DOE, Environmental Management (EM-50), Efficient Separation and Processing Crosscutting Programs and the Mixed Waste Focus Area for supporting the mercury separations work; the Los Alamos Waste Management group for the lead in soils and mercury debris work; the Dual Axis Radiographic Hydrotest Program for supporting the studies on firing site soils; and Los Alamos National Laboratory Directed Research and Development (LDRD) and program development (LDRD-PD) for supporting other portions of the project. The work was performed under DOE contract #W-7405-ENG-36.

There are a number of people who have contributed to these projects over the years, including Yvonne Rogers, Brandy Duran, Trudi Foreman, Deborah Ehler, James Brainard, Laura Vanderberg, Deborah Apodaca, Jason Jarvinen, Christopher Lubeck, Bryan Carlson, Omar Hernandez and Raj Jain.

REFERENCES

1 B. F. Smith, T. W. Robison, and J. W. Gohdes, "Water-Soluble Polymers and Compositions Thereof," No. 5,891,956, USA: The Regents of the University of California, 1999.

2 R. R. Grinstead, "Process and Composition for Removal of Metal Ions from Aqueous Solutions," No. 4,741,831, USA: Dow Chemical Company, 1988.

3 B. F. Smith and T. W. Robison, "Water-Soluble Polymers for Recovery of Metals Ions from Aqueous Streams," No. 5,766,478, USA: The Regents of the University of California, 1998.

4 A. S. Michaels, "Ultrafiltration," in *Progress in Separations and Purification,* vol. 1, E. S. Perry, Ed. New York: Interscience, 1968, pp. 297–334.

5 K. E. Geckeler and K. Volchek, "Removal of Hazardous Substances from Water Using Ultrafiltration in Conjunction with Soluble Polymers," *Environmental Science and Technology,* vol. 30, pp. 725–734, 1996.

6 B. F. Smith, T. W. Robison and G. D. Jarvinen, "Water-Soluble Metal-Binding Polymers with Ultrafiltration: A Technology for the Removal, Concentration, and Recovery of Metal Ions from Aqueous Streams," in *ACS Symposium Series 716, Metal-Ion Separations and Preconcentration: Progress and Opportunities,* A. H. Bond, M. L. Dietz, and R. D. Rogers, Eds. Washington, DC: American Chemical Society, pp. 294–330, 1999.

7 W. S. Winston and K. K. Sirkar, *Membrane Handbook.* New York: Van Nostrand Reinhold, 1992.

8 R. A. Leonard, D. B. Chamberlain and C. Conner, "Centrifugal Contactors for Laboratory-Scale Solvent Extraction," *Separation Science and Technology,* vol. 32, pp. 193–210, 1997.

9 M. Cheryan, *Ultrafiltration Handbook.* Lancaster, PA: Technomic, 1986.

10 H. Strathmann, "Selective Removal of Heavy-Metal Ions from Aqueous-Solutions by Diafiltration of Macromolecular Complexes," *Separation Science and Technology,* vol. 15, pp. 1135, 1980.

11 M. Tuncay, S. D. Christian, E. E. Tucker, and R. W. S. Taylor, "Ligand-Modified Polyelectrolyte-Enhanced Ultrafiltration with Electrostatic Attachment of Ligands. 1. Removal of Cu(II) and Pb(II) with Expulsion of Ca(II)," *Langmir,* vol. 10, pp. 4688–4692, 1994.

12 U.S.EPA, *Laboratory Manual Physical/Chemical Methods. SW-846,* vol. 1C: U.S. Environmental Protection Agency, 1986.

13 J. A. Roach, "Mixed Waste Focus Area Department of Energy Complex Needs Report," INEL, INEL-95/0555, November 1995.

14 J. J. Perona and C. H. Brown, "Mixed Waste Integrated Program: A Technology Assessment for Mercury-Containing Mixed Wastes," Oak Ridge National Laboratory, Oak Ridge, DOE/MWIP-9, March 1993.

15 D. T. Reed, I. R. Tasher, J. C. Cunane, and G. F. Vandegrift, *Environmental Restoration and Separation Science.* Washington, DC: American Chemical Society, 1992.

16 U.S.EPA, "Superfund. Superfund Engineering Issue: Treatment of Lead-Contaminated Soils," U.S.EPA Office of Solid Waste and Emergency Response, Washington, DC, EPA 540/2-91/009, 1991.

17 S. Krishnamurthy, "Extraction and Recovery of Lead Species from Soil," *Environmental Progress,* vol. 11, pp. 256–260, 1992.

18 M. Peterson, "DOE Plume Focus Area Overview," in *Environmental Science Program Review,* Gaithersburg, Maryland, 1996.

19 S. Warren, "Waste Retrieval, Treatment and Processing," presented at DOE Information Exchange Meeting, Houston, Texas, 1993.

20 D. L. Feldman, L. Miller and M. E. Castle, "Evaluating the Acceptability of Soil Treatment Methods in the Department of Energy's Formerly Utilized Sites Remedial Action Program: A Stakeholder Analysis," DOE Ref #FC05-920R22056, September 1994.

21 M. E. Watanabe, "Phytoremediation on the Brink of Commercialization," *Environmental Science and Technology,* vol. 31, pp. 182A–186A, 1997.

22 D. B. Sogorka, H. Gabert, D. Goswami, and B. Sorgorka, "Emerging Technologies for Soils Contaminated with Metals," *Contaminated Soils,* vol. 3, pp. 451–473, 1998.

23 I. Twardowska, S. Schulte-Hostede and A. A. F. Kettrup, *Fate and Transport of Heavy Metals in the Vadose Zone,* Boca Raton, Florida: Lewis Publishers, 1999.

24 S. K. Sikdar, D. Grosse, and I. Rogut, "Membrane Technologies for Remediating Contaminated Soils: A Critical Review," *Journal of Membrane Science,* vol. 151, pp. 75–85.

25 G. A. Torma, P. C. Hsu, and A. E. Torma, "Remediation Processes for Heavy Metals Contaminated Soils," in *Extraction and Processing for Treatment, Minimization Wastes, 1 Proc. 2nd Int. Symp. Minerals, Metals and Materials Society,* V. Ramachandram and C. C. Nesbitt, Eds. Warrendale, PA, 1996, pp. 289–304.

26 B. Rubin, R. Gaire, P. Cardenas, and H. Masters, "U.S. EPA's Mobile Volume Reduction Unit for Soil Washing," presented at Superfund '90—Proceedings of the 11th National Conference, 1990.

27 W. S. Richardson, T. B. Hudson, J. G. Wood, and C. R. Phillips, "Characterization and Washing Studies on Radionuclide Contaminated Soils," presented at Superfund '89—Proceedings of the 10th National Conference, 1989.

28 T. J. Nunno, J. A. Hyman, and T. Pheiffer, "Assessment of Site Remediation Technologies in European Countries," presented at Superfund '88—Proceedings of the 9th National Conference, 1988.

29 R. W. Peters and L. Shem, "Use of Chelating Agents for Remediation of Heavy Metal Contaminated Soil," in *ACS Symposium Series, Environmental Remediation,* vol. 509. Washington, DC: American Chemical Society, pp. 70–84, 1992.

30 B. F. Smith, T. W. Robison, N. N. Sauer and D. S. Ehler, "Water-Soluble Polymers for Recovery of Metals from Solids," No. 5,928,517, USA: The Regents of the University of California, 1999.

31 C. A. Cicero, "Summary of Pilot-Scale Activities with Mercury Contaminated Sludges," Savanna River Technology Center, WSRC-TR-95-0404, September 30, 1995.

32 E. Marshall, "The Lost Mercury At Oak-Ridge," *Science,* vol. 221, pp. 130–132, 1983.

33 K. R. Campbell, C. J. Ford, and D. A. Levine, "Mercury Distribution in Poplar Creek; Oak Ridge, Tennessee; USA," *Environmental Toxicology and Chemistry,* vol. 17, pp. 1191–1198, 1998.

34 K. T. Klasson, L. J. K. Jr., D. D. Gates and P. A. Cameron, "Removal of Mercury from Solids Using the Potassium Iodide/Iodine Leaching Process," ORNL, ORNL/TM-13137, December 1997.

35 G. A. Hulet, "U.S. DOE, Mixed Waste Focus Area Program, Idaho Falls, ID," 1998.

36 W. D. Bostick, D. E. Beck, K. T. Bower, D. H. Bunch, R. L. Fellows, and G. F. Sellers, "Treatability Study for Removal of Leachable Mercury in Crushed Fluorescent Lamps," Oak Ridge K-25 Site, Oak Ridge K/TSO-6, February 1996.

37 E. S. Rubin, "Toxic Release from Power Plants," *Environmental Science and Technology,* vol. 33, pp. 3062–3067, 1999.

38 R. Ferrara, B. E. Maserti, A. D. Liso, H. Edner, P. Ragnarson, S. Svanberg and E. Wallinder, "Could the Geothermal Power Plant of Mt. Amiata (Italy) Be a Source of Mercury Contamination?," in *Mercury Pollution, Integration and Synthesis,* C. J. Watras and J. W. Huckabee, Eds. Boca Raton, FL: CRC Press, Inc., pp. 601–607, 1994.

39 C. C. Lee and G. L. Huffman, "Metal Behavior during Medical Waste Incineration," *ACS Symposium Series,* vol. 515, pp. 189–198, 1993.

40 R. Allan, "Introduction: Mining and Metals in the Environment," *Journal of Geochemical Exploration,* vol. 58, pp. 95–100, 1997.

41 I. Thornton, "Impacts of Mining on the Environment; Some Local, Regional and Global Issues," *Applied Geochemistry,* vol. 11, pp. 355–361, 1996.

42 L. A. Ruth, "Energy from Municipal Solid Waste: A Comparison with Coal Combustion Technology," *Progress in Energy and Combustion Science,* vol. 24, pp. 545–564, 1998.

43 C. S. Krivanek, "Mercury Control Technologies for MWC's: The Unanswered Questions," *Journal of Hazardous Materials,* vol. 47, pp. 119–136, 1996.

44 J. Werther and T. Ogada, "Sewage Sludge Combustion," *Progress in Energy and Combustion Science,* vol. 25, pp. 55–116, 1999.

45 R. C. Kistler and F. Widmer, "Behavior of Chromium, Nickel, Copper, Zinc, Cadmium, Mercury, and Lead during Pyrolysis of Sewage Sludge," *Environmental Science and Technology,* vol. 21, pp. 704–708, 1987.

46 D. S. Charlton, J. A. Harju, D. J. Stepan, V. Kuhnel, C. R. Schmit, R. D. Butler, K. R. Henke, F. W. Beaver and J. M. Evans, "Natural Gas Industry Sites Contaminated with Elemental Mercury: An Interdisciplinary Research Approach," in *Mercury Pollution, Integration and Synthesis,* C. J. Watras and J. W. Huckabee, Eds. Boca Raton, FL: CRC Press, Inc., pp. 595–600, 1994.

47 P. B. Queneau and L. A. Smith, "U.S. Mercury Recyclers Provide Expanded Process Capabilities," *Hazmat World,* vol. 7, pp. 31–34, 1994.

48 C. Hanisch, "Where Is Mercury Deposition Coming From?," *Environmental Science & Technology,* vol. 32, pp. A176–A179, 1998.

49 C. J. Lin and S. O. Pehkonen, "The Chemistry of Atmospheric Mercury: A Review," *Atmospheric Environment,* vol. 33, pp. 2067–2079, 1999.

50 W. V. Farrar and A. R. Williams, "A History of Mercury," in *The Chemistry of Mercury,* C. A. McAuliffe, Ed. London: The Macmillan Press Ltd., pp. 3–45, 1977.

51 U. S. G. Survey, "Mercury in the Environment," U.S. Department of the Interior, Washington, DC, p. 173, 1970.

52 F. Baldi, "Mercury Pollution in the Soil and Mosses around a Geothermal Plant," *Water Air and Soil Pollution,* vol. 38, pp. 111–119, 1988.

53 B. S. Panov, A. M. Dudik, O. A. Shevchenko, and E. S. Matlak, "On Pollution of the Biosphere in Industrial Areas: The Example of the Donets Coal Basin," *International Journal of Coal Geology,* vol. 40, pp. 199–210, 1999.

54 G. Kaiser and G. Tolg, "Mercury," in *The Handbook of Environmental Chemistry,* vol. 3, Part A, O. Hutzinger, Ed. Berlin: Springer-Verlag, pp. 1–58, 1980.

55 P. O'Neill, "Mercury," in *Environmental Chemistry,* 2nd ed. New York: Chapman and Hall, 1993.

56 W. Baeyens, R. Ebinghaus, and O. Vasiliev, "Global and Regional Mercury Cycles: Sources, Fluxes, and Mass Balances," Dordrecht: Kluwer Academic Publishers, 1999.

57 D. F. Foust, "Extraction of Mercury and Mercury Compounds from Contaminated Materials and Solutions, No. 5,226,545," USA: General Electric Corp., 1993.

58 B. A. Weir, N. K. Chung, J. E. Litz, D. W. Whisenhunt, and B. M. Frankhouser, "Mercury Removal from DOE Solid Mixed Waste Using the GEMEP Technology," presented at Waste Management-'99 Conference, Tucson, AZ, published on CD-ROM, W. M. Symposia Inc., 1999.

59 N. V. Jarvis and J. M. Wagener, "Mechanistic Studies of Metal Ion Binding to Water-Soluble Polymers Using Potentiometry," *Talanta,* vol. 42, pp. 219–226, 1995.

60 K. E. Geckeler and E. Bayer, "Water-Soluble Quinolin-8-ol Polymer for Liquid-Phase Separation of Elements," *Analytica Chemica Acta,* vol. 230, pp. 171–174, 1990.

61 B. L. Rivas and K. E. Geckeler, "Synthesis and Metal Complexation of Poly(ethyleneimine) and Derivatives," *Advances in Polymer Science*, vol. 102, pp. 173–186, 1992.

62 B. F. Smith, T. W. Robison, Y. C. Rogers, and C. Lubeck, "Polyelectrolyte Enhanced Removal of Mercury from Mixed Waste Debris," presented at Waste Management-'99 Conference, Tucson, AZ, published on CD-ROM, W. M. Symposia Inc., 1999.

63 Y. Uludag, H. O. Ozbelge, and L. Yilmaz, "Removal of Mercury from Aqueous Solutions via Polymer-Enhanced Ultrafiltration," *Journal of Membrane Science*, vol. 129, pp. 93–99, 1997.

64 R. R. Navarro, K. Sumi, N. Fujii, and M. Matsumura, "Mercury Removal from Wastewater Using Porous Cellulose Carrier Modified with Polyethyleneimine," *Water Research*, vol. 30, pp. 2488–2494, 1996.

65 T. H. Wirth and N. Davidson, "Mercury (II) Complexes of Guanidine and Ammonia, and a General Discussion of the Complexing of Mercury (II) by Nitrogen Bases," *Journal American Chemical Society*, vol. 86, pp. 4325–4329, 1964.

66 T. H. Wirth and N. Davidson, "Studies of the Chemistry of Mercury in Aqueous Solutions. I. Mercury(I) and Mercury (II) Complexes of Aniline," *Journal American Chemical Society*, vol. 86, pp. 4314–4329, 1964.

67 P. Brooks and N. Davidson, "Mercury (II) Complexes of Imidazole and Histidine," *Journal American Chemical Society*, vol. 82, pp. 2118–2123, 1959.

68 Z. Stojek and J. Osteryoung, "The Mechanism of Oxidation of Mercury in the Presence of EDTA," *Journal of Electroanalytical Chemistry*, vol. 127, pp. 67–74, 1981.

69 B. F. Smith, T. W. Robison, Y. C. Rogers, and C. Lubeck, "Polyelectrolyte Enhanced Removal of Mercury from Mixed Waste Debris," presented at Waste Management-'99 Conference, Tucson, AZ, published on CD-ROM, W. M. Symposia Inc., 1999.

70 A. J. Bard, R. Parsons, and J. Jordan, "Standard Potentials in Aqueous Solution," New York: Marcel Dekker, Inc., 1985, pp. 129.

71 A. J. Bard, R. Parsons, and J. Jordan, "Standard Potentials in Aqueous Solution," New York: Marcel Dekker, Inc., 1985, pp. 282.

72 A. J. Bard, R. Parsons, and J. Jordan, "Standard Potentials in Aqueous Solution," New York: Marcel Dekker, Inc., 1985, pp. 57.

73 D. C. Wigfield and S. L. Perkins, "Oxidation of Elemental Mercury by Hydroperoxides in Aqueous-Solution," *Canadian Journal of Chemistry-Revue Canadienne De Chimie*, vol. 63, pp. 275–277, 1985.

74 W. H. Schroeder, G. Yarwood, and H. Niki, "Transformation Processes Involving Mercury Species in the Atmosphere: Results from a Literature Survey (Vol 56; p. 653; 1991)," *Water Air and Soil Pollution*, vol. 66, p. 203, 1993.

75 L. G. Hepler and G. Olofsson, "Mercury: Thermodynamic Properties; Chemical-Equilibria; and Standard Potentials," *Chemical Reviews*, vol. 75, pp. 585–602, 1975.

76 L. Grondahl, A. Hammershoi, A. M. Sargeson, and V. J. Thom, "Stability and Kinetics of Acid- and Anion-Assisted Dissociation Reactions of Hexaamine Macrobicyclic Mercury(II) Complexes," *Inorganic Chemistry*, vol. 36, pp. 5396–5403, 1997.

77 B. F. Smith, Y. C. Rogers, T. W. Robison, C. R. Lubeck, N. N. Sauer, D. S. Ehler, B. L. Duran, M. Grigorova, P. Nelson and B. J. Carlson, "Selective Removal/Recovery of RCRA Metals from Waste and Process Solutions Using Polymer Filtration Technology," Los Alamos National Laboratory, LA-UR 99-2878, 1999.

78 C. Galindo and A. Kalt, "UV-H$_2$O$_2$ Oxidation of Monoazo Dyes in Aqueous Media: A Kinetic Study," *Dyes and Pigments*, vol. 40, pp. 27–35, 1998.

79 G. Mark, A. Tauber, L. A. Rudiger, H. P. Schuchmann, D. Schulz, A. Mues, and C. von Sonntag, "OH-Radical Formation by Ultrasound in Aqueous Solution: Part II: Terephthalate and Fricke Dosimetry and the Influence of Various Conditions on the Sonolytic Yield," *Ultrasonics Sonochemistry*, vol. 5, pp. 41–52, 1998.

80 H. Yanagida, Y. Masubuchi, K. Minagawa, T. Ogata, J. Takimoto, and K. Koyama, "A Reaction Kinetics Model of Water Sonolysis in the Presence of a Spin-Trap," *Ultrasonics Sonochemistry*, vol. 5, pp. 133–139, 1999.

81 S. W. Paff, B. Bosilovich, and N. J. Kardos, "Acid Extraction Treatment System for Treatment of Metal Contaminated Soils," Environmental Protection Agency, Cincinnati, Ohio, EPA/540/R-94/613, 1994.

82 S. A. Wasay, P. Arnfalk, and S. Tokunaga, "Remediation of a Soil Polluted by Mercury with Acidic Potassium-Iodide," *Journal of Hazardous Materials,* vol. 44, pp. 93–102, 1995.

83 E. Schuster, "The Behavior of Mercury in the Soil with Special Emphasis On Complexation and Adsorption Processes: a Review of the Literature," *Water Air and Soil Pollution,* vol. 56, pp. 667–680, 1991.

84 K. Geckeler, G. Lange, H. Eberhardt, and E. Bayer, "Preparation and Application of Water-Soluble Polymer-Metal Complexes," *Pure and Applied Chemistry,* vol. 52, pp. 1883–1905, 1980.

85 M. C. Steele and J. Pichtel, "Ex-situ Remediation of a Metal Contaminated Superfund Soil Using Selective Extractants," *Journal of Environmental Engineering,* vol. 124, pp. 639–645, 1998.

86 B. E. Reed, P. C. Carriere, and R. Moore, "Flushing of a Pb(II) Contaminated Soil Using HCl, EDTA, and $CaCl_2$," *Journal of Environmental Engineering,* vol. 122, pp. 48–50, 1996.

87 C. Kim and S. K. Ong, "Effects of Soil Properties on Lead Extraction from Lead-Contaminated Soils Using EDTA," *Hazard. Ind. Wastes,* vol. 28, pp. 425–431, 1996.

88 R. Raghavan, E. Coles, and D. Dietz, "Cleaning Excavated Soils Using Extraction Agents: A State-of-the-Art Review," Final Report to the Superfund Technology Demonstration Division on Contract 68-03-3255; Foster Wheeler Enviresponse, Inc., EPA/600/2-89/034, 1989.

89 R. M. Smith and A. E. Martell, *Critical Stability Constants,* vol. 1–6. New York: Plenum Press, 1976.

90 L. A. Vanderberg, T. M. Foreman, M. J. Attrep, J. R. Brainard, and N. N. Sauer, "Treatment of Heterogeneous Mixed Wastes: Enzyme Degradation of Cellulosic Materials Contaminated with Hazardous Organics and Toxic and Radioactive Metals," *Environmental Science and Technology,* vol. 33, p. 1256, 1999.

91 N. F. Gray, "Environmental Impact and Remediation of Acid Mine Drainage: A Management Problem," *Environmental Geology,* vol. 30, pp. 62–71, 1997.

92 D. W. Boening and C. M. Chew, "A Critical Review: General Toxicity and Environmental Fate of Three Aqueous Cyanide Ions and Associated Ligands," *Water Air and Soil Pollution,* vol. 109, pp. 67–79, 1999.

93 C. A. Pristos and J. Ma, "Biochemical Assessment of Cyanide-Induced Toxicity in Migratory Birds from Gold Mining Hazardous Waste Ponds," *Toxicology and Industrial Health,* vol. 13, pp. 203–209, 1997.

94 M. Scoble and L. K. Daneshmend, "Mine of the Year 2020: Technology and Human Resources," *Cim Bulletin,* vol. 91, pp. 51–60, 1998.

95 S. A. Shuey, "Mining Technology for the 21st Century," *Engineering and Mining Journal,* vol. 200, p. 18, 1999.

96 N. Vagenas, "Advanced Mining Technologies: The Productivity Tool for Mining in the 21st Century," *Cim Bulletin,* vol. 91, p. 16, 1998.

97 R. W. Bartlett, *Solution Mining, Leaching, and Fluid Recovery of Materials,* 2nd ed. Amsterdam: Gordon and Breach Science Publishers, 1998.

98 R. W. Bartlett, "Metal Extraction from Ores by Heap Leaching," *Metallurgical and Materials Transactions B,* vol. 28B, pp. 529–545, 1997.

99 S. M. Committee, "Solution Mining: A Review of 1992 Activities," *Mining Engineering,* pp. 610–611, 1993.

100 D. J. Millenacker, "In-Situ Mining," *Engineering and Mining Journal,* pp. 56–58, 1989.

101 G. A. Kordosky, J. M. Sierakoski, M. J. Virnig, and P. L. Mattison, "Gold Solvent-Extraction from Typical Cyanide Leach Solutions," *Hydrometallurgy,* vol. 30, pp. 291–305, 1992.

102 G. A. Kordosky, "Copper Solvent-Extraction: The State-of-the-Art," *Journal of the Minerals Metals and Materials Society,* vol. 44, pp. 40–45, 1992.

CHAPTER 6

Case Studies for Immobilizing Toxic Metals with Iron Coprecipitation and Adsorption

KASHI BANERJEE[1]

INTRODUCTION

THE toxic effects of heavy metals in water, even in trace quantities, are well known. Trace elements, such as copper, lead, zinc, arsenic, etc., found in wastewaters are designated as priority pollutants by the U.S.EPA. Industrial sources of heavy metals include metal processing and refining, metal plating, petroleum refining, chloroalkali production, battery manufacturing, steel production, pigment manufacturing, tanning, anodizing, photographic film manufacturing, automotive production, etc. Every year, in the United States, these industries produce billions of gallons of wastewater containing heavy metals. The release of these substances into the environment needs to be controlled to minimize effects on aquatic life and downstream uses. Increasingly stringent environmental regulations pertaining to discharges containing heavy metals necessitate the development of technically and economically feasible processes for the removal of these metals from wastewater.

Literature Review of Potential Treatment Processes

The standard treatment method for removal of heavy metals from wastewater is chemical precipitation using hydroxide, carbonate or sulfide, or some combination of these chemicals. The most common technique is the hydroxide precipitation process, because it is simple, easy to operate and economical if lime is used. The theoretical minimum solubility values of metal hydroxide are low; however, these levels are seldom achieved in practice because of poor solid/liquid separation, slow reaction rates, pH fluctuation, and the presence of other cations and complexing agents in the wastewater. Existing literature data [1–7] indicate that the concentrations of heavy metals, such as copper, lead and zinc, in the treated effluent can be consistently reduced to about 0.5 mg/L by the hydroxide precipitation process with proper pH control, clarification and filtration. Sulfide precipitation has

[1]USFilter Corporation, North American Technology Center, 600 Clubhouse Drive, Pittsburgh, PA 15108, U.S.A., banerjeek@usfilter.com

been demonstrated to be an effective alternative to hydroxide precipitation for removal of heavy metals from wastewater, regardless of whether a solution of sodium sulfide, hydrosulfide or a slightly soluble ferrous sulfide slurry is used as the reagent [8–11]. Bhattacharyya et al. [12] found that sulfide precipitation is a highly effective process for removal of cadmium, copper, lead, zinc, arsenic and selenium. The results of their studies revealed that this process is capable of reducing the concentrations of heavy metals to less than 0.1 mg/L. However, because of the small particle size, filtration may be required following clarification for effective metals removal. The potential for hydrogen sulfide gas evolution, sulfide toxicity and odor makes the inorganic sulfide precipitation process less attractive. Some of these problems with the inorganic sulfide precipitation process can be eliminated through the use of organic sulfide compounds, such as Trimercapto-S-Triazine (TMT), but the high cost of TMT makes the process economically unfeasible.

Ion exchange is primarily a volume reduction process. The ions are removed from the wastewater and concentrated on the exchange resin. During regeneration, the ions are released from the resin and are solubilized in the regenerant solution. Treatment of this concentrated waste is required prior to discharge. Ion exchange is capable of reducing the concentrations of heavy metals to nondetectable levels. However, the high capital and operating costs sometimes make this process economically unfeasible, unless the treated effluent is recycled and reused.

Iron coprecipitation/adsorption is a well-accepted process for removal of heavy metals from wastewater. Freshly prepared oxides of iron are quite amorphous and have high binding capacity (moles of sites per kilogram of solid). Published literature [13–16] indicates that amorphous iron oxyhydroxide is capable of removing various cations, such as copper, lead, zinc, chromium, etc. The concentrations of metals can be reduced to ppb levels [17]. Depending on the pH condition, this process is also capable of removing oxyanions of metals (such as arsenate, chromate, selenite, etc.) from water and wastewater.

A variety of methods have been developed for removal of arsenic compounds from water and wastewater. Commonly used treatment methods include pH adjustment and lime precipitation [18], coagulation and coprecipitation/adsorption onto metal hydroxide [19–25], precipitative softening [26,27], sulfide treatment [28,29], adsorption onto activated carbon [30] and alumina [31–34], and ion exchange [32,35].

The reactive hydroxyl surface sites and the high surface area of iron oxyhydroxide make the iron coprecipitation/adsorption process a very effective and economical method of reducing heavy metals to the ppb level in wastewater. The theoretical aspects and the engineering design considerations for the iron coprecipitation/adsorption process are discussed in this chapter with other pertinent information, including results from treatability studies. Results from two full-scale wastewater treatment plants for the removal of copper, lead, zinc and arsenic are also presented and discussed.

THEORETICAL CONSIDERATION OF THE IRON COPRECIPITATION/ADSORPTION PROCESS

Copper, Lead and Zinc Removal

In the iron coprecipitation/adsorption process, ferric salt (e.g., ferric chloride or ferric sulfate) is added to water and forms an amorphous precipitate of iron oxyhydroxide ($Fe_2O_3 \cdot H_2O$). The trace elements (both dissolved and suspended) are adsorbed onto and trapped within the precipitate. The settled precipitate is then separated from the water, leaving a purified effluent. Binding of a contaminant is a function of reaction pH, adsorbent (iron) concentration, adsorbate (contaminant) concentration and the presence of competing cations and anions. Depending upon the solution pH, the oxide surface can act as a weak acid or base and gain or lose a proton (i.e., it can undergo protonation or deprotonation). The following reactions are expected to occur at the surface [14]:

$$\overline{SO^-} + H^+ = \overline{SOH} \tag{1}$$

$$\overline{SOH} + H^+ = \overline{SOH_2^+} \quad (\overline{SOH} \text{ represents a singly protonated oxide site.}) \tag{2}$$

Cation adsorption reaction:

$$\overline{SOH} + M^{n+} + mH_2O = \overline{SO \cdot M(OH)_m^{n-m-1}} + (m+1)H^+ \tag{3}$$

Anion adsorption reaction:

$$\overline{SOH} + A^{n-} + mH^+ = \overline{SOH_{m+1}A^{m-n}} \tag{4}$$

M and A represent cationic and anionic adsorbate, respectively. The above equations reveal that protons are released into the system when an uncomplexed metal is adsorbed and are removed from the system during the adsorption of anions. The above equations also reveal that increasing the concentration of protonated surface oxide sites increases the adsorption capacity for cations and anions. Figure 6.1 illustrates pH-adsorption edge, the relationship between pH and the rapid change in adsorption behavior. The figure indicates that increasing the pH increases the adsorption of cations and decreases the adsorption of an-

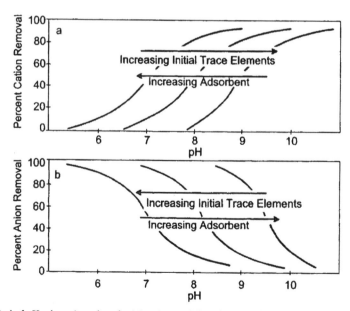

Figure 6.1 Typical pH-adsorption edges for (a) cations and (b) anions as a function of adsorbent concentration. Source: EPRI Report No. CS-4087, 1985 [39].

TABLE 6.1. pK Values at 25°C.

H$_3$AsO$_4$	H$_3$AsO$_3$
2.2	9.22
6.98	12.13
11.55	13.40

ions. Conversely, adsorption of anions increases as the pH decreases. However, the location of the pH-adsorption edge primarily depends on the concentration of the adsorbent. Increasing the adsorbent dosage shifts the pH-adsorption edge to the left during the adsorption of cations and to the right during the adsorption of anions.

Arsenic Removal

The chemistry of arsenic in an aquatic system is complex and consists of oxidation-reduction, ligand exchange, precipitation and adsorption. Arsenic is stable in four oxidation states (+5, +3, 0 and –3) under different redox potential (*Eh*) conditions. However, the occurrence of arsenic metal (As^0) in an aquatic system is very rare, and –3 arsenic is found only at an extremely low redox potential (*Eh*) value. The principal aqueous forms of inorganic arsenic are arsenite [As(III)] and arsenate [As(V)]. In an oxidizing environment, the arsenate species predominates; under moderately reducing conditions, the arsenite species becomes predominant. Compounds of arsenic sulfide, such as orpiment (As$_2$S$_3$), become stable at pH values below 5.5 and at a redox potential (*Eh*) of approximately zero volts. Because of the slow redox transformation, both arsenite and arsenate exist in either redox environment. However, arsenite is more toxic than arsenate and more mobile in the environment [36].

As discussed above, arsenic can be present at different oxidation states, and some forms of arsenic are more easily removed from water than others. Oxidation state is one of the major factors that influence arsenic removal treatment efficiency. The reaction between arsenate and iron (Fe^{3+}) forms an insoluble precipitate known as ferric arsenate (FeAsO$_4$), whereas ferric arsenite (FeAsO$_3$) is soluble in water and does not precipitate. The solubility of ferric arsenate is about 20 mg/L (8 mg/L as As). However, at low concentrations, arsenic can be removed by the adsorption process, which involves surface complexation and electrostatic interaction. As shown in Table 6.1, the pK value of arsenious acid (H$_3$AsO$_3$) is greater than that of arsenic acid (H$_3$AsO$_4$). Deprotonation, or loss of hydrogen ion from H$_3$AsO$_3$, is more difficult than loss of hydrogen ion from H$_3$AsO$_4$ at a neutral pH condition. The surface complexation efficiency is significantly influenced by the constituent's deprotonation ability, and surface complexation will occur if undissociated arsenic acid donates a proton to the hydroxyl group present in hydrous ferric oxide. Increasing deprotonation ability increases the potential for surface complexation. Because H$_3$AsO$_4$ deprotonates much more efficiently than H$_3$AsO$_3$ at a neutral pH, oxidation of As(III) to As(V) would increase the treatment efficiency at this pH.

CASE STUDY: COPPER, LEAD AND ZINC TREATMENT

Selection and Optimization of Process Parameters

Selection and optimization of process variables are important aspects of water/wastewater treatment plant design. The applicability and optimum operating conditions of a selected technology are determined by conducting laboratory or pilot-scale treatability studies. The following case study describes the development of a treatment scheme for metals removal.

Because of stringent discharge limitations, a steel-making facility implemented a wastewater management plan that included segregation of process wastewater from storm water to minimize the flow requiring treatment. Under this plan, the concentrations of copper, lead and zinc in the treated process water were to be reduced to ppb levels prior to discharging to the creek. The target concentrations of the selected metals in the treated water are presented in Table 6.2. The objective of the wastewater treatment plant was to achieve copper and zinc concentrations of <10 µg/L, as well as a lead concentration of <5 µg/L in the treated effluent.

During the initial phase of the project, various treatment options were developed, and a feasibility study was conducted. The study results revealed that iron coprecipitation/adsorption process was the most technically and economically sound method for achieving the desired target limits for copper, lead and zinc from this wastewater.

Composite samples were collected to characterize the wastewater and to conduct treatability studies. From the time of sample collection to analysis, all samples were preserved and handled in accordance with the U.S.EPA approved procedure. All analyses were performed in accordance with the procedure described in Standard Methods for the Examination of Water and Wastewater [37]. Throughout the laboratory-scale study, the concentrations of copper, lead and zinc in each of the untreated and treated samples were measured after filtration using 0.45-micron filter paper, so that the analytical results would approximately correspond to the dissolved concentrations of these metals. The Graphite Furnace Atomic Absorption (GFAA) spectrometric method was used to determine the metals concentrations.

The wastewater characterization results presented in Table 6.3 indicate that the initial concentrations of total copper, lead and zinc in the filtered sample were 510, 33 and 160 µg/L, respectively. The concentration of total iron in the sample was about 2 mg/L. The sample had a pH of about 7 S.U.

During the treatability study, the effects of reaction pH, iron (Fe^{3+}) dosage and reaction time were investigated, and the optimum operating conditions were determined.

Effect of Reaction pH on Treatment Efficiency

The experiment was performed at three different pH conditions ranging between 8.0 and 10.0 S.U., at a constant iron concentration of 10 mg/L as Fe^{3+}. After about 20 minutes of reaction time, a solution of polymeric flocculant was added to each sample. The samples were then settled for 30 minutes, filtered and analyzed for the selected metals.

The effect of changing pH on the removal of copper, lead and zinc during the iron

TABLE 6.2. Target Concentration of Selected Metals.

Metal	Target Concentration in Treated Effluent (µg/L)
Copper	<10
Zinc	<10
Lead	<5

TABLE 6.3. Wastewater Characterization Results.

Parameter	Concentration
pH (S.U.)	7.0
TSS (mg/L)	118
TDS (mg/L)	516
Iron (mg/L)	2
Lead (µg/L)	33
Copper (µg/L)	510
Zinc (µg/L)	160

coprecipitation/adsorption process is shown in Figure 6.2. As expected, increasing the pH increased the fractional adsorption of metals. The concentrations of copper, lead and zinc in the treated effluent were reduced to less than their target values. The results revealed that the most efficient pH for the removal of copper, lead and zinc with this process is approximately 8.0 S.U.

Determination of Optimum Iron (Fe^{3+}) Dosage

Various dosages of ferric ion ranging between 2 and 20 mg/L were tried. These dosages represent iron-to-copper weight ratios of 4:1, 10:1, 20:1, 30:1 and 40:1. The reaction pH was maintained at 8.0 S.U., as determined from the study described above. After about 20 minutes of reaction time, followed by coagulation/flocculation, settling and filtration, the concentrations of the selected metals in the treated effluent were measured.

Figure 6.3 shows the effect of iron dosage on metals removal. Increasing the dosage of iron decreased the concentrations of metals in the treated water. Initially, the effect of iron dosage on removal efficiency is significant, and then it gradually approaches a plateau. At a

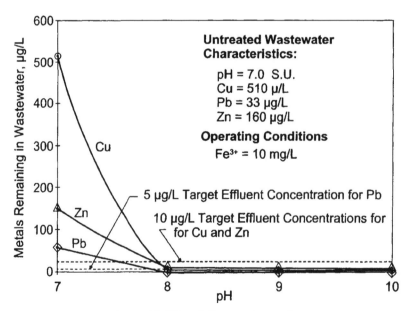

Figure 6.2 Effect of pH on metal removal.

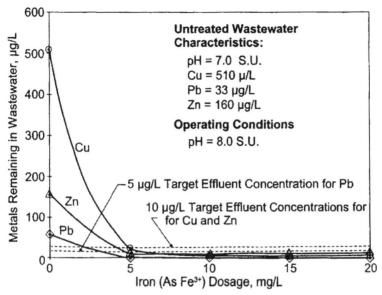

Figure 6.3 Effect of iron dosage on metal removal.

pH of 8.0 S.U. and an iron dosage of 5 mg/L (as Fe^{3+}), the concentrations of lead and zinc were reduced to less than their target concentrations of 5 and 10 µg/L. A 10 mg/L (as Fe^{3+}) dosage of iron reduced the concentration of copper from 510 to µg/L. Therefore, 10 mg/L of iron was selected as the optimum dosage for the removal of copper, lead and zinc from the wastewater.

Determination of Optimum Reaction Time

To determine if any benefit would be achieved by providing a longer reaction time for the iron coprecipitation/adsorption process, a study was conducted in which about 10 mg/L of iron (as Fe^{3+}) were added to the wastewater sample. The reaction pH was maintained at 8.0 S.U. The iron dosage and pH were determined from the above studies. Periodically between 1 and 60 minutes of reaction, samples were withdrawn from the reactor and were filtered and analyzed for copper. The above experiments demonstrated that copper was the rate-limiting adsorbate, because the treatment process removed lead and zinc from wastewater prior to removing copper. Therefore, an adsorption rate study was conducted to establish the equilibrium reaction time for copper. The data from the kinetic studies for copper adsorption onto iron oxyhydroxide are presented in Figure 6.4. The decreasing concentration of copper remaining in the solution indicates that the metal was adsorbed onto iron oxyhydroxide. It is also evident from the figure that, initially, the rate of copper adsorption was significantly high, with much slower subsequent removal rates that gradually approached an equilibrium condition. About 98% of the copper was removed during the first 10 minutes of reaction, while only 1 to 1.5 percent of additional removal occurred during the following 40 minutes of contact. A near straight-line fit can be seen in Figure 6.5, where the value of the logarithmic ratio of the copper concentration at a specific time to the initial copper concentration is plotted as a function of reaction time (linear regression coefficient, $R^2 = 0.892$). This figure indicates that the reaction can be approximated to first-order reversible kinetics with an overall reaction rate constant value of about 0.1154 per minute.

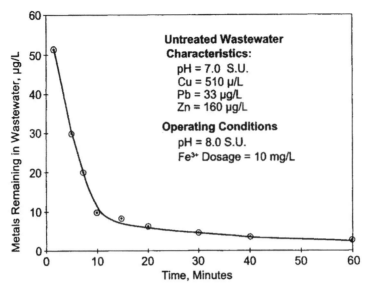

Figure 6.4 Effect of reaction time on copper removal.

Additional Study to Predict Effluent and Sludge Quality

In order to determine the overall effectiveness of the treatment process at the optimum operating conditions as well as to investigate the characteristics of the sludge generated from the iron coprecipitation/adsorption process, an additional laboratory-scale treatability study was performed prior to the actual engineering design work. The pH of the sample was adjusted to about 8.0 S.U. by the addition of sodium hydroxide. Approximately 10 mg/L of iron (as Fe^{3+}) was added to the sample, and the reaction was allowed to progress for 10 minutes. The concentrations of copper, lead and zinc in the filtered sample were then measured.

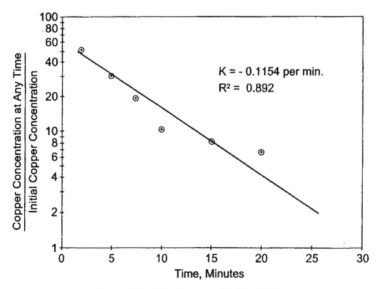

Figure 6.5 First-order reversible kinetic fit.

TABLE 6.4. TCLP Results: Concentration of
Lead in TCLP Extract.

Parameter	Concentration
pH (S.U.)	6.7
Lead (mg/L)	<0.10

The results show that the iron coprecipitation/adsorption process reduced the copper and zinc concentrations to <10 µg/L and the lead concentration to <5 µg/L.

By following the procedure outlined in 40 CFR Part 261, a Toxic Characteristics Leaching Procedure (TCLP) test was performed on the dewatered sludge. Lead was the only TCLP toxic metal that the wastewater contained; therefore, the TCLP extract was analyzed only for this metal. The TCLP test results presented in Table 6.4 indicate that the concentration of lead in the extract was less than 0.1 mg/L, which is below the toxicity limit of 5 mg/L; therefore, the sludge can be disposed of as a nonhazardous material.

Design and Implementation of Treatment System

Based on the laboratory-scale study results, the process design conditions were selected and are summarized in Table 6.5. As the table illustrates, the full-scale treatment system was designed with the following process parameters:

- Iron coprecipitation/adsorption will occur at a pH between 8 and 8.5 S.U.
- The weight ratio of iron (as Fe^{3+}) to copper will be maintained between 20:1 and 30:1.
- About 10 minutes of reaction time will be provided.

Using the above criteria, a mass balance calculation was performed for a wastewater flow of 60 gpm. The conceptual flow diagram and the mass balance results for this process are depicted in Figure 6.6. The treatment system consists of flow equalization, iron coprecipitation/adsorption reaction, coagulation/flocculation, clarification, filtration and sludge dewatering. The mass balance results predict that this treatment system is capable of producing effluent containing about 2 lb of total suspended solids (TSS) per day; <0.5 lb of total iron; and insignificant amounts of total copper, lead and zinc. The treatment system is expected to generate about 48 gallons of nonhazardous sludge per day, containing approximately 24–25% (by weight) dry solids.

The preliminary engineering flow diagram, including the size of the major equipment, is depicted in Figure 6.7. Wastewater is pumped to the treatment facility at a rate of approxi-

TABLE 6.5. Selected Process Conditions for
Copper, Lead and Zinc Treatment.

Process Parameter	Preferred Condition
pH	8 to 8.5 S.U.
Fe^{3+}	10 mg/L
Reaction Time	10 minutes
Fe^{3+} to Copper Ratio	20:1 to 30:1

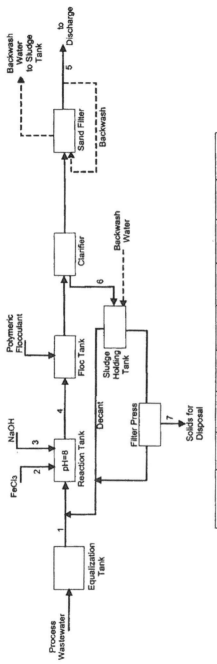

Stream No.		1	2	3	4	5	6	7
Parameters	Unit							
Flow	lb/hr	30,024	8.5(a)	2(b)	30,035	26,020	4,015	20.5
	gpm	60	1 GPH	<0.5GPH	60	52	8	2GPH
TSS	mg/L	85	-	-	100	2	98	98
	lb/day	118	-	-	139	<5	1020	24%
Iron	mg/L	1.5	7.5	-	9	<0.5	9	9
	lb/day	2	3.92%	-	12.5	<1	94	2.5%
Copper	lb/day	0.367	-	-	0.367	neg	0.367	0.367
	mg/L	0.51	-	-	0.51	<0.01	4.00	900
Lead	lb/day	0.024	-	-	0.024	neg	0.024	0.024
	mg/L	0.033	-	-	0.033	<0.005	0.250	60
Zinc	lb/day	0.116	-	-	0.116	neg	0.116	0.116
	mg/L	0.160	-	-	0.160	<0.01	1.21	285

a) 10% Solution of $FeCl_3$
b) 15% Solution of NaOH
neg = Negligible

Figure 6.6 Conceptual flow diagram for metal treatment.

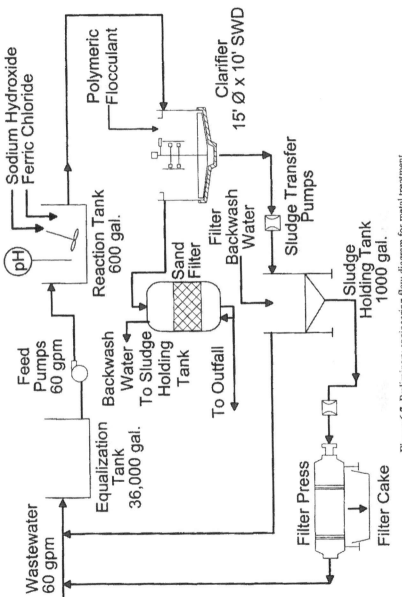

Figure 6.7 Preliminary engineering flow diagram for metal treatment.

191

mately 60 gpm and is received by a 36,000-gallon equalization tank. At this influent rate, the equalization tank will have a hydraulic retention time of about 10 hours. This retention time should be sufficient to minimize wastewater quality variability and thus enable the wastewater treatment facility to function efficiently and produce an effluent with consistently acceptable pollutant concentrations. Following equalization, the wastewater is pumped to a 600-gallon reaction tank where sodium hydroxide is added to maintain the pH between 8.0 and 8.5 S.U. In addition to sodium hydroxide, ferric chloride is added in this tank at the rate needed to achieve a ferric ion concentration of 10 mg/L. From the reaction tank, the wastewater flows by gravity to the center well of the solids contact clarifier, where polymeric flocculant is added. The clarifier, which has a diameter of 15 feet and a side wall depth of 10 feet, serves two purposes. First, it gently mixes the wastewater/polymeric flocculant to promote floc growth. Following flocculation, the clarifier provides sufficient surface area and hydraulic retention time for the newly formed floc to settle. The clarifier effluent flows by gravity to a sand filter prior to discharging to the creek.

The accumulated solids from the clarifier bottom are collected in a 1,000-gallon sludge holding tank. A 5-cubic-foot filter press is used to dewater the sludge before off-site disposal.

The full-scale wastewater treatment plant was installed in late 1997. The treatment plant is fully automated, designed to operate 24 hours per day, 7 days per week. Pertinent data collected from January 1998 through May 1999 are presented in Figure 6.8. During this period, the influent concentrations ranged between 400 and 600 µg/L copper, between 20 and 50 µg/L lead and between 100 and 200 µg/L zinc. The average wastewater flow to the treatment plant varied between 50–60 gpm. The analytical results reveal that the effluent concentrations of these metals are consistently <5 µg/L, which is below the target effluent concentrations.

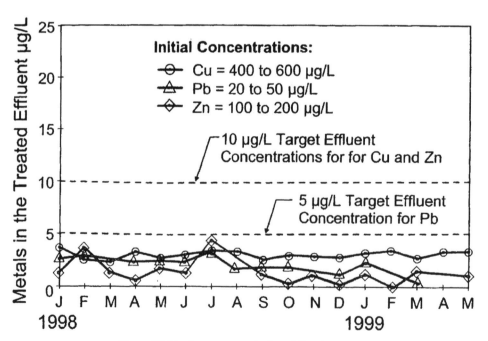

Figure 6.8 Metal treatment results from full-scale system.

CASE STUDY: ARSENIC TREATMENT

Selection and Optimization of Process Parameters

As discussed in the section entitled "Theoretical Consideration," the ability of H_3AsO_4 to deprotonate is much greater than that of H_3AsO_3 at a neutral pH condition, and the oxidation of As(III) to As(V) would increase the treatment efficiency of the iron coprecipitation/adsorption process at this pH. Common oxidizing agents include chlorine, sodium/calcium hypochlorite, ozone, hydrogen peroxide, potassium permanganate, hydroxyl radicals, etc. Hydroxyl radicals are generated by an advanced oxidation process that utilizes UV/H_2O_2, UV/O_3, O_3/H_2O_2 or Fenton's reagent. The selection of an oxidizing agent depends on the wastewater characteristics, contaminant concentration, treated water quality requirements and economic feasibility. The arsenic removal case study that is discussed below used Fenton's reagent. The technical and economic feasibility evaluation for this project indicated that the free hydroxide radicals generated by the Fenton's reaction provided the most economically feasible process for wastewater containing a high concentration of arsenic.

The fundamental aspects of Fenton chemistry are well documented. Fenton's reagent is characterized by the catalytic generation of hydroxyl free radicals that result from the chain reaction between ferrous ion (Fe^{2+}) and hydrogen peroxide. The hydroxyl radicals generated by this reaction efficiently oxidize a variety of organic and inorganic compounds. The oxidation potential values of various oxidizing agents are presented in Table 6.6, which reveals that the oxidation power of the hydroxyl radical is close to that of fluorine. The following mechanism for Fenton chemistry has been proven and accepted:

$$H_2O_2 + Fe^{2+} \rightarrow OH^{\cdot} + OH^- + Fe^{3+} \tag{5}$$

$$Fe^{3+} + H_2O_2 \rightarrow Fe^{2+} + HO_2^* + H^+ \tag{6}$$

In the above reactions, $OH^{\cdot}$ represents a hydroxyl radical, and HO_2^* is a superoxide radical. The advantage of Fenton's reagent is that it first oxidizes As(III) to As(V). The ferric ions present in the oxidation process then remove As(V) as an insoluble ferric-arsenate precipitate. The remaining oxyanions of arsenic are adsorbed onto the surface of hydrous ferric oxide ($Fe_2O_3 \cdot H_2O$).

In this case study, groundwater at an industrial facility in the northeastern United States was contaminated with arsenic compounds comprised of mixed inorganic and organic ar-

TABLE 6.6. Oxidation Potentials of Radical Species Compared to Common Molecular Oxidants.

Oxidants	Potentials (volts)
F_2	3.06
HO^*	2.80
O_3	2.07
H_2O_2	1.77
HOO^*	1.70
HOCl	1.49
Cl_2	1.39

TABLE 6.7. Groundwater Characteristics.

Parameter	Analytical Result
pH (S.U.)	7.2
TDS (mg/L)	1,400
TSS (mg/L)	50
Alkalinity to pH 8.3 (mg/L)	425
Chloride (mg/L)	30
Sulfate (mg/L)	150
Phosphate (mg/L)	26
Iron (mg/L)	5
Calcium (mg/L)	120
Magnesium (mg/L)	50
Aniline (mg/L)	2
Total Arsenic (mg/L)	30

senic species. The initial objectives of the project were to develop a cost-effective process for treating the contaminated groundwater to the current U.S. drinking water standard of less than 0.05 mg/L arsenic, to determine the optimum process conditions and to develop a conceptual process design. Treatability studies were conducted to verify the process efficiency, as well as to determine the optimum process design conditions. Based on the process design information, a full-scale treatment system was designed and installed.

During the study, all analyses were performed in accordance with the procedures described in *Standard Methods for the Examination of Water and Wastewater* [37]. Throughout the study, the concentrations of the contaminants in the untreated and treated samples were measured after filtration using a 0.45-micron filter paper. As mentioned before, the analytical results of the filtered samples approximately correspond to the dissolved concentrations of these contaminants. The total arsenic concentrations were determined using the Inductively Coupled Plasma (ICP) method and the Graphite Furnace Atomic Absorption (GFAA) Spectrometric method. Ion chromatography was used for the determination of inorganic arsenic species, such as As(III) and As(V). The concentration of organic arsenic species was determined by subtracting the sum of As(III) and As(V) from the total arsenic concentration.

The groundwater characterization results presented in Table 6.7 indicate that the total arsenic concentration in the sample was approximately 30 mg/L. About 50% of the total arsenic was As(V), 40% was As(III) and about 10% was organic arsenic. To define the design parameters for the process, the treatability study addressed the amount of iron required to achieve an arsenic concentration of less than 0.05 mg/L in the treated effluent, the amount of hydrogen peroxide required, the pH condition of the reaction and the required reaction time. The effects of each of the parameters are briefly described below.

Effect of Hydrogen Peroxide Dosage

To determine the optimum amount of hydrogen peroxide required in the oxidation step, 300 mg/L of iron (as Fe^{2+}) was added to the groundwater sample. Because Fenton's reagent is efficient under acidic conditions, the pH of the samples was adjusted to 4 S.U. with hydrochloric acid. Various dosages of hydrogen peroxide ranging between 25 and 300 mg/L were added separately to each sample. After thorough mixing, the pH of the samples was adjusted to 8 S.U. The samples were then aerated, and a solution of polymeric flocculant (1 mg/L) was added to each sample. Following a thorough mixing, the samples were allowed to settle for 30 minutes. Catalase, a reducing agent, was then added to the samples to destroy any excess hydrogen peroxide. The filtered samples were then analyzed for total arsenic.

The effect of hydrogen peroxide dosage on arsenic removal is presented in Figure 6.9. The results reveal that increasing the hydrogen peroxide dosage decreased the arsenic concentration in the treated effluent up to a hydrogen peroxide-to-iron weight ratio of about 0.5 to 1. At a ratio higher than that, the arsenic concentration did not appreciably decrease. In the presence of 300 mg/L of iron and at a pH of 4.0 S.U., a hydrogen peroxide dosage of 150 mg/L resulted in the maximum arsenic removal efficiency. The total arsenic concentration in the sample was reduced from 30 to less than 0.05 mg/L.

Determination of Optimum Iron (Fe^{2+}) Dosage

Figure 6.10 shows the effect of iron dosage on arsenic treatment. Five different dosages of iron, including 50, 90, 150, 200 and 300 mg/L (as Fe^{2+}), were added. These dosages represent iron-to-arsenic ratios of 1.67:1, 3:1, 5:1, 6.67:1 and 10:1. The data demonstrate that as the iron-to-arsenic ratio increases, the arsenic concentration in the treated water decreases. Initially, the effect of iron dosage on the removal efficiency is significant, and then it gradually approaches a plateau. In order to reduce the arsenic concentration to less than 0.05 mg/L in the treated water, an iron-to-arsenic weight ratio of about 10 to 1 is required for this type of wastewater.

Determination of Optimum Reaction pH

This treatability study was conducted to investigate the effect of pH on Fenton reaction. About 300 mg/L of iron (Fe^{2+}) and 150 mg/L of hydrogen peroxide were added to the groundwater sample. After thorough mixing, the samples were split into four portions.

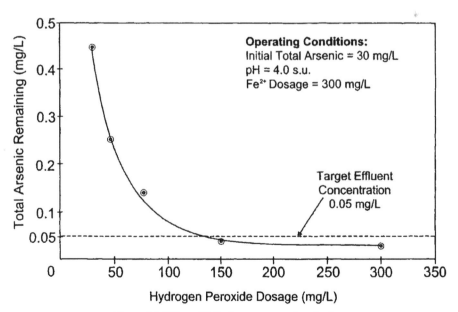

Figure 6.9 Effect of H_2O_2 dosage on arsenic reduction.

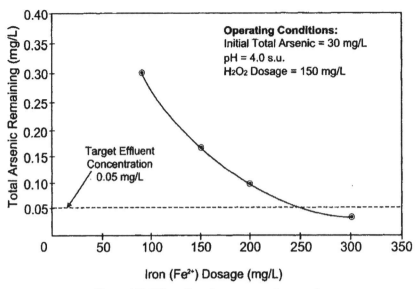

Figure 6.10 Effect of iron dosage on arsenic removal.

Using hydrochloric acid, the pH of each sample was adjusted to the range of 2.5 to 6.5. After about 30 minutes of mixing, the pH of each sample was adjusted to 8 S.U. with sodium hydroxide. The samples were then aerated, settled and filtered, prior to analyzing for total arsenic.

As shown in Figure 6.11, the arsenic treatment efficiency increases with the decreasing pH condition of the system. These results illustrate that the most effective pH for arsenic treatment, using Fenton's reagent, is about 2.5 S.U. At this pH condition, the arsenic concentration in the treated effluent was reduced to approximately 0.02 mg/L. At the low pH,

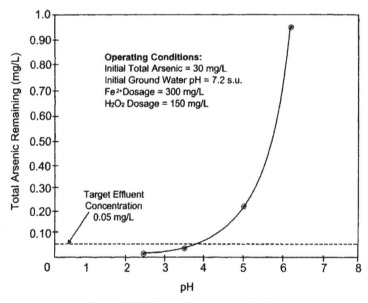

Figure 6.11 Effect of pH on arsenic reduction.

arsenite and organoarsenical compounds are completely oxidized to arsenate. These anions react with the ferric ions produced during the oxidation process and precipitate as a ferric-arsenate complex at a pH between 6 and 8 S.U. The remaining arsenate ions are then adsorbed onto ferrihydride.

Determination of Optimum Reaction Time

About 300 mg/L iron (Fe^{2+}) and 150 mg/L hydrogen peroxide were added to the groundwater sample. The pH of the mixture was adjusted to 2.5 S.U. with hydrochloric acid. Periodically between 1 and 60 minutes of reaction, samples were withdrawn from the reactor, and catalase was added to stop the oxidation reaction. The pH value of each sample was raised to 8 S.U. The samples were then aerated, settled and filtered, and the arsenic concentrations were measured.

Figure 6.12 shows the effect of reaction time on arsenic removal at a pH value of 2.5 S.U. with hydrogen peroxide and iron dosages of 150 and 300 mg/L, respectively. The data reveal that the oxidation reaction is very fast. The arsenic concentration was reduced from 30 to less than 0.02 mg/L within 5 minutes of reaction.

Additional Study to Predict Effluent and Sludge Quality

The overall effectiveness of the treatment system at the optimum operating conditions and the sludge characteristics were determined by conducting an additional process verification study.

Approximately 300 mg/L of iron (Fe^{2+}) and 150 mg/L of hydrogen peroxide were added to the groundwater sample. The pH of the sample was adjusted to 2.5. Although the reaction kinetics study indicated that the oxidation reaction is very fast, and nearly complete oxidation can be achieved within a few minutes of the reaction, the sample was allowed to react

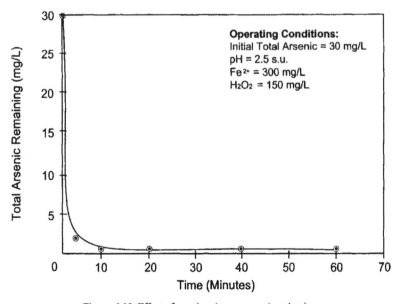

Figure 6.12 Effect of reaction time on arsenic reduction.

TABLE 6.8. Selected Process Conditions for Arsenic Removal.

Parameter	Preferred Condition
pH (Oxidation Tank)	2.5 S.U.
Fe^{2+}	300 mg/L
H_2O_2	150 mg/L
Oxidation Time	20 minutes
Aeration Time	30 minutes
pH (Coprecipitation/Adsorption Tank)	7.5 to 8.0 S.U.
Fe^{2+} to As Ratio	10:1
H_2O_2 to Fe^{2+} Ratio	0.5:1

for about 20 minutes to achieve complete oxidation. After the oxidation reaction, pH of the sample was adjusted to 8. S.U., and the sample was aerated, settled and filtered. The treated sample was then analyzed for arsenic. Because arsenic is a TCLP toxic metal, a TCLP test was conducted on the filter cake by following the procedure outlined in 40 CFR Part 261.

The results demonstrate that oxidation followed by the iron coprecipitation/adsorption process reduced the concentration of arsenic in the treated water from 30 to less than 0.05 mg/L. In addition, the TCLP results indicate that the sludge does not exhibit hazardous waste characteristics [38].

Design and Implementation of the Arsenic Treatment System

Based on the laboratory-scale treatability study results, the design conditions for the arsenic treatment process were selected and are summarized in Table 6.8. In order to achieve complete oxidation of arsenite and organoarsenical compounds, 20 minutes of reaction time were provided. An iron-to-arsenic ratio of 10 to 1 and a hydrogen peroxide-to-iron ratio of 0.5 to 1 is recommended in the oxidation tank. It is also recommended that the oxidation reaction be conducted at a pH of 2.5 S.U., and that the coprecipitation/adsorption reaction be conducted at a pH between 7.5 and 8.0 S.U. The conceptual flow diagram, including the mass balance calculation for this process, is depicted in Figure 6.13. As the figure illustrates, the treatment system consists of flow equalization, oxidation followed by coprecipitation/adsorption, coagulation and flocculation, clarification, filtration and sludge dewatering. The treatment plant was designed to handle about 150 gpm of groundwater. The mass balance results indicate that the treatment system is capable of producing an effluent containing less than 10 lb of TSS per day, less than 2 lb of iron, and an insignificant amount of arsenic. Approximately 575 gpd of nonhazardous sludge containing about 30% dry solids (by weight) is anticipated from the treatment plant. It is also expected that approximately 104 lb of sludge will be produced to treat 1 lb of arsenic.

The preliminary engineering flow diagram for the full-scale treatment is presented in Figure 6.14. Groundwater is pumped to a 20,000-gallon holding tank at the rate of approximately 150 gpm. The water is then pumped to a 3,000-gallon closed tank. As indicated above, the optimal dosages of ferrous chloride and hydrogen peroxide were added to this tank. The reaction pH was maintained at 2.5 S.U. by the addition of hydrochloric acid. In this tank, arsenite and organoarsenical compounds are oxidized to arsenate. Additionally, some organic compounds present in the groundwater are also oxidized. After about 20 minutes of reaction, the water flows by gravity to a 5,000-gallon tank. The pH of water in this

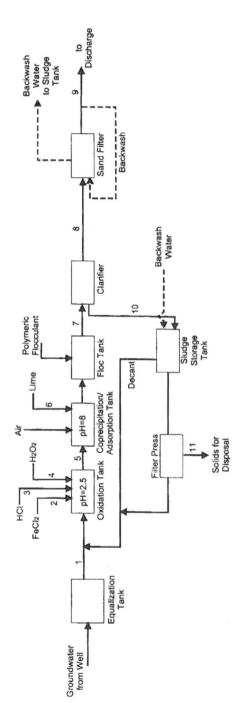

Stream No.	Unit	1	2	3	4	5	6	7	8	9	10	11
Parameters												
Flow	lb/hr	75,060	255(a)	240(b)	23(c)	75,578	458	76,036	67,150	67,150	8,886	232
	gpm	150	<0.5	0.5	2 GPH	153	1	154	136	136	18	24 GPH
TSS	lb/day	90	neg	-	-	-	550	1740	65	<10	1,675	1,675
	mg/L	50	neg	-	-	-	4.6%	940	40	<5	7,750	30%
Iron	lb/day	9	540	-	-	549	neg	549	<5	<2	549	549
	mg/L	5	10.5%	-	-	300	neg	300	<2	<1	2,540	11.5%
Arsenic	lb/day	54	neg	-	-	54	neg	54	0.10	<0.1	54	54
	mg/L	30	neg	-	-	30	neg	30	0.06	<0.05	250	1.1%

a) 20% Solution of FeCl₂
b) 5% Solution of HCl
c) 50% Solution of H₂O₂
neg = Negilible

Figure 6.13 Conceptual flow diagram for arsenic treatment.

199

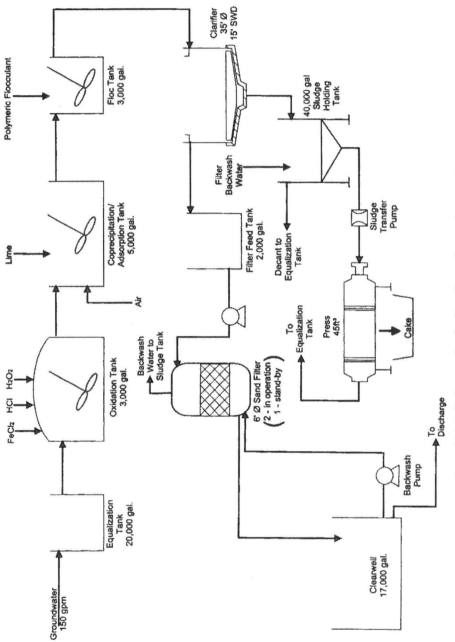

Figure 6.14 Preliminary engineering flow diagram for arsenic treatment.

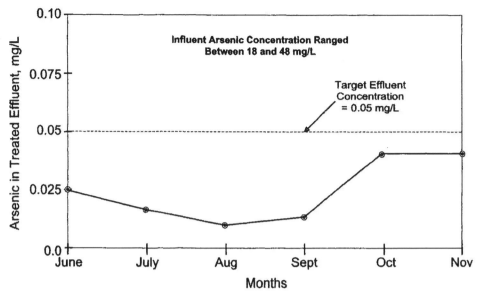

Figure 6.15 Arsenic treatment results from full-scale system.

tank is maintained between 7.5 and 8.0 S.U. A centrifugal blower is used to aerate the water, thus oxidizing the unreacted ferrous ions to ferric ions. A solution of polymeric flocculant (1 mg/L) is added to the flocculation tank. From this tank, the water flows by gravity to a 35-foot-diameter clarifier where solids are separated from the liquid. The clarifier effluent is filtered prior to discharging to the creek. Two sand filters are operated in parallel, with one in standby mode. The filters are backwashed once a day.

The accumulated solids from the clarifier are collected in a 40,000-gallon sludge storage tank. Using a 45-cubic-foot filter press, the sludge is dewatered before off-site disposal.

The full-scale treatment plant has been in operation for about six months. The plant operates in the automatic mode on a 24-hour, 7-day schedule. Daily composite samples were collected from the treatment plant influent and effluent streams for a period of six months. The influent concentration of total arsenic ranged between 18 and 48 mg/L. The average groundwater flow to the treatment plant varied between 145,000 and 200,000 gpd. The groundwater is treated using an iron (Fe^{2+})-to-arsenic weight ratio of 10 to 1 and a hydrogen peroxide-to-iron weight ratio of 0.5 to 1. The chemical usage rates are controlled based on the groundwater flow to the treatment plant. Results from the full-scale operation presented in Figure 6.15 reveal that the concentration of arsenic in the treated effluent is consistently less than the target concentration of 0.05 mg/L.

SUMMARY AND CONCLUSIONS

The iron coprecipitation/adsorption process is a reliable, effective and economical method of reducing heavy metals to ppb levels. However, a feasibility study is recommended prior to selecting this process. Important design parameters for the iron coprecipitation/adsorption process include reaction pH, iron-to-heavy metal weight ratio, reaction time and wastewater flow and characteristics. The concentrations of metals decrease with an increasing iron dosage. Usually, the weight ratio of iron to heavy metal is maintained between 20:1 and 30:1; however, the presence of a significant amount of competitive cations and anions in the wastewater is likely to increase the iron demand. At a constant iron dosage, increasing the wastewater pH increases the adsorption of cations onto iron oxyhydroxide. The desired pH for copper, lead, and zinc removal is typically between 8 and 8.5 S.U. It is advisable to conduct a laboratory-scale treatability study on the actual wastewater sample to determine the optimum operating conditions for the key process parameters.

The oxidation process followed by iron coprecipitation/adsorption is capable of reducing the arsenic concentration below the current U.S. drinking water standard of 0.05 mg/L. The key process design parameters for arsenic treatment include the type and dosage of oxidizing agent, the reaction pH, iron dosage, reaction time and wastewater flow and characteristics. The selection of an oxidizing agent should be determined by the wastewater characteristics, concentration of arsenic, desired quality of the treated water and economic feasibility. Free hydroxyl radicals generated by the Fenton reaction provide a potentially economical process for wastewater containing high concentrations of arsenite and organoarsenical compounds. To achieve less than 0.05 mg/L of arsenic in the treated effluent, an iron-to-arsenic weight ratio of 10 to 1 is recommended for this type of wastewater. The case study results showed that increasing the hydrogen peroxide dosage increases arsenic removal efficiency until a hydrogen peroxide-to-iron weight ratio of 0.5 to 1 is achieved. However, the iron and hydrogen peroxide requirements depend on the initial concentrations of arsenic (V), arsenic (III), organoarsenical compounds and the presence of other competitive anions and organic compounds. This study also concluded that the maximum oxidizing capability of Fenton's reagent is achieved at a pH between 2.5 and 3.0 S.U.

The mechanism for removing high concentrations of dissolved arsenic from wastewater can be described as a three-step process, including solids formation and chemical precipitation, surface complexation and electrostatic interaction. As (III) and organoarsenical compounds are first oxidized to As (V), which are soluble and ionized in water at a pH between 6 and 8. The ferric ions generated in the oxidation process as well as in the coprecipitation/adsorption tank precipitate As (V) as an insoluble ferric-arsenate complex. The remaining oxyanions of arsenic are then adsorbed onto iron oxyhydroxide.

ACKNOWLEDGEMENTS

The author wishes to thank Carla G. Robinson for her contribution to the preparation of this chapter, as well as Keith A. Benson and Terence J. Smith for their review.

REFERENCES

1 Nuquist, O. W. and H. R. Carroll. 1959. "Design and Treatment of Metal Processing Wastewaters," *Sew. Ind. Wastes,* 31:941–948.

2 Stone, E. H. F. 1972. "Treatment of Non-Ferrous Metal Processing Waste," *Metal Finishing J.,* 18:280–290.

3 Stone, E. H. F. 1967. "Treatment of Non-Ferrous Metal Process Waste of Kynoch Works," Proceedings of the 25th Industrial Waste Conference, May 1967, Purdue University, West Lafayette, Indiana, pp. 848–865.

4 Watson, K. S. 1953. "Treatment of Complex Metal-Finishing Wastes," Proceedings of the 26th Annual Meeting, *Fed. Sew. Ind. Wastes,* Oct. 1953, Miami, FL, pp. 182–194.

5 Knapp, R. and E. Paulson. 1982. "Gravity Filtration Reduces Suspended Metals in a Lime Precipi-

tation System," Proceedings of the 37th Industrial Waste Conference, May 1982, Purdue University, West Lafayette, Indiana, pp. 95–104.

6 Osantowski, R. and A. Ruppersberger. 1977. "Upgrading Foundry Wastewater Treatment," Proceedings of the 32nd Industrial Waste Conference, May 1977, Purdue University, West Lafayette, Indiana, pp. 102–115.

7 McVaugh, J. and W. T. Wall. 1976. "Optimization of Heavy Metals Wastewater Treatment Effluent Quality versus Sludge Treatment System," Proceedings of the 31st Industrial Waste Conference, May 1976, Purdue University, West Lafayette, Indiana, pp. 17–25.

8 Bhattacharyya, D., A. Jumawan, G. Sun, and K. Schwitzebel. 1981. "Precipitation of Sulfide: Bench Scale and Full Scale Experimental Results," AIChE Symposium Series, 77(209):31–32.

9 Kim, B. M. and P. A. Amodeo. 1983. "Calcium Sulfide Process for Treatment of Metal Containing Wastes," *Environ. Prog.*, 2(3):175–180.

10 Peters, R. W., Y. Ku, and T. K. Chang. 1984. "Heavy Metals Crystallization Kinetics in an MSMPR Crystallizer Employing Sulfide Precipitation," AIChE Symposium Series, 80(240):55–75.

11 Peters, R. W. and Y. Ku. 1984. "Removal of Heavy Metals from Industrial Plating Wastewaters by Sulfide Precipitation," Proceedings of the Industrial Wastes Symposia, 57th Water Poll. Control Fed. Annual Conference, pp. 279–311.

12 Bhattacharyya, D., A. B. Jumawan, and R. B. Grieves. 1979. "Separation of Toxic Heavy Metals by Sulfide Precipitation," *Sep. Sc. Tech.*, 14:441–452.

13 Gradde, R. and H. Laitenen. 1974. "Studies of Heavy Metal Adsorption by Hydrous Iron and Manganese Oxides," *Analytical Chemistry*, 46:20–22.

14 Benjamin, M. M., K. F. Hayes, and J. O. Leckie. 1982. "Removal of Toxic Metals from Power Generation Waste Streams by Adsorption and Coprecipitation," *J. Water Poll. Control Fed.*, 54(11):1472–1481.

15 Davis, J. A. and O. J. Leckie. 1978. "Surface Ionization and Complexation at the Oxide-Water Interface II: Surface Properties of Amorphous Iron Oxyhydroxide and Adsorption of Metal Ions," *J. Colloid Interface Sc.*, 67:90.

16 Appleton, A. R., C. Papelis, and J. O. Leckie. 1988. "Adsorptive Removal of Trace Elements from Coal Fly-Ash Wastewater onto Iron Oxyhydroxide," Proceedings of the 43rd Industrial Waste Conference, Purdue University, West Lafayette, Indiana, pp. 375–387.

17 Banerjee, K. and C. D. Blumenschein. 2000. "An Innovative Process for Removal of Copper, Lead, and Zinc from a Steel Mill Effluent," *AISE Steel Technology*, April: 27–32.

18 Magnusen, L. M., T. C. Waugh, O. K. Galle, and J. Bredfeldt. 1970. "Arsenic in Detergents; Possible Danger and Pollution Hazards," *Science*, 168:389–390.

19 Rosehart, R. and J. Lee. 1972. "Effective Methods of Arsenic Removal from Gold Mine Wastes," *Canad. Min. J.*, June, 53–57.

20 Shen, Y. S. 1973. "Study of Arsenic Removal from Drinking Water," *J. Amer. Water Works Assn.*, 65(8):543–548.

21 Banerjee, K. 2000. "Removal of Soluble Arsenic Species from Flyash Leachate by Iron Coprecipitation/Adsorption Process," Proceedings of the 3rd International R&D Conference, March 2000, Jabalpur, India, pp. 439–448.

22 NcNeill, L. S. and M. Edwards, 1995. "Soluble Arsenic Removal at Water Treatment Plants," *J. Amer. Water Works Assn.*, 4:105–113.

23 Scott, K. N., J. F. Green, H. D. Do and S. J. McLean. 1995. "Arsenic Removal by Coagulation," *J. Amer. Water Works Assn.*, 4:114–126.

24 Hering, J. G., P. Y. Chen, J. A. Wilkie, and M. Elimelech. 1997. "Arsenic Removal from Drinking Water during Coagulation," *J. Environ. Engineering*, 123(8):800–807.

25 Gulledge, J. H. and J. T. O'Connor. 1973. "Removal of Arsenic (V) from Water by Adsorption on Aluminum and Ferric Hydroxide," *J. Amer. Water Works Assn.*, 8:548–552.

26 McNeill, L. S. and M. Edwards. 1997. "Arsenic Removal during Precipitative Softening," *J. Environ. Engineering*, 123(5):453–460.

27 Sorg, T. J. and G. S. Logsdon. 1978. "Treatment Technology to Meet the Interim Primary Drinking Water Regulations for Inorganics: Part 2," *J. Amer. Water Works Assn.*, 70(7):379–393.

28 Shen, Y. S. and C. S. Chen. 1964. "Relation between Blackfoot Disease and the Pollution of Drinking Water by Arsenic in Taiwan," Proceedings of the 2nd International Conference on Water Pollution Research, Tokyo, 1:173–190.

29 Lee, J. Y. and R. G. Rosehart. 1972. "Arsenic Removal by Sorption Process from Wastewaters," *Canad. Min. Met. (CIM) Bull.*, 65(11):33–37.

30 Huang, C. P. and P. L. K. Fu. 1984. "Treatment of Arsenic (V) Containing Water by the Activated Carbon Process," *J. Water Poll. Control Fed.*, 56(3):233–242.

31 Gupta, S. K. and K. Y. Chen. 1978. "Arsenic Removal by Adsorption," *J. Water Poll. Control Fed.*, 3:493–506.

32 Fox, K. R. 1989. "Field Experience with Point-of-Use Treatment Systems for Arsenic Removal," *J. Amer. Water Works Assn.*, 2:94–101.

33 Ghosh, M. M. and J. R. Yuan. 1987. "Adsorption of Inorganic Arsenic and Organoarsenicals on Hydrous Oxides," *Environ. Progress*, 6:150–157.

34 Hathaway, S. W. and F. Rubel Jr. 1987. "Removing Arsenic from Drinking Water," *J. Amer. Water Works Assn.*, 8:61–65.

35 Schlicher, R. J. and M. M. Ghosh. 1985. "Removal of Arsenic from Water by Physical-Chemical Processes," *AIChE Symposium Series*, 81(243):152–164.

36 Korte, N. E. and Q. Fernando. 1991. Review of Arsenic(III) in Groundwater, *Crit. Rev. Environ. Control*, 21:1–39.

37 Greenburg, A. E., L. S. Clesceri, and A. D. Eaton, ed. 18th Edition, 1992. *Standard Methods for the Examination of Water and Wastewater*, Washington, DC: Published jointly by APHA, AWWA, WEF.

38 Benerjee, K., R. P. Helwick, and S. K. Gupta. 1999. "A Treatment Process for Removal of Mixed Inorganic and Organic Arsenic Species from Groundwater," *Environ. Prog.*, 18(4):280–284.

39 Electric Power Research Institute. 1985. "Trace Element Removal by Coprecipitation with Amorphous Iron Oxyhydroxide: Engineering Evaluation," EPRI Report No. CS-4087, Palo Alto, California.

CHAPTER 7

Removal of Heavy Metals by Activated Carbon

BRIAN E. REED[1]

INTRODUCTION

THE presence of heavy metals in the environment is of major concern because of their toxicity and threat to human life and the environment. Primary drinking water standards for metals are regulated by the Clean Water Act. Values of maximum concentration limit (MCL) for metals are presented in Table 7.1. Metal discharge limits for industrial wastewater discharges are a function of the industry and can vary from state to state. While it is not possible to list general wastewater discharge limits for a particular metal, it can be said that there is substantial regulatory pressure for reducing the metal limits in drinking water and in wastewater discharges. As heavy metal regulations become more stringent, new technologies must be developed.

Anthropogenic sources of heavy metals include wastes from the electroplating and metal finishing industries, metallurgical industry, tannery operations, chemical manufacturing, mine drainage, battery manufacturing, leachates from landfills and contaminated groundwater from hazardous waste sites. For wastes with high metal concentrations, precipitation processes (e.g., hydroxide, sulfide) are often the most economical. However, given the tendency for stricter heavy metal limits, additional treatment processes, down line from the precipitation process, may be required to "polish" the effluent prior to discharge. These tertiary processes may also be the primary metal removal process for waste streams with low concentrations of metals or for drinking water. Examples of tertiary processes include ion exchange, reverse osmosis and adsorption. Ion exchange and reverse osmosis, while effective in producing an effluent low in metals, have high operation and maintenance costs and are subject to fouling. Adsorption by activated carbon is an established treatment method for organic contaminants but has been rarely used in an actual treatment setting for inorganic adsorbates, despite the fact that the ability of activated carbons to remove heavy metals has been established by numerous researchers. It is hoped that as engi-

[1]Department of Civil and Environmental Engineering, University of Missouri at Columbia, Columbia, MO 65211, U.S.A., reedb@missouri.edu

TABLE 7.1. Primary Drinking Water Standards for Metals.

Metal	MCL[1], mg/L
Antimony (Sb)	0.006
Arsenic (As)	$0.05/(0.005-0.020)^2$
Barium (Ba)	2
Beryllium (Be)	0.004
Cadmium (Cd)	0.005
Chromium (Cr)	0.1
Copper (Cu)	1.3
Lead (Pb)	$0.015/0^2$
Mercury (Hg)	0.002
Selenium (Se)	0.05
Thallium (Tl)	$0.002/0.0005^2$

[1]Maximum Concentration Limit (MCL).
[2]MCL Goal.

neers, scientists and regulators become aware of activated carbon's ability to remove heavy metals from the aqueous phase, the use of activated carbon technology for metal removal will become more common.

Granular activated carbon (GAC) for metals removal has not had widespread commercial use but may have economic benefits for certain applications. With this in mind, the purpose of this chapter is to introduce the reader to the important topics regarding heavy metal removal by activated carbon. In many instances, research results are in conflict and, concrete statements regarding certain phenomena cannot be made. When this has occurred, examples of opposing results are presented. In the second section of this chapter, an overview of activated carbon surface chemistry is presented. Heavy metal removal mechanisms are presented in the third chapter. Factors affecting metal removal are presented and discussed in the fourth section, and results from batch studies are presented to illustrate basic concepts. In the fifth section, acid-base behavior and heavy metal removal are modeled using the Surface Complex Formation (SCF) approach. If activated carbon is to find widespread use in the treatment of metal-bearing wastewaters, it will most likely be through the use of GAC columns. Accordingly, emphasis is placed on providing the reader with experimental results from kinetic experiments (section six) and GAC column studies (section seven).

CARBON SURFACE CHEMISTRY

The type and number of activated carbon surface groups will influence the extent and rate at which organic and inorganic compounds are adsorbed. The presence of functional groups on the surfaces of activated carbons has been speculated for over a century. In 1863, Smith [1] hypothesized that when oxygen is allowed to react with the surface of carbon, a chemical change occurs. While other gases, such as nitrogen, are easily removed from the surface of carbon, oxygen can only be removed (as CO_2) by heating the carbon. Rhead and Wheeler [2,3] reported that some sort of oxygen-carbon complex(s) was formed when oxygen contacts carbon surfaces. It is not known what surface functional groups are formed during the carbon activation process. Whether the surface oxides are acidic, basic or amphoteric in nature depends on the method of activation. Steenberg [4] suggested that carbons capable of adsorbing a strong base be called L-type (acidic carbons) and those capable of adsorbing a strong acid be called H-type (basic carbons). L-type carbons are generally exposed to oxygen at temperatures of 200°C to 500°C or solution oxidants during the acti-

TABLE 7.2. pH_{zpc} for Various Commercial
Activated Carbons.

Carbon	pH_{zpc}
Nuchar SA	4.0[1]
Nuchar SN	5.35[2], 5.84[3]
Filtrasorb 200	8.2[1]
Filtrasorb 400	10.4[1]
Hydrodarco 3000	8.5[1]
Hydrodarco B	7.45[1]

[1]Corapcioglu and Huang [12].
[2]Reed and Matsumoto [13].
[3]Huang [9].

vation process. H-type carbons are formed using activation methods that remove indigenous surface oxide groups. This is accomplished by heating the carbon in the presence of an inert gas or vacuum and cooling to low temperatures in a similar environment. Thus, H-type carbons are exposed to oxygen only at low temperatures, and the oxidizing conditions necessary for acidic functional groups to form are minimized. As the activation temperature increases, the number of H-type surface groups increases (amount of acid adsorbed by the carbon increases). The reverse is observed at low activation temperatures. Steenberg [4] suggested that the transition temperature from L- to H-type carbons is approximately 500°C. Puri and Bansal [5,6] reported that the majority of CO_2 is evolved at temperatures less than 600°C ($\approx$ maximum activation temperature of L-type carbons), while CO evolution occurs between 800° and 1,000°C (typical activation temperatures for H-type carbons). These researchers hypothesized that the surface groups that evolved CO_2 were acidic, and those evolving CO were basic. Numerous surface functional groups have been suggested as being responsible for the acidic nature of activated carbons. The most frequently mentioned acidic groups are the carboxylic, fluorescein-type lactones, quinone and phenolic groups. Proposed basic surface groups include chromene [7] and pyrone [8] structures.

Activated carbons when added to water develop a surface charge and exhibit acid-base properties. In the absence of specifically adsorbed anions or cations, the pH at which the proton excess on the surface is zero is defined as the zero point of charge (pH_{zpc}). At solution pH values above the pH_{zpc}, the surface has a net negative charge, while at pH values below the pH_{zpc}, the surface has a net positive charge. L-type carbons have a low pH_{zpc} and tend to carry negative surface charge over a wide pH range. The opposite is true for H-type carbons: high pH_{zpc} and positive surface charge. Experimentally, the pH_{zpc} can be determined from the intersection of the net acid-base titration curves at low ionic strengths [9]. Low ionic strength curves are used because as the ionic strength is increased, the adsorption of background electrolytes can become important. The pH_{zpc} can also be determined by titrating the carbon with an inert salt solution and determining the change in pH with increasing ionic strength [10,11]. The addition of an inert electrolyte to a carbon suspension changes the pH in the direction of the pH_{zpc}. In Table 7.2, the pH_{zpc} for several activated carbons is presented.

REMOVAL MECHANISMS

Metals can be removed via the following phenomena: physical adsorption, chemisorption, hydrogen bonding, ion exchange, surface precipitation and filtration. Physical adsorption is nonspecific in nature and is recognized as the primary removal mechanism for organic adsorbates. Chemisorption is specific in nature and involves the formation of a covalent bond (sharing of electrons) between the adsorbate and a specific carbon surface site. Chemisorption is generally considered to be irreversible, while physical adsorption is considered to be reversible. In hydrogen bonding, a long-range attractive force exists between the hydrogen atom of hydrated metal ions and carbon surface oxygen groups. Hydrogen bonding can be classified under chemisorption, but the hydrogen bond is much weaker than a covalent bond. With covalent bonds, the much stronger inner-sphere complex is formed, while an outer-sphere complex is formed for a hydrogen bond. Ion exchange is the sorption phenomenon that occurs when the adsorbate and adsorbent are oppositely charged. Precipitation of a metal on a surface is easier than the formation of the same solid in solution. For this reason, given the proper chemical conditions, heavy metals will preferentially precipitate on the carbon surface. Localized high concentrations of the metal and OH^- in an activated carbon's extensive pore volume can enhance metal removal. Finally, if the metal exists as a solid prior to entering the GAC column (e.g., if the GAC column is located after a coagulation-flocculation-solid/liquid separation system) or is associated with other solid particles (e.g., sediment-bound metal), the metal can be removed by filtration. Additional discussions on removal mechanisms will be provided throughout this chapter in conjunction with experimental results.

FACTORS AFFECTING METAL REMOVAL BY ACTIVATED CARBON

Factors affecting heavy metal removal by activated carbon include: solution pH, metal type and concentration, surface loading, presence of complexing ligands and competing adsorbates, ionic strength, temperature and carbon type.

Solution pH

Solution pH is the primary variable governing metal adsorption by all hydrous solids (activated carbon, metal oxides, clays, etc.). Sigworth and Smith [14] first observed that metal removal by activated carbon was inversely proportional to the solubility of the metal. For uncomplexed cationic heavy metals, removal increases with increasing pH (the fraction of metal removed increases from zero to one over a relatively narrow pH range). The metal removal versus pH curve is referred to in the literature as a "pH-adsorption edge." Examples of pH-adsorption edges for a cationic heavy metal (10 mg/L Pb) are presented in Figure 7.1 [15]. When the heavy metal exists as an anion, removal generally increases with decreasing pH. An example of this phenomenon is presented in Figure 7.2 for Cr(VI) [16]. An example of an exception to the first rule is presented in Figure 7.3 for Hg(II) [17]. Mercury removal was highest in the acidic region and decreased with increasing pH. Regeneration of activated carbon that has been exhausted with a heavy metal is possible using the pH-adsorption relationships demonstrated in Figures 7.1 through 7.3. For example, a carbon that has been exhausted with Pb can be regenerated by exposure to an acid. Regeneration techniques will be discussed further in the section on the use of GAC columns for metal removal.

Specifically, the pH affects the status of the outer hydration sheaths of the metal ion, aqueous metal speciation, complexation and solubility and the electrochemical behavior of the carbon surface. Netzer and Hughes [18] reported that the onset of cationic metal adsorp-

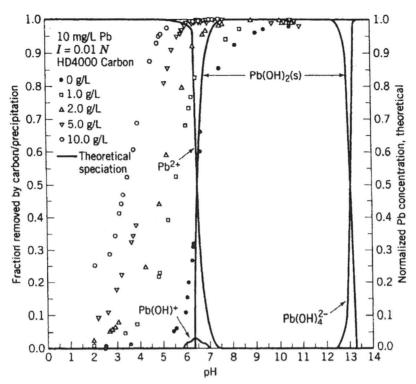

Figure 7.1 Pb pH-adsorption edges for Norit Americas Hydrodarco 4000 and the Pb aqueous phase speciation diagram [15].

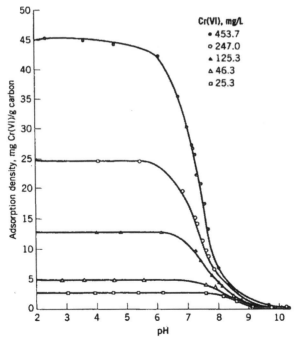

Figure 7.2 Cr(VI) pH-adsorption edges for Calgon Filtrasorb 400. Experimental conditions: carbon concentration = 10 g/L, ionic strength = 0.1 M and Cr(VI) = 25.3 to 453 mg/L [16]. Reprinted with permission of the Purdue Industirial Waste Conference and the Purdue Research Foundation.

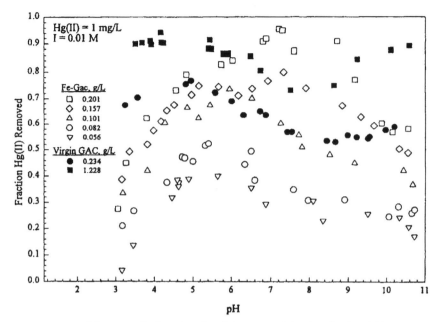

Figure 7.3 Hg(II) pH-adsorption edges for a commercial activated carbon [17].

tion coincided with the loss of the outer hydration sheath of the metal ions. The outer hydration sheath is removed at a pH slightly lower than the pH at which $Me(OH)^-_{(aq)}$ forms. Reed [19] discussed possible Pb removal mechanisms for the activated carbon HD4000 in relationship to the pH that solution precipitation begins (pH_{prec}). At pH values below pH_{prec}, adsorption was hypothesized to be the primary removal mechanism. As the solution pH approached pH_{prec}, surface precipitation (either on the external surface or in the carbon pores) became important. Based on the shape of experimental pH-adsorption edges, Reed and Matsumoto [20] reported that surface precipitation could occur at pH values 1/2 to 1 pH units lower than the pH at which solution precipitation occurs. Reasons for this behavior include the following: (1) the carbon surface pH may be higher than the solution pH; (2) the surface behaves as a "nucleus" for $Me(OH)_{2(s)}$ formation; and (3) a locally high concentration of metal may exist on the carbon surface, increasing the opportunity for precipitation to occur. A possible removal scenario is that a metal is adsorbed by the carbon and creates a condition on the surface whereby additional metal removal can occur via surface precipitation (i.e., the first layer of metal is by adsorption and additional layers are due to the formation of a metal solid). This phenomenon can be thought of as "multilayer sorption" or enhancement of precipitation by providing a nucleus (i.e., heterogeneous nucleation) and a locally high metal concentration. Similar behavior has been observed for metal oxide adsorbents. Farley et al. [21] proposed a surface precipitation model for metal oxide adsorbents that predicts the gradual increase in the surface concentration observed at very high adsorbate/adsorbent ratios. Following single-layer coverage, a new surface is formed that is comprised of the adsorbate and the metal hydroxide adsorbent.

In Figure 7.1, the theoretical Pb speciation diagram is included with the Pb-HD4000 pH-adsorption edges [15]. Based on thermodynamic calculations, the formation of $Pb(OH)_2(s)$ begins at about pH 6.25 (pH_{prec}) for 10 mg/L Pb. There was significant Pb removal at pH values less than pH_{prec}, and for several carbon concentrations, removal was completed prior to the start of solution precipitation, indicating that adsorption was important. $Pb(OH)_{2(s)}$ resolubilized at higher pH values, but in the presence of carbon, Pb desorption/resolubilization did not occur. At higher pH values, both adsorption and surface precipitation may be important. However, several researchers have reported a decrease in metal removal at high pH values through the formation of soluble metal complexes [e.g., $Pb(OH)_4^{-2}$] [22,23].

The effect of pH on heavy metals that exist as anions [e.g., Cr(VI), As(III) and As(VI)] is not as well understood. The pH may affect the carbon surface chemistry more than the metal's aqueous chemistry. For example, at low pH values, most carbon surfaces will be positively charged and attract anionic metals. As the pH increases, the magnitude and eventually the sign of the surface charge will change, resulting in a decrease in adsorption. In addition, there may be specific surface sites that are more reactive at lower pH values.

Metal Type and Concentration

In general, the lower the pH at which cationic metals form aqueous hydroxide complexes [$Me^x(OH)_n^{(x-n)}$] and solids [$Me(OH)_{2(s)}$], the better the metal will be removed. The formation of the first metal hydroxide species can be represented by the following chemical reaction:

$$Me^{2+} + H_2O \leftrightarrow Me(OH)^+ + H^+; \quad K_1$$

Pb, Cu, Zn and Ni form hydroxide complexes in the following order based on increasing pH:

$$Pb\ (K_1 = 6.45) > Cu\ (7.73) > Zn\ (8.96) > Ni\ (10.0)$$

Values of K_1 are from Smith and Martell [24]. Corapcioglu and Huang [23] reported that the initiation of metal pH-adsorption edges followed this type of trend for a variety of commercial activated carbons. While this may be an oversimplification of a complex phenomenon, it demonstrates the effect metal chemistry has on cationic metal removal. For anionic and select cationic metals the removal phenomenon is more complex. For example, Huang and Blankenship [25] reported that Hg(II) is adsorbed by activated carbon and is reduced to Hg^o. Once in the Hg^o form, mercury volatilized.

The concentration of the metal ion affects the removal mechanism as well as the metal surface loading (discussed in next section). If the concentration of the cationic metal is large enough such that $Me(OH)_{2(s)}$ forms, then surface precipitation can occur. If the metal concentration is less than the metal's solubility, then the primary removal mechanisms are sorptive in nature (physical/chemical adsorption, ion exchange). The calculation of the metal surface concentration (X/M, mg metal/g carbon) at pH values that equal or exceed the pH_{prec} is difficult because theoretically no carbon is required for metal removal (see no-carbon removal curves in Figure 7.1). Metal surface concentrations under these conditions increase rapidly because the denominator of the surface concentration expression approaches zero while removal approaches 100%.

Surface Loading

As with organic adsorbates, increasing the adsorbate surface loading (initial mass of metal per gram of carbon) results in an increase in the adsorbates' aqueous and surface phase concentrations. An example of this phenomenon is demonstrated in Figure 7.4 where the Cd surface concentration ($\Gamma \equiv X/M$, mg Cd/g carbon) versus the aqueous phase Cd concentration is presented for Hydrodarco B (Norit Americas, Inc., Atlanta, GA, USA) [26]. The solid lines in Figure 7.4 represent results from the Freundlich isotherm adsorption model. The effect of pH on metal removal is also evident from Figure 7.4.

Complexing Ligands, Competing Adsorbates, Ionic Strength and Temperature

The presence of complexing ligands and competing adsorbates can alter metal removal from that observed in the metal-only system. A metal ion that has been complexed in solution may adsorb more strongly, weakly or the same as an uncomplexed metal species. Factors determining the effect ligands have on metal adsorption include type and concentration of ligand and metal, carbon type and solution pH. In systems with more than one adsorbate, competition between the adsorbates for surface sites may occur. The degree of competition is dependent on the type and concentration of the competing ions, number of surface sites and the affinity of the surface for adsorbate. Changes in ionic strength affect metal adsorption by altering aqueous metal chemistry and the structure of the electric double layer surrounding the carbon surface. Similarly, temperature influences the degree to which both aqueous and solid-phase reactions proceed. In the remaining portion of this section, four scenarios are forwarded that demonstrate the effect influent characteristics have on the pH-adsorption edge. The general shape of the pH-adsorption edge for each scenario, including one representing a ligand-free (i.e., metal-only) system are presented in Figure 7.5 [27]. For each scenario, experimental results will be presented. While useful information can be gleaned from the experimental results presented in this section, the effectiveness of activated carbon in removing heavy metals must be ascertained on a case-by-case basis by performing treatability studies on the actual wastewater (as is done for organic adsorbates).

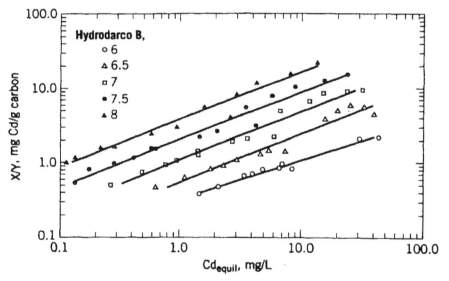

Figure 7.4 Cd surface concentration ($\Gamma = X/M$) versus equilibrium aqueous phase Cd concentration for Norit Americas Hydrodarco B. The solid line represents Freundlich isotherm model [26].

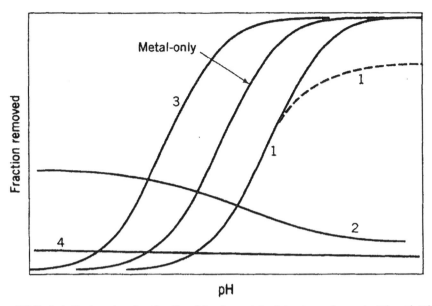

Figure 7.5 Typical pH-adsorption edges for a ligand-free (i.e., metal-only) system and scenarios 1 through 4 [27].

SCENARIO 1

For Scenario 1, the general shape of the pH-adsorption edge remains relatively unchanged compared to the metal-only/ligand-free system, except that the adsorption edges move to a different pH region. A leveling off of the pH-adsorption edge at less than 100% metal removal may also be observed (dotted line in Figure 7.5). Altering the adsorbate surface loading (initial mass of metal per gram of carbon), the presence of complexing ligands and competing adsorbates, and changes in the ionic strength and temperature can cause this behavior. Increasing the metal surface loading decreases the number of surface sites per unit of metal ion. This behavior was illustrated in Figures 7.1 and 7.2 where the pH-adsorption edges shifted as the adsorbate/adsorbent ratio was changed. Organic and inorganic ligands can form complexes with the metal ion decreasing the amount of free metal ions available for interaction with the solid surface. In Figure 7.6, Ni pH-adsorption edges are presented for F400 in the absence and presence of orthophosphate [27]. The addition of orthophosphate shifted the pH-adsorption edge to higher pH values (i.e., decreased adsorption). In Figure 7.7, the Cd pH-adsorption edge is presented for Hydrodarco (HD) KB (Norit Americas, Inc., Atlanta, GA, USA) in the absence and presence of EDTA [27]. The presence of EDTA caused pH-adsorption edge to shift slightly and level off at higher pH values. Kuennen et al. [28] reported that the presence of CO_3^{2-} can reduce Pb adsorption through the formation of the less adsorbable $PbCO_{3(aq)}$. The presence of a competing adsorbate ("secondary adsorbate") can also affect the removal of the primary metal. Netzer and Hughes [18] studied the effect of competing metal ions (Cu and Co) on Pb adsorption onto ten activated carbons. Pb was two and ten times more adsorbable than Cu and Co, respectively. Pb adsorption was affected more by the presence of Cu than by Co because Cu competed more strongly for adsorption sites than Co. When Co and Cu were both present, Pb adsorption decreased to a greater extent than when Cu and Co were present separately. The

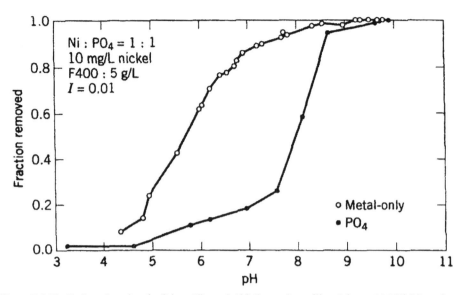

Figure 7.6 Ni pH-adsorption edges for Calgon Filtrasorb 400. Comparison of ligand-free and 1:1 Ni:PO$_4$ molar ratio systems [27].

presence of a second adsorbate does not always decrease metal removal. In Figure 7.8, pH-adsorption edges for Hydrodarco KB and 10 mg/L Ni in the presence of 0, 5 and 10 mg/L Cd are presented [27]. Ni removal was not affected by the presence of Cd. Similar behavior was observed for Cd removal in the presence of Ni. Boomhower (29) reported that the presence of catechol and sulfosalicylic acid shifted the Cd-F400 pH-adsorption to a lower pH region (increased removal). Huang [30] reported that the presence of organic adsorbates did not adversely affect the removal of several heavy metals.

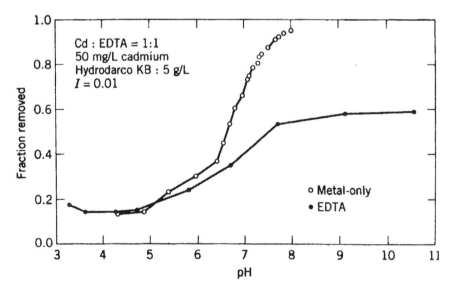

Figure 7.7 Cd pH-adsorption edges for Norit Americas Hydrodarco KB. Comparison of ligand-free and 1:1 Cd:EDTA molar ratio systems [27].

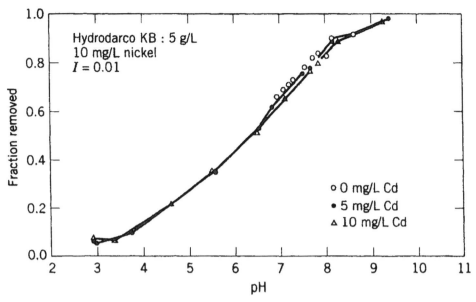

Figure 7.8 Ni pH-adsorption edges for Norit Americas Hydrodarco KB at 10 mg/L Ni and 0, 5 and 0.1 mg/L Cd. Effect of competing metals [27].

Changes in the ionic strength will alter the metal's aqueous chemistry and the electric double layer surrounding the carbon surface. At high ionic strengths, the carbon surface is "swamped" by the electrolyte, and access to the surface is made more difficult. In Figure 7.9, the Cd pH-adsorption edges for Hydrodarco KB at ionic strengths of 0.01 and 0.1 are presented [27]. As the ionic strength increased, metal removal decreased. Similar results were reported by Huang and Smith [31] for Cd on Nuchar SN and SA. The effect of ionic strength is important because a number of metal-bearing waste streams (e.g., landfill

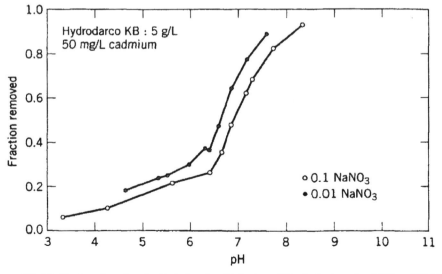

Figure 7.9 Cd pH-adsorption edges for Norit Americas Hydrodarco KB at ionic strengths of 0.01 and 0.1 [27].

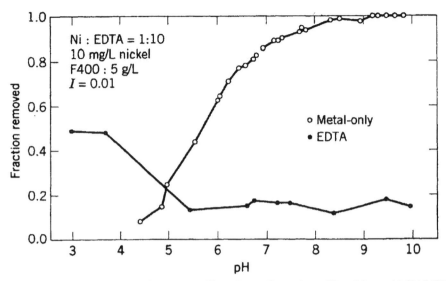

Figure 7.10 Ni pH-adsorption edges for 1 Calgon Filtrasorb 400. Comparison of ligand-free and 1:10 Ni:EDTA molar ratio systems [27].

leachates, metal-plating wastes) contain high amounts of total dissolved solids. The effect of temperature on metal removal by activated carbon has not been studied in great detail. Several researchers [23,28] reported that unlike organic adsorbates, removal does not decrease with increasing temperature. If adsorption is an exothermic reaction, as most researchers believe, then other removal mechanisms must be operative. For example, precipitation is often an endothermic reaction; thus, removal by this mechanism would increase with increasing temperature. Corapcioglu and Huang [23] reported that the Cu pH-adsorption edges shifted to lower pH values (increased removal) as temperature was increased from 25° to 102°C. Removal was 100% for all temperatures at pH = 6 (pH_{prec}), so it is unlikely that surface precipitation was the predominant removal mechanism. The shedding of the secondary hydration sheaths and the formation of the more adsorbable $Cu(OH)^-$ may occur to a greater extent at higher temperatures causing the pH-edges to shift toward low pH values.

SCENARIO 2

In Scenario 2, the metal-ligand complex adsorbs via a ligand-solid interaction (or equivalently, the ligand adsorbs on the solid, and the metal ion reacts with the solid-ligand complex). The shape of the pH-adsorption edge is radically different from that observed in the metal-only/ligand-free system. Generally, the adsorption of complexed cationic metals increases at lower pH values and decreases at higher pH values compared with the metal-only/ligand-free case. An example of this phenomenon is provided in Figure 7.10 where the Ni pH-adsorption edge is presented for F400 at a Ni:EDTA molar ratio of 1:10 [27]. The pH_{zpc} for F400 was reported to be 10.4 [12]. Below a pH of 10.4, the surface has a net positive charge, and above a pH of 10.4, the surface has a net negative charge. The Ni-EDTA complex is negatively charged over a wide pH range. Thus, at lower pH values, the electrostatic force is attractive. As the pH increases, this attractive force decreases until it becomes repulsive at pH values greater than 10.4. As mentioned previously, metal adsorption cannot be explained solely by electrostatic theory. Chemical interactions between the carbon surface and a particular metal-ligand complex can also vary with pH. Rubin and Mercer [32] studied the effect of EDTA and 1,10 phenanthroline on Cd adsorption. At low Cd:carbon ratios, EDTA enhanced adsorption, but as the amount of EDTA was increased, removal was adversely affected. The presence of 1,10 phenanthroline promoted Cd removal.

SCENARIOS 3 AND 4

In Scenario 3, unique complexation between the metal, ligand and carbon surface causes an increase in metal removal at all pH values compared with the metal-only system. The general shape of the pH-adsorption edge is similar to the metal-only/ligand-free system. Scenario 3 behavior is rarely observed. In Scenario 4, adsorption is suppressed at all pH values. The presence of a strong metal complexing agent (e.g., EDTA) or a large amount of a weak complexing agent or competing adsorbate can cause this behavior. The presence of a complexing agent reacts with the metal ion causing a decrease in removal provided the chelated metal complex is not adsorbed. An example of this phenomenon is presented in Figure 7.11 for Hydrodarco KB-Cd-EDTA system [27]. The high concentration of EDTA caused a decrease in metal removal compared to the metal-only/ligand-free system. At lower pH values, there is a small amount of removal, indicating limited Scenario 2 behavior. Huang et al. [33] reported that at EDTA:Co(II) molar ratios greater than 1:1 cobalt adsorption was almost completely suppressed.

Carbon Type

Netzer and Hughes [18] and Corapcioglu and Huang [23] conducted cationic metal removal studies on 10 and 14 commercially available activated carbons, respectively. In Fig-

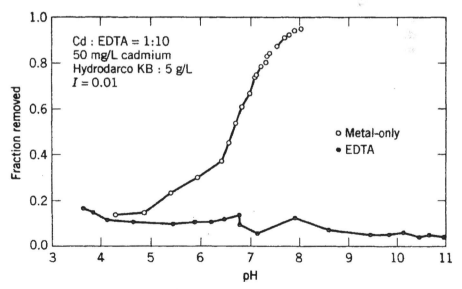

Figure 7.11 Cd pH-adsorption edges for Norit Americas Hydrodarco KB. Comparison of ligand-free and 1:10 Cd:EDTA molar ratio systems [27].

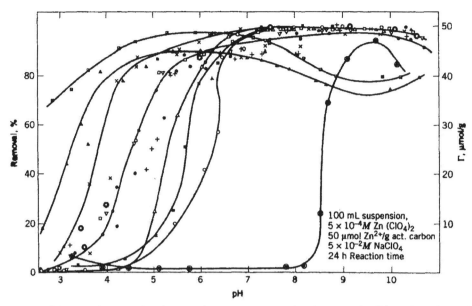

Figure 7.12 Zn pH-adsorption edges for several commercial activated carbons. Reprinted from *Water Research* 9(9), M. O. Corapcioglu and C. P. Huang, "The Adsorption of Heavy Metals onto Hydrous Activated Carbon," 1031–1044, copyright 1987 [23], with kind permission from Elsevier Science Ltd., The Boulevard, Langford Lane, Kidlington OX5 1GB, UK.

ure 7.12, Zn pH-adsorption edges for the carbons investigated by Corapcioglu and Huang [23] are presented. Each symbol represents a different carbon type. Both research groups concluded that there was a large variability in the ability of the carbons to adsorb metals, but removal could not be correlated to any one carbon property. Several researchers have proposed that metal removal can be quantitatively predicted based on whether a carbon is L-type (acidic nature) or H-type (basic nature). The more acidic the carbon, the lower the pH at which it deprotonates and takes on a negative surface charge (i.e., lower the pH_{zpc}). A negatively charged surface favors cation removal. The opposite behavior occurs for H-type carbons—deprotonation occurs at alkaline pH values, and removal of anions [e.g., Cr(VI), Se(IV), Pb-EDTA-2, etc.] is favored. While these rules-of-thumb are useful, adsorption studies are necessary if the optimum carbon for a given water/wastewater is to be selected.

SCF MODELING OF ACID-BASE BEHAVIOR AND HEAVY METAL REMOVAL

Acid-Base Behavior

The acid-base and metal removal behavior of several commercial-activated carbons has been modeled successfully by the Surface Complex Formation (SCF) model [9,13,23]. The basic premise of the surface complex formation model is the use of mass action laws to describe ion interactions with the hydrous solid surface. The surface of the hydrous solid acquires a surface charge, due to various surface groups or sites, and an electric double layer (EDL) develops around the charged particle. Thus, the total free energy of interaction is defined by the sum of the free energy of the chemical interaction at the solid surface and an electrostatic component. In the past, it has been assumed that all surface sites are capable of binding and releasing protons equally, thereby, the solid can be modeled as a single, weak diprotic acid. With this assumption, the surface functional groups can be represented by the following surface reactions:

$$\equiv SOH_2^+ \leftrightarrow \ \equiv SOH + H_s^+ \quad K_{a1} \tag{1a}$$

$$\equiv SOH \leftrightarrow \ \equiv SO^- + H_s^+ \quad K_{a2} \tag{1b}$$

where the symbol ($\equiv$) distinguishes surface complexes from soluble complexes. K_{a1} and K_{a2} are the first and second surface acidity constants defined by the reactions (1a) and (1b), respectively. H_s^+ is the activity of the proton at the solid surface. The Boltzman equation is used to relate the activity of a solute in the bulk solution to the activity at the surface.

$$\{H_s^+\} = \{H_b^+\} \ \exp(-F\Phi/RT) \tag{2}$$

where R, F, and T are the gas constant, Faraday constant, and absolute temperature, respectively. Φ is the potential difference across the electric double layer, measured in volts. Substituting the Boltzman equation into the mass action laws and assuming that the activity coefficients for the surface complexes are equal [34], and the activity coefficients for the bulk and surface activities are equal [35] yields,

$$K_{a1} = \frac{[\equiv SOH]\{H_b^+\}}{[\equiv SOH_2^+]} \ \exp(-F\Phi/RT) \tag{3a}$$

$$K_{a2} = \frac{[\equiv SO^-]\{H_b^+\}}{[\equiv SOH]} \ \exp(-F\Phi/RT) \tag{3b}$$

where [ ] and { } represent molar concentrations and activities, respectively, and the exponential term can be considered as a correction term included in the mass action laws to account for the electrostatic interactions. The material balance equation for surface sites is

$$N_s = [\equiv SOH_2^+] + [\equiv SOH] + [\equiv SO^-] \tag{4}$$

where N_s is the total concentration of a site in moles/L. N_s can be expressed in moles/g or moles/cm^2 if the concentration and specific surface area of the solid are known.

Snoeyink and Weber [36] have reported that phenolic and lactone functional groups may be responsible for the amphoteric behavior of activated carbon, while Mattson and Mark [37] have suggested carboxyl and quinone groups. Regardless of the specific functional group(s) present on the surface of the carbon, it would seem plausible to model the carbon surface as a number of weak monoprotic acids rather than a single diprotic acid. There are two types of monoprotic sites to consider if the charge of the carbon surface is to be properly modeled, a positively charged site,

$$\equiv P^i OH_2^+ \leftrightarrow \ \equiv P^i OH^0 + H_s^+ \quad K_a^{p_i} \tag{5a}$$

and a negatively charged site,

$$\equiv N^i OH^0 \leftrightarrow \ \equiv N^i O^- + H_s^+ \quad K_a^{n_i} \tag{5b}$$

where i is an index to differentiate between sites. Each of these reactions will have associated with it a K_a and N_s. The acidity constants for the two types of reactions are as follows:

$$K_a^{p_i} = \frac{[\equiv P^i OH^0]\{H_b^+\}}{[\equiv P^i OH_2^+]} \ \exp(-F\Phi/RT) \tag{6a}$$

$$K_a^{n_i} = \frac{[\equiv N^i O^-]\{H_b^+\}}{[\equiv N^i OH^0]} \ \exp(-F\Phi/RT) \tag{6b}$$

For a two-monoprotic site representation, there is one of each type of site. For the three and four monoprotic representations, there can be several combinations of Equations (6a) and (6b); however, there must be at least one of each type of site to account for the change in sign of the surface charge with pH.

The potential can be related to the surface charge through a number of different electric double-layer models, such as the constant capacitance model, the Gouy-Chapman diffuse layer model and the triple-layer model [38]. The constant capacitance model requires one capacitance, while the triple-layer model requires two. While capacitances have been estimated experimentally for metal oxides, there has been no such work using PAC. Westall and Hohl [38] have reported that all of these electrostatic models represent data equally well but with different corresponding parameters. Thus, the Gouy-Chapman model, which does not require a specified capacitance, is often chosen to represent the EDL of activated carbons. The Gouy-Chapman diffuse-layer model is represented by the following equation:

$$\sigma = 0.1172 I^{1/2} \sinh \ (zF\Phi_o/2RT) \tag{7}$$

where σ is the surface charge in coulombs per surface area (C/m^2), Φ_o is the potential at the surface of the solid, I is the ionic strength of the medium and z is the valence of the background electrolyte. If Φ_o is approximately equal to the potential at the plane of the closest approach of adsorbed ions, then Φ_o is also the potential difference across the electric double layer in Equations (3) and (6).

The pH-σ-Φ relationship is determined experimentally by performing acidimetric-alkalimetric titrations as described by Huang [9]. The surface charge-pH relationship is calculated experimentally using the following expression:

$$\sigma = \frac{(\Delta eq)(96,500 \ C/equivalent)}{S_a C_s V} \tag{8}$$

where S_a is the specific surface area (m^2/g), C_s is the concentration of the solid (g/L), V is the volume titrated (L) and (Δeq) is the amount of acid/base added in equivalents, at each pH value. (Δeq) is taken from the net titration curve corrected for the pH$_{zpc}$ (see Reference [9] for details of the pH$_{zpc}$ correction procedure).

Various mathematical methods exist by which values of the surface acidity parameters can be determined. Huang [9] presented a graphical method for the single diprotic representation, while Westall [39] developed a statistical parameter optimization procedure entitled FITEQL, by which surface parameters could be determined for various surface site representations. FITEQL determines the values of the surface acidity constants and N_s based on minimizing the sum of squares of errors in the titration data. Only the FITEQL results will

TABLE 7.3. Properties of Activated Carbons Used in Acid-Base Modeling Exercise.

Carbon	Ash, %	S_a, m^2/g	Density, kg/m^3	pH$_{zpc}$	Particle Size Distribution		
					100 mesh	200 mesh	325 mesh
Darco HDB	NA	600–650	500	7.45	99	95	90
Nuchar SN	3–5	1,400–1,800	350	5.35	95–100	85–95	65–85

be presented here. The reader is referred to Reed and Matsumoto [13] for a comparison of the results from the graphical and FITEQL approaches.

The acid-base behavior of two commercially available carbons [Darco HDB (Norit Americas, Inc., Atlanta, GA, USA) and Nuchar SN (Westvaco Co., New York, NY, USA)] and the applicability of the SCF modeling approach using the multiple monoprotic site approach will be presented. Selected properties of the two carbons are given in Table 7.3. Average values for surface area were used in model calculations. The values of pH$_{zpc}$ (see Figure 7.13) were determined using the salt titration method described by Yates and Healy [10]. Both carbons underwent pretreatment as described by Huang [9] and were then titrated as described in Reed and Matsumoto [13].

The net titration curves corrected for pH$_{zpc}$ are presented in Figure 7.14 for both carbons at $I = 10^{-1}$ [40]. The surface of Nuchar SN will have a net negative charge at pH values >5.35, while the surface of Darco HDB will be negative at pH values >7.45. Complicating the selection of which acid-base representation to use is the fact that as the number of parameters used to describe the data increases, a better fit may result regardless of whether the representation is physically correct. Thus, the following surface site representations, with the number of adjustable parameters indicated in parentheses, were investigated: single diprotic (three), two diprotic (six), two monoprotic (four), and three monoprotic (six).

For Darco HDB, the two diprotic and three monoprotic (two positive sites, one negative site) representations are practically identical and describe the titration data adequately. Each of these two representations has six adjustable parameters. Thus, on a purely stochastic basis, their ability to fit the data should be similar. The single diprotic and two monoprotic representations did not reproduce the titration curve at pH values below 7.

For Nuchar SN, only the three-monoprotic representation (one positive site, two negative sites) adequately described the titration data. The two-diprotic representation, which has the same number of adjustable parameters as the three-monoprotic representation, failed to simulate the data at lower pH values. For a given diprotic site, N_s for the positive and negatively charged species is, by theory, equal. For Nuchar SN, a large value of N_s for the second diprotic site was required for the surface charge (i.e., titration curve) at higher pH values to be modeled. The higher values of N_s resulted in the overprediction of the amount of acid adsorbed at the lower pH values. Thus, when dealing with charged surfaces, it is possible for models having the same number of adjustable parameters to give radically different fits to the data.

The choice of which surface site representation to use cannot be based strictly on the goodness of the model fit to experimental data. Other criteria such as the coherence of the chosen representation with literature results must be considered. For Darco HDB, the three monoprotic and two diprotic gave similar results. For Nuchar SN, only the three monoprotic representation adequately simulated the titration data. Based on literature results, the acid-base behavior of activated carbons can be attributed to a number of monoprotic sites

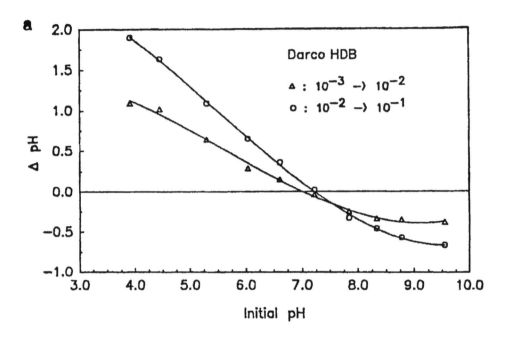

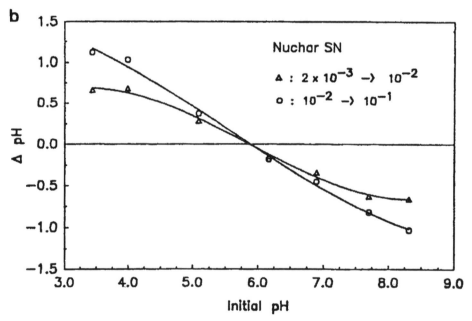

Figure 7.13 NaNO₃ titration curves for (a) Darco HDB and (b) Nuchar SN [13].

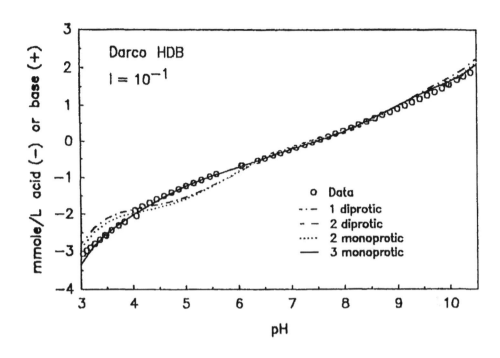

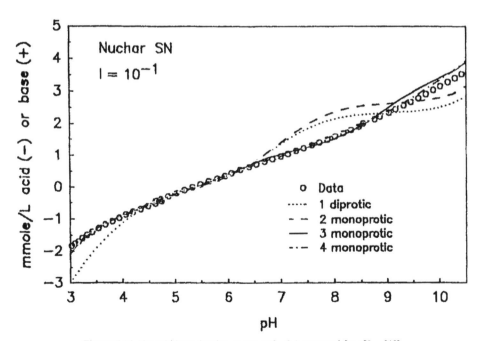

Figure 7.14 Net acid-base titration curves at $I = 0.1$ corrected for pH_{zpc} [40].

223

[36,37]. Thus, the three monoprotic site representation would seem to be the most plausible surface representation for the two PACs studied.

A summary of pK_a and N_s values for the three monoprotic representation for both carbons studied is given in Table 7. 4. In Figure 7.15, speciation diagrams for the two carbons at $I = 0.1$ are presented [20]. The speciation diagram for an individual surface group is similar to that of a weak acid in solution. The values of N_s vary with ionic strength, but the total number of surface sites (ΣN_s) for each carbon were approximately 4×10^{-4} mole/L/g Darco and 3.6×10^{-4} mole/L/g Nuchar. If the triple layer model is used to represent the electric double layer, then theoretically a single set of pK_a values across differing ionic strengths can be determined. Also, it should be pointed out that the discrete monoprotic sites represent a class of surface sites with the calculated pK_a acting as a typical acidity constant. Nondiscrete models, such as those used for humic acid, could also be used to describe the titration data.

SCF Modeling of Heavy Metal Removal

SCF modeling of heavy metals by activated carbon will be demonstrated using Cd and the two activated carbons used in the acid-base modeling exercise. The surface groups that are responsible for the coordination and release of H+ are also assumed to be responsible for the coordination of heavy metals. Development of the SCF model to account for heavy metal adsorption is relatively straightforward. If only monodentate surface reactions are considered the cadmium-surface reactions can be written as follows:

$$\equiv P^i OH_2^+ + Cd_s^+ + mH_2O \leftrightarrow [\equiv P^i OH - Cd(OH)_m]^{2-m} + (m+1)H_s^+ \quad K_{i,m+1}^p \quad (9a)$$

$$\equiv N^i OH^0 + Cd_s^{2+} + mH_2O \leftrightarrow [\equiv N^i O - Cd(OH)_m]^{1-m} + (m+1)H_s^+ \quad K_{i,m+1}^n \quad (9b)$$

for $m = 0, 1, 2, 3$. Adsorption of heavy metals by hydrous solids is strongly dependent on the pH of the solution, and pH has been identified as the variable governing metal adsorption onto hydrous solids [41]. Equations (9a) and (9b) reflect this pH dependence. Writing the cadmium-surface complexation constants in terms of $\equiv P^i OH^0$, $\equiv N^i OH^0$ and Cd_s^{2+} yields

$$K_{i,m+1}^p = \frac{[\equiv P^i OH - Cd(OH)_m]^{2-m}\{H_b^+\}^m}{[\equiv P^i OH^o][\{Cd_b^{2+}\}} \exp(-\Delta z \Phi F / RT) \quad (10a)$$

$$K_{i,m+1}^n = \frac{[\equiv N^i O - Cd(OH)_m]^{1-m}\{H_b^+\}^{m+1}}{[\equiv N^i OH^0]\{Cd_b^{2+}\}} \exp(-\Delta z \Phi F / RT) \quad (10b)$$

TABLE 7.4. Values of pK_a and N_s for the Three-Monoprotic Site Model.

Carbon	I	$pK_a^{p_1}$	$pK_a^{p_2}$	$pK_a^{n_1}$	$pK_a^{n_2}$	$N_s^{p_1}$	$N_s^{p_2}$	$N_s^{n_1}$	$N_s^{n_2}$
						10^{-4} mole/g Carbon			
Darco HDB	10^{-1}	4.94	6.73	8.47	—	1.57	0.89	1.64	—
Darco HDB	10^{-2}	5.51	7.62	8.04	—	0.87	0.86	21.7	—
Darco HDB	10^{-3}	6.00	8.40	7.75	—	0.49	0.81	2.84	—
Nuchar SN	10^{-1}	4.61	—	6.9	8.56	1.08	—	1.22	2.20
Nuchar SN	10^{-2}	4.64	—	6.24	8.41	0.81	—	1.13	1.98
Nuchar SN	2×10^{-3}	5.63	—	5.99	7.76	0.22	—	0.70	1.44

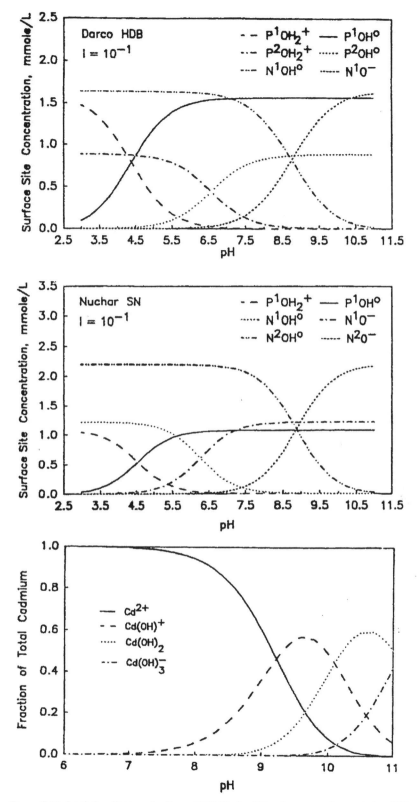

Figure 7.15 Speciation diagrams for Darco HDB. Nuchar SN and aqueous-phase cadmium [20].

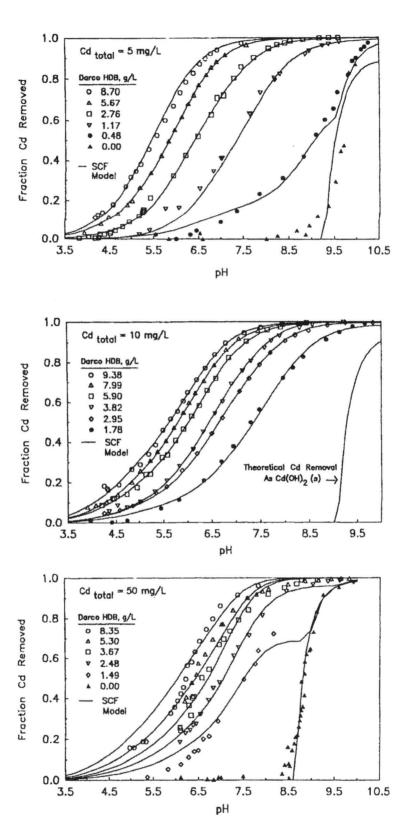

Figure 7.16 Darco HDB cadmium pH-adsorption edges. Solid lines are SCF model predictions [20]

226

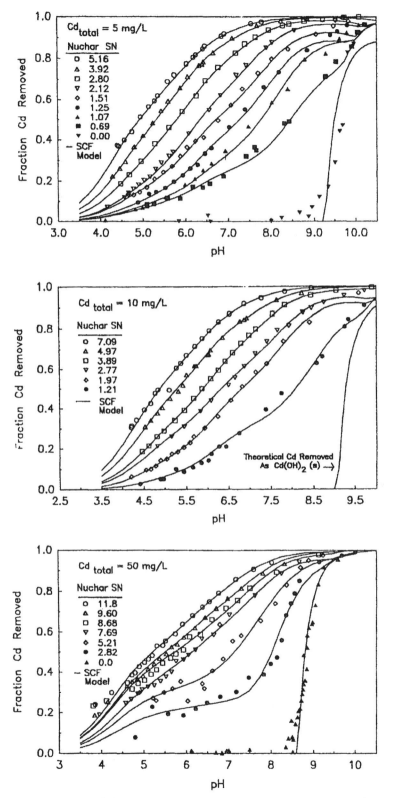

Figure 7.17 Nuchar SN cadmium pH-adsorption edges. Solid lines are SCF model predictions [20].

for $m = 0, 1, 2, 3$. The value of Δz is the net change in the charge number of the surface species determined from Equations (9a) and (9b). Polydentate reactions are also possible. The overall stoichiometric coefficient (N), defined as the moles of H^+ released per mole of metal adsorbed, is an indicator of whether polydentate reactions are important. Results from the literature for metal oxides [42] indicate that values of N are between 1 and 2, indicating that polydentate reactions may not be operative for these solids. The value of N at $I = 0.1$ for Darco and Nuchar carbons varied between 0.89–1.24 and 0.9–1.1, respectively. Because a value of N close to 1.0 indicates that monodentate reactions predominate, polydentate reactions were not considered for this modeling exercise.

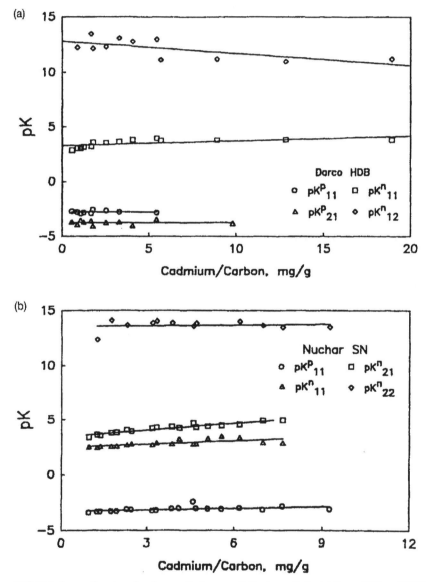

Figure 7.18 Cadmium-surface complexation constants versus Cd/carbon ratio for (a) Darco HDB and (b) Nuchar SN [20].

Cadmium pH-adsorption edges for Darco HDB and Nuchar SN are presented in Figures 7.16 and 7.17, respectively [20]. Removal of cadmium from solution by precipitation (i.e., absence of carbon) is also presented in Figures 7.16 and 7.17. Removal by precipitation was determined experimentally for cadmium concentrations of 5 and 50 mg/L and theoretically for 10 mg/L Cd using constants from Westall et al. [43]. Cadmium precipitation begins at about pH 9.2 for a 5 mg/L Cd solution, 9.0 for a 10 mg/L Cd solution and at about 8.6 for 50 mg/L Cd solution. For the majority of the cadmium/carbon ratios studied, the fraction of cadmium removed was close to 1.0 prior to the solution pH reaching the pH at which solution precipitation occurred.

The cadmium-surface reactions responsible for cadmium adsorption can be deduced by comparing the speciation diagrams of soluble cadmium and the carbon surface (see Figure 7.15) and the cadmium removal curves (see Figures 7.16 and 7.17). Reactions involving soluble cadmium species or surface species when either is present in small amounts are eliminated. Reactions corresponding to ($i = 1$, $m = 0$) and ($i = 2$, $m = 0$) in Equation (9a) and ($i = 3$, $m = 0,1$) in Equation (9b) were used in the SCF model to describe cadmium adsorption by Darco HDB. For Nuchar SN reactions corresponding to ($i = 1$, $m = 0$) in Equation (9a) and ($i = 2$, $m = 0$) and ($i = 3$, $m = 0,1$) in Equation (9b) were used. The solid lines in Figures 7.16 and 7.17 are SCF model simulations of the cadmium pH-adsorption edges using these surface reactions. Cadmium-surface complexation constants are presented in Figure 7.18 versus the cadmium/carbon ratio [20]. The SCF model simulated the experimental data fairly well. Precipitation of cadmium as cadmium hydroxide was allowed to occur in the SCF model and the predicted amounts of cadmium removed by precipitation and adsorption were added together for the total fraction of cadmium removed. The 5 mg/L Cd–0.48 g/L Darco HDB, 5 mg/L–0.69 g/L and 10 mg/L–1.21 g/L Nuchar SN were the only systems in which precipitation definitely played a major role in removal of cadmium from solution.

METAL REMOVAL KINETICS

The consensus of the literature is that metal removal kinetics are rapid relative to the adsorption of organic compounds. Factors affecting metal removal kinetics include metal and carbon type, adsorbate/adsorbent ratio (i.e., surface loading), pH and carbon particle size. Netzer and Hughes [18] reported that it took 30 minutes for complete removal for Cu and Pb and 2 hours for the complete removal of Co. Time to equilibrium was increased to 2 hours when all three metals were present. Bhattacharys and Venkobachar [44] reported that Cd removal by two bituminous coals was complete within 24 hours and followed a first-order reversible kinetic relationship. Reed and Matsumoto [20] reported that the adsorption of Cd by Hydrodarco KB and Nuchar SN was rapid for low adsorbate/adsorbent ratios, but the time to equilibrium increased as the adsorbate/adsorbent ratio increased. Netzer and Hughes [18] reported similar results, but Huang [45] reported that the adsorbate/adsorbent ratio had no effect on kinetics for Cd and several activated carbons. Reed and Matsumoto [20] also reported that the time to equilibrium increased as the pH increased, most likely because surface precipitation became more important (precipitation reactions are slower than adsorption reactions). Huang and Blankenship [25] reported that the Hg(II) adsorption was very fast (within minutes), but Hg(II) reduction/volatilization followed a linear $t^{1/2}$ relationship. Pandey and Chaudhuri [46] also reported that Hg(II) removal kinetics were rapid. Rao et al. [47] reported that Cr(VI) removal was complete within 70 minutes and followed a first-order rate law. Rajakovic [48] reported that As(III) and As(V) adsorption was 50%

complete within 30 minutes and was 100% complete after about 4 hours. Wilczak and Keinath [49] reported that the rate of Cu and Pb adsorption occurred in two distinct steps. The first step was rapid and was followed by a prolonged second step that lasted several days. Because most metal-carbon kinetic studies have been conducted over a short time period, the equilibrium capacity of carbons may have been underestimated. These same authors reported that desorption of carbon-bound metals was very rapid—essentially complete after the first few minutes. There are contradictory results on whether film or pore diffusion is the rate-limiting step. Arulanantham et al. [50] reported that adsorption of Cd and Pb by a coconut shell carbon and a commercial carbon was film diffusion controlled. Bhattacharya and Venkobachar [44] reported that film diffusion was the rate limiting step for the removal of Cd by two bituminous coal carbons. Rao et al. [47] reported that the rate limiting step in Cr(VI) removal was pore diffusion. Huang and Wirth [51] investigated the effect of particle aggregation on removal kinetics and reported that as the size of the carbon particle increased, interparticle diffusion became more important.

HEAVY METAL REMOVAL BY GAC COLUMNS

The majority of heavy metal-activated carbon research has been conducted in the batch mode (equilibrium and kinetic studies). While results from batch studies are useful in understanding the basic phenomena that occur when metals are in contact with activated carbon, it is difficult to determine how activated carbon will perform in a dynamic process such as a GAC column. The use of GAC columns for metal removal represents the most likely treatment scenario for actual waters and wastewaters. The focus of this section will be to present results from heavy metal-GAC column studies. The majority of the results are from treatability studies that employed synthetic wastewaters. While there have been several successful treatability studies using actual wastewaters, the companies that performed these studies have not allowed the results to be made public. The design of a GAC column for the treatment of heavy metal-bearing wastewater is similar to the process used for organic adsorbates (a detailed description of the design process is presented by Clark and Lykins [52]). Following a general overview of how a GAC column treatability study is conducted, literature results will be presented based on metal type.

Conducting a GAC Column Treatability Study

Because metal removal is highly dependent on pH, the dependence of removal on pH must be determined prior to column testing. The optimum pH for metal removal can be determined from experimental pH-adsorption edges. The pH of the influent should be adjusted so as to maximize the efficiency of the column. Care should be taken to account for any precipitates that may form as these can foul the column.

The most reliable method for determining if GAC columns are a feasible treatment option for a particular wastewater is to conduct column breakthrough studies. Water or wastewater is introduced to the GAC column, and effluent contaminant(s) concentration is monitored versus the volume of liquid treated (or time). The resulting graph is referred to as a "breakthrough" curve. Wastewater volume can be presented by a true volume or a normalized volume (number of bed volumes treated = volume of wastewater/empty bed volume of column). When the effluent contaminant concentration exceeds a predetermined value, breakthrough has occurred. The breakthrough concentration is determined by regulatory or down-line treatment concerns. When the effluent concentration is equal to the influent concentration, the column is said to be exhausted. Operationally, exhaustion is often taken as $C_e = 0.95 C_o$. Process performance parameters used to ascertain the effectiveness of the GAC treatment process include volume of liquid treated at breakthrough (V_b) and exhaustion (V_{exh}); contaminant surface concentration (mg contaminant/g carbon) at breakthrough (X/M_b) and exhaustion (X/M_{exh}); carbon usage rate (CUR = mass of carbon/V_b), degree of column utilization (DoCU = $X/M_b \div X/M_{exh} \times 100\%$), waste concentration factor (WCF = V_b/V_{res}, where V_{res} is the volume of residuals produced) and volume reduction factor [VRF = $(V_b - V_{res})/V_b$].

When designing GAC columns for organic contaminant removal, the empty bed contact time (EBCT) is a critical design parameter that influences the effectiveness and life of a GAC column. The EBCT is used by the activated carbon industry to represent the length of time a liquid stream is in contact with a granular activated carbon bed and thus, is related to the system's removal kinetics. The EBCT is defined as the time required for a fluid to pass through the volume equivalent of the media bed, without the media being present. Typically, the void space in a GAC column is about 45%, so the EBCT is about twice the true contact time between the liquid being treated and the GAC particles. For a given activated carbon, wastewater and effluent requirements (i.e., breakthrough concentration), a minimum EBCT (EBCT$_{min}$) exists that must be exceeded if the GAC column is to produce any volume of liquid that meets discharge standards. As the EBCT is increased, V_b, X/M_b and DoCU increase while CUR decreases. For systems where adsorption is the dominant removal mechanism, X/M_{exh} should not vary with EBCT because X/M_{exh} is determined by the influent contaminant concentration. At the optimal EBCT, additional increases in the EBCT will not cause further significant changes in V_b, X/M_b, DoCU and CUR. Thus, there is a trade-off economically between increases in capital costs (function of EBCT, column size) and decreases in operating costs (CUR, regeneration schedule). The optimal EBCT is a function of the type and concentration of the adsorbate, number of contaminants present, effluent requirements and carbon type. The hydraulic loading rate (HLR = volumetric flow rate/column cross-sectional area) affects the thickness of the hydrodynamic boundary layer and the headloss through the column. As the HLR is increased, the boundary layer thickness decreases, decreasing film resistance. Snoeyink [53] reported that for most applications, adsorbate transport through the liquid film is not the rate-limiting step. Cover [54] reported that the performance of columns operated at different HLRs but the same EBCT, was identical, provided that the EBCT was greater than EBCT$_{min}$.

Cadmium

Huang and Wirth [51] investigated the use of GAC columns and a continuous mixed flow PAC process for the removal of Cd(II). Only the results from the GAC column study will be presented here. A synthetic waste stream that mimicked a cadmium fluoborate plating wastewater was used. Characteristics of the synthetic wastewater were as follows: 10^{-4} M Cd (112 mg/L), 7×10^{-5} M NH$_4$BF$_4$, 5×10^{-5} M H$_3$BO$_3$. The wastewater pH was adjusted to 7 prior to its introduction to the column. The powdered activated carbon Nuchar SN (Westvaco) was binded into beads having a particle size that ranged from 0.425 mm to 0.833 mm and averaged 0.629 mm. Forty grams of the carbon beads were placed in a 2.54 cm inner diameter Plexiglas column. After packing, the column had a bed depth of 21.5 cm and an empty bed volume of 109 cm^3. The flow rate to the column was 43 cm^3/min, resulting in a HLR of 4.9 m/h (2 gal/min-ft^2) and an EBCT of about 2.5 minutes. Following exhaustion, the carbon was removed from the column and regenerated with 4.2 L of 1 M H$_2$SO$_4$

for 6 hours and then contacted with a NaOH solution until the pH of the carbon slurry was 7. Four treatment cycles were conducted using the same carbon. In Figure 7.19, Cd breakthrough curves are presented [51]. In Table 7.5, V_b and DoCU are presented. The DoCU is a measure of how much of the total capacity of the column is being utilized. Prior to breakthrough, the Cd concentration was less than the detection limit of 10^{-6} M (0.112 mg/L). About a 20% loss in treatment capacity occurred between column runs and may have been due to incomplete regeneration. The carbon was regenerated in the batch mode, and the Cd concentration gradient may not have been favorable. Performing a series of batch acid rinses or contacting the carbon with the acid in the column mode should provide more favorable concentration gradients. Even with the decrease in treatment capacity, 1,586 bed volumes (BVs) of Cd wastewater were treated to moderately low Cd levels. The authors performed a preliminary cost analysis on a single-stage adsorber and reported that the process was not cost effective if the carbon was discarded after a single pass through. Thus, the development of regeneration methods that do not adversely affect subsequent column performance are required. For example, if there was no loss in treatment capacity between runs, a total of 1,980 BVs of wastewater would be treated prior to breakthrough for the four runs. This represents an increase of about 25% over what actually was observed. In addition, the use of multiple-staged columns may improve column performance because the first column can be operated to exhaustion rather than breakthrough.

Chromium(VI)

Bowers and Huang [55] investigated the removal of Cr(VI) by Filtrasorb 400 (Calgon Co., Pittsburgh, PA, USA) as a function of carbon mass (bed depth or EBCT), influent Cr(VI) concentration and pH. For all column studies, the hydraulic loading rate was 2 gal/min-ft^2 (4.9 m/h). The influent Cr(VI) concentration and pH were 10^{-3} M (52 mg/L) and 2.5, respectively. In Figure 7.20, Cr(VI) breakthrough and effluent pH curves are presented

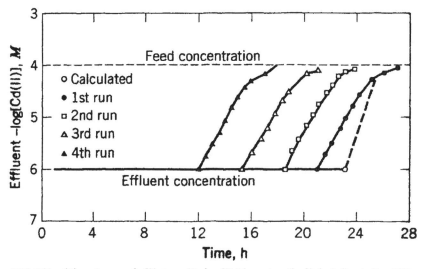

Figure 7.19 Cd breakthrough curves for Westvaco Nuchar SN. Flow rate = 43 ml/min. Influent pH and Cd concentration were 7 and 112 mg/L, respectively [51]. Reprinted from *ASCE J. of Environmental Engineering* 108(6), C. P. Huang and P. K. Wirth, "Activated Carbon for the Treatment of Cadmium Wastewater," 1280–1299, copyright 1982, with kind permission from ASCE.

TABLE 7.5. Bed Volumes Treated at Breakthrough and
the Degree of Column Utilization for Studies [51].

Run #	V_b (BV)	DoCU (%)
1	495	64
2	440	57
3	367	48
4	284	37
Total	1,586	—

for carbon masses of 10, 30 and 50 g [55]. The effluent column pH was initially elevated because F400 is an H-type carbon with a high pH_{zpc}. The elevated pH over about the first 100 bed volumes caused the effluent Cr(VI) concentration to be high (recall from earlier discussions that removal of anionic heavy metals decreases with increasing pH). As the basicity of the carbon was neutralized by the influent acidity, the effluent column pH and Cr(VI) concentration decreased. Cr(VI) removal was a strong function of carbon mass (note that for a constant column diameter and HLR, carbon mass, bed depth and EBCT are interchangeable). At 10 g F400, the lowest Cr(VI) effluent concentration observed was 10^{-4} M (≈ 5 mg/L). At higher bed depths, a significant volume of wastewater was treated to low concentrations of Cr(VI), although column exhaustion did not occur. To eliminate the high concentration of Cr(VI) at the beginning of the column run, the carbon was rinsed with 150 BVs of a pH = 2.5 liquid. In Figure 7.21, results from column experiments that employed acid rinsing are presented [55]. The effluent pH was constant at about 3, and there was no breakthrough of Cr(VI). However, Cr(III) was present in the effluent at a concentration of about

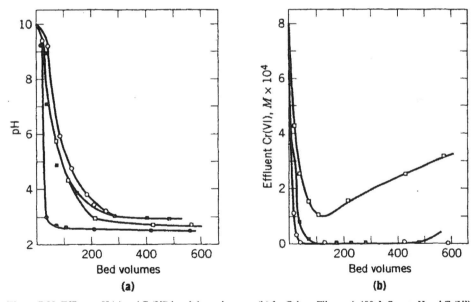

Figure 7.20 Effluent pH (a) and Cr(VI) breakthrough curves (b) for Calgon Filtrasorb 400. Influent pH and Cr(VI) concentrations were 2.5 and 52 mg/L, respectively. ● 0 mg/L (blank) Cr(VI), □ 10 g F400, ■ 30 g F400, ○ 50 g F400. Reprinted from *Prog. Water Science & Technology* 12, A. R. Bowers and C. P. Huang, "Activated Carbon Processes for the Treatment of Chromium (VI) Containing Industrial Wastewaters," 629–650, copyright 1980, with kind permission from Elsevier Science Ltd., The Boulevard, Langford Lane, Kidlington OX5 1GB, UK [55].

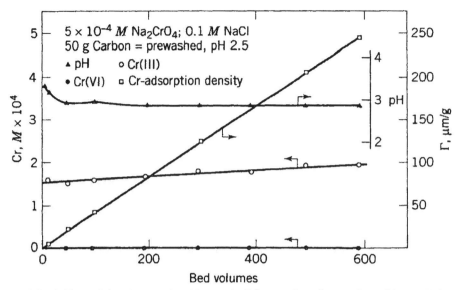

Figure 7.21 Cr(VI) breakthrough curves for acid prewashed Filtrasorb 400 Column. Influent pH and Cr(VI) concentrations were 2.5 and 26 mg/L, respectively. Reprinted from *Prog. Water Science & Technology 12*, A. R. Bowers and C. P. Huang, "Activated Carbon Processes for the Treatment of Chromium (VI) Containing Industrial Wastewaters," 629–650, copyright 1980, with kind permission from Elsevier Science Ltd., The Boulevard, Langford Lane, Kidlington OX5 1GB, UK [55].

1.5 to 2×10^{-4} M for the duration of the column experiment. Carbon-bound Cr(VI) is easily reduced to Cr(III) in an acidic environment and in the presence of activated carbon [56]. The production of Cr(III) was similar in the experiments for which acid rinsing was not employed. In an actual treatment system, Cr(III) would have to be removed using an additional process (e.g., precipitation, ion exchange or activated carbon with pH adjustment).

Chelated Heavy Metals

Shay and Etzel [57] investigated the removal of several heavy metals (Cu, Ni, Zn) that were chelated with citrate or EDTA. The effect of influent pH, metal:chelate molar ratio, EBCT, presence of hardness and heavy metal competition were investigated. The authors studied the effect of chelating agents on metal removal to determine if the use of activated carbon to treat metal plating wastewaters was feasible. An unnamed activated carbon ($pH_{zpc} = 8.5$) was added to a 50 ml burette such that the carbon bed depth was about 30.5 cm. The HLR was varied by adjusting the burette's stopcock so that the EBCT ranged between 5 and 25 minutes. Influent pH varied between 3 and 9. Influent metal concentrations were 1 mg/L, and metal:chelate molar ratios between 1:1 and 1:6 were employed. For citrate (and EDTA), increasing the metal:chelate molar ratio improved column performance for all three studied heavy metals. This is contrary to what has been generally reported in the literature—the presence of chelating agents reduces metal removal by activated carbon. Possible reasons for the improved removal include the following: (1) the activated carbon had specific surface sites that had a strong attraction for the chelated metal and (2) the electrostatic attraction component of the overall adsorption force was attractive. The carbon had a basic pH_{zpc} (at pH pH_{zpc} the surface is positively charged), and the metal-citrate complex is negatively charged over a wide pH range (e.g., at pH > 5.5 the predominant Cu species is

$CitCu^{-2}$). When the influent pH was lowered to 3, Cu removal decreased. The predominant metal complex at pH = 3 is Cu^{2+}, thus, the electrostatic force was repulsive. Removal increased with increasing EBCT (5 to 25 minutes), however, the column studies were not conducted to exhaustion, so it was difficult to determine the optimum EBCT.

For the three metals investigated, better adsorption was observed when the metals were chelated with EDTA compared to citrate. However, metal removal was not improved by increasing the metal:EDTA ratio above 1:1. EDTA is a strong metal complexing agent, and at a 1:1 metal:EDTA ratio, all the metal is chelated except at very low pH values. Thus, for this particular activated carbon, a high EDTA concentration should not affect column performance unless enough EDTA was added so that competition between EDTA and the heavy metal for the carbon surface became important. Column performance was poorest at pH = 3 and the following phenomena may have contributed to the decrease in removal: (1) H^+ competes for the EDTA, reducing the amount of EDTA-metal complex and (2) EDTA-metal species and the carbon surface can be positively charged, resulting in a repulsive force. Metal removal was similar at pH values of 7 and 9. The effect of multiple metals (Cu and Zn at a metal:EDTA molar ratio of 1:1) was insignificant at pH values of 3, 7 and 9. Finally, the presence of hardness (360 mg/L as $CaCO_3$, two-thirds calcium, one-third magnesium) did not adversely affect the removal of the Cu-EDTA complexes at the three pH values investigated.

Lead

Reed and Arunachalam [58] investigated the efficacy of using GAC columns to treat several synthetic lead-bearing waste streams. Hydrodarco (HD) 4000 (Norit Americas, Inc., Atlanta, GA, USA) was selected as the activated carbon. Two Pb concentrations, 10 and 50 mg/L, were used for each of the following three synthetic wastewaters: (1) Pb-only at influent pH values of 4 and 5.4, (2) 10^{-3} M acetic acid at an influent pH = 4.7 and (3) EDTA at Pb:EDTA molar ratios of 1:0.1 and 1:1 and an influent pH = 5.4. Wastewaters 2 and 3 were used to simulate wastes containing weak organic acids and strong complexing agents, respectively. Approximately 50 g (dry) of the carbon was weighed, sieved through a U.S. No. 50 sieve and washed with distilled water to remove fines. The carbon was placed in 38 cm long, 2.85 cm inner diameter acrylic columns such that the formation of air voids was minimized. Rounded stones and glass wool were used as the column support media. The length of the adsorptive bed was approximately 19 cm, and the bed volume was 120 cm^3. A flow rate of 40 ml/min (HLR = 3.67 m/h, 1.5 gal/ft^2-min) was maintained using a Cole-Parmer peristaltic pump. Columns were operated in the up-flow mode. Effluent samples were collected either every 30 or 60 minutes, measured for pH, and acidified. Breakthrough was defined as $C_e = 0.03C_o$. During the course of the column cycle (operation + regeneration), the volume and Pb concentration of the influent, effluent and regenerants were measured and used to calculate Pb surface concentrations (*X/M*) and desorption efficiencies. Following each column run, the majority of the carbon columns were regenerated using an acid-base rinse procedure. Approximately 1 L of 0.1 N HNO_3 was pumped through the column at a flow rate of 10 ml/min. Samples of the acid rinse were taken periodically and measured for pH and Pb content. This procedure was repeated using 1 L of 0.1 N NaOH. The 0.1 N NaOH remaining in the column after the base rinse was allowed to contact the carbon for approximately five days. The five-day contact time was chosen strictly for logistical reasons (i.e., it was the minimum amount of time before the next column run could begin). Because Pb removal increased following the first regeneration step, the regeneration procedure was used as a carbon pretreatment step for virgin carbon in later experiments. Two other regeneration schemes were investigated and will be discussed later.

In Table 7.6, the X/M_b and V_b are presented for the Pb-only column studies. Breakthrough occurred at about 50 BV for Run 1. Pb breakthrough and effluent pH curves for the 10 mg/L Pb-only experiments at an influent pH = 5.4 are presented in Figure 7.22 [58]. Pb removal increased by approximately 600% following the 0.1 N HNO_3-0.1 NaOH N regeneration step. Based on the Pb-only results, the regeneration procedure was used as a pretreatment step for virgin carbon for the remainder of the column experiments. Pb breakthrough and effluent pH curves for 50 mg/L Pb and an influent pH of 5.4 are presented in Figure 7.23 [58]. The effect of pH on the effluent Pb concentration was dramatic—the sudden increase in effluent Pb concentration coincides directly with the rapid decrease in column pH. Based on the column effluent pH and Pb concentration and the pH-adsorption edges and Pb-only speciation diagram that were presented in Figure 7.1, the following removal hypotheses were forwarded. For Run 1, removal was a combination of adsorption and surface precipitation. At the pH values observed in Run 1, precipitation can occur although all the Pb may not be in the $Pb(OH)_{2(s)}$ form. Given the low concentration of lead in the effluent, some other removal mechanism in addition to precipitation must be operative. For Runs 2 through 4, it is hypothesized that the dominant removal mechanisms were surface and pore liquid precipitation. The NaOH conditioning step of the regeneration process deposited OH^- on the surface and raised the pH of the pore liquid, creating a sink of OH^- for $Pb(OH)_{2(s)}$ formation. The observation that column performance was not affected by incomplete Pb desorption (discussed later) affirms this hypothesis. If adsorption was a dominant removal mechanism, then the presence of residual lead would decrease the number of sites available for adsorption, leading to a deterioration of column performance with col-

TABLE 7.6. Summary of Column Performance for
Pb-Only Experiments [58].

Influent	V_b (BV)	X/M_b (mg/g)
10 mg/L—pH = 4		
Run 1[1]	40	0.97
Run 2	250	6.11
Run 3	240	5.43
Run 4	235	5.00
10 mg/L—pH = 5.4		
Run 1[1]	50	1.21
Run 2	325	7.01
Run 3	315	7.18
Run 4	300	7.36
50 mg/L—pH = 4		
Run 1	20	5.64
Run 2	110	25.6
Run 3	110	26.4
Run 4	110	27.4
50 mg/L—pH = 5.4		
Run 1	25	6.81
Run 2	125	26.8
Run 3	120	28.7
Run 4	120	31.0

[1]Virgin carbon.

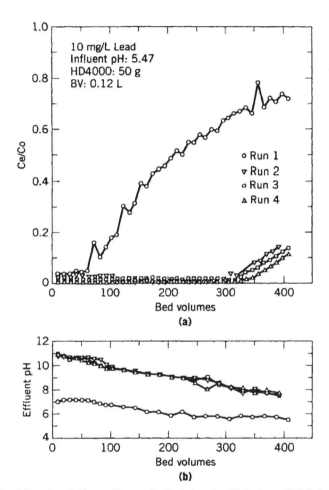

Figure 7.22 Pb breakthrough and effluent pH curves for Norit Americas Hydrodarco 4000. Influent Pb concentration and pH were 10 mg/L and 5.47, respectively [58].

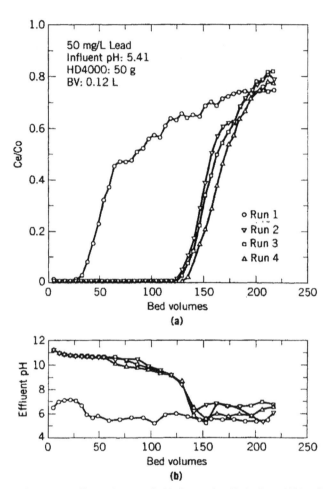

Figure 7.23 Pb breakthrough and effluent pH curves for Norit Americas Hydrodarco 4000. Influent Pb concentration and pH were 50 mg/L and 5.4, respectively [58].

umn run number. Surface or pore liquid precipitation should not be adversely affected by the residual lead.

Pb breakthrough and effluent pH curves for 10 mg/L Pb and 10^{-3} M acetic acid are presented in Figure 7.24 [58]. Acetic acid was used to simulate a moderately acidic wastewater, such as a landfill leachate. Results for 50 mg/L followed a similar trend. V_b, V_{exh}, X/M_b, X/M_{exh} and the DoCU are presented in Table 7.7 for both 10 and 50 mg/L Pb experiments. The presence of acetic acid caused a deterioration of the column performance compared to the Pb-only experiments. Because of differences in influent pH, it is not possible to directly compare the two systems. However, if results for the lead-only system are interpolated to pH = 4.7, the presence of acetic acid decreased V_b by about two-thirds for 10 mg/L and one-third for 50 mg/L. As in the Pb-only column studies, the increase in effluent Pb concentration coincided with the rapid drop in column pH. The increase in waste acidity accelerated this phenomenon and caused breakthrough to occur earlier than in the Pb-only experiments. Eventually, the pH in the column decreased such that Pb desorption occurred ($C_e / C_o > 1$). Remembering that Pb removal decreases with decreasing pH, this was ex-

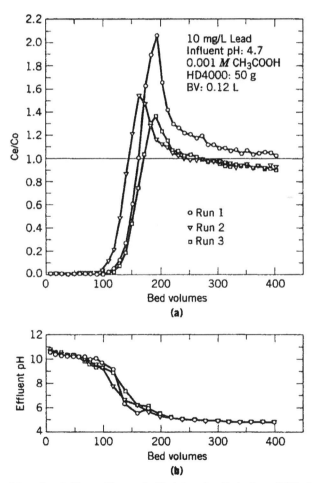

Figure 7.24 Pb breakthrough and effluent pH curves for Norit Americas Hydrodarco 4000 in the presence of 0.001 M acetic acid. Influent Pb concentration and pH were 10 mg/L and 4.7, respectively [58].

TABLE 7.7. Summary of Column Performance for Pb-Acetic Acid Experiments [58].

Influent	V_b (BV)	V_{exh} (BV)	X/M_b (mg/G)	X/M_{exh} (mg/g)	DoCU (%)
10 mg/L—pH = 4.7					
Run 1	120	155	2.55	3.23	79
Run 2	100	145	2.57	3.06	84
Run 3	120	170	2.97	3.58	83
50 mg/L—pH = 4.7					
Run 1	75	120	93.66	12.5	77
Run 2	70	110	8.42	10.4	81
Run 3	70	110	8.32	10.5	79

pected. In fact, the use of an acid to regenerate the column is based on this phenomenon. In an actual treatment scenario, the performance of the column would be improved by neutralizing the acidity of the wastewater prior to its entering the GAC column. Regenerating the carbon column did not significantly affect subsequent column performance.

Pb breakthrough and effluent curves for 10 mg/L Pb at a Pb:EDTA molar ratio of 1:0.1 are presented in Figure 7.25 [58]. For all experiments, C_e was greater than $0.03C_o$ at the be-

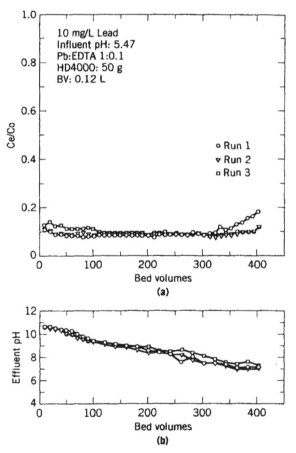

Figure 7.25 Pb breakthrough and effluent pH curves for Norit Americas Hydrodarco 4000 at a Pb:EDTA molar ratio of 0.1:1. Influent Pb concentration and pH were 10 mg/L and 5.47, respectively [58].

ginning of the column run. Thus, based on the definition of breakthrough, no wastewater was adequately treated. The Pb X/Ms, presented in Table 7.8, were taken from the end of the column run (400 and 160 BV for 1:0.1 and 1:1 Pb:EDTA molar ratio, respectively). Except for V_b, results from 1:0.1 Pb:EDTA were similar to the Pb-only results. According to chemical thermodynamic calculations, at a 1:0.1 Pb:EDTA molar ratio, 10% of the Pb is complexed by EDTA. The Pb-EDTA complexes were not removed effectively by this brand of carbon. Similar behavior was observed for a Pb:EDTA molar ratio of 1:1. At 1:1 Pb:EDTA, theoretically all of the Pb is complexed by EDTA, and for column experiments, C_e/C_o was always greater than 0.9. These results are in contrast to those presented by Shay and Etzel [57] who reported that their study carbon removed Pb-EDTA complexes at Pb:EDTA molar ratios greater than 1:1. The results from these two studies underscore the effect carbon type has on metal removal.

Regeneration, in the Reed and Arunachalam [58] study, was defined as the sum of the processes used to prepare spent or exhausted carbon for the next column run. With this definition, the removal of carbon-bound Pb (referred to as "desorption" for the remainder of this discussion) is the first step in the regeneration process. Initial regeneration procedures consisted of using 0.01 N or 0.1 N HNO_3 followed by a tap water rinse. Lead desorption using 0.01 HNO_3 produced poor results, only 14–40% of the carbon-bound lead was removed. Increasing the HNO_3 concentration to 0.1 N greatly increased desorption efficiencies. However, the tap water rinse was not able to significantly raise the pH of the column after the acid rinse. The low column pH (pH 1.7) caused poor removal in subsequent column runs. In addition, a large volume of rinse water was produced, which in turn would require treatment. Based on these results, a desorption step consisting of pumping 1 L of 0.1 N HNO_3 through the column for 1 hour was selected. In response to the low column pH following the desorption step, a "reconditioning" step was added. The reconditioning step consisted of circulating 0.1 N NaOH through the column in the same manner as was done for the nitric acid rinse. The reconditioning step rid the column of residual acid, deposited OH^- on the carbon surface and raised the pore liquid pH.

Desorption efficiencies, presented in Table 7.9, were determined using as a benchmark either the mass of lead removed from solution during the run in question or the total mass of lead on the carbon at the start of the desorption [i.e., includes undesorbed lead from previ-

TABLE 7.8. Summary of Column Performance for
Pb-EDTA Experiments [58].

Influent	X/M^1 (mg/g)
10 mg/L–1:0.1 Pb:EDTA	
Run1	8.69
Run 2	8.71
Run 3	8.22
50 mg/L–1:0.1 Pb:EDTA	
Run 1	24.5
Run 2	28.4
Run 3	26.0
10 mg/L–1:1 Pb:EDTA	
Run 1	3.25
50 mg/L–1:1 Pb:EDTA	
Run 1	17.9

[1]Determined at the end of the column run.

TABLE 7.9. Desorption Efficiencies for Pb Experiments [58].

Influent	Desorption Efficiencies, %			
	Run 1	Run 2	Run 3	Run 4
10 mg/L Pb–pH = 4	68	101 (81)	94 (76)	112 (84)
10 mg/L Pb–pH = 5.4	77	83 (69)	100 (73)	94 (68)
50 mg/L Pb–pH = 4	73	95 (83)	83 (70)	110 (78)
50 mg/L Pb–pH = 5.4	50	99 (73)	88 (66)	97 (66)
10 mg/L Pb–10^{-3} M HAc	98	98 (97)	93 (90)	NA
50 mg/L Pb–10^{-3} M HAc	90	96 (86)	89 (77)	NA
10 mg/L Pb–1:0.1 Pb:EDTA	99	97 (96)	98 (94)	NA
50 mg/L Pb–1:0.1 Pb:EDTA	92	82 (77)	86 (68)	NA

Number in ( ) is based on total carbon-bound Pb; NA: Not applicable.

ous run(s)]. The value in parentheses represents results when the Pb from previous runs was accounted for. Desorption efficiencies based on the mass of lead removed during an individual column run were greater than 80% for all but the first run of the Pb-only systems. For several experiments, the desorption of Pb from previous runs produced efficiencies greater than 100%. When the total amount of the Pb on the carbon is used as the desorption efficiency benchmark, it becomes obvious that a significant amount of Pb remained on the carbon from run to run. However, as discussed earlier, the presence of Pb on the carbon from previous runs did not appear to adversely affect column performance.

In a subsequent study, Reed et al. [59] investigated the effect of different regeneration schemes for GAC columns loaded with either 10 mg/L or 50 mg/L Pb. Acid type (HCl, HNO_3 and NaOH), regenerant concentration (0.1 and 1 N), acid reuse and a base-only (NaOH) rinse were studied. Two Pb concentrations (10 and 50 mg/L) were investigated. Columns were operated in the same manner as described earlier [58]. For the experiments where acid was reused, the normality and volume of the regenerant were adjusted to the initial values before being reused, and desorption efficiencies were corrected to account for the Pb present from the previous regeneration step. In Table 7.10, X/M_{exh}, V_b, V_{exh} and desorption efficiencies are presented for acid type and regenerant concentration experiments. For six of the 12 experiments, desorption efficiencies were less than 90% however, incomplete regeneration did not adversely affect column performance. If the majority of the Pb was removed by precipitation on the carbon surface or in the carbon pore liquid, then incomplete regeneration would not adversely affect column performance unless the carbon pores were blocked with $Pb(OH)_{2(s)}$. Given the relatively low surface concentrations of Pb, blockage of the pore passages was unlikely. For both 10 and 50 mg/L Pb, column performance and desorption efficiencies were not significantly affected by the type of acid used. Thus, the less expensive HCl was used for the remainder of the column experiments.

In Figure 7.26 [59], the Pb concentration and pH of the column effluent during the acid rinse are presented for the 10 mg/L Pb–0.1 N HCl-NaOH experiments. During the acid rinse, the pH dropped to about 1, and the carbon-bound Pb was almost completely desorbed after five bed volumes of acid were pumped through the column. Ten BVs of acid appear to be more than sufficient for lead desorption, and a reduction in the amount of acid used during regeneration is possible. There was little additional Pb removal during the 0.1 N NaOH rinse because the majority of the Pb was desorbed during the acid rinse. The pH of the column increased to above 12 after about 3 BVs of base addition, and it remained at this level

TABLE 7.10. Summary of Results from the Effect of Acid Type and Regenerant Concentration [59].

Pb (mg/L)	Regeneration Scheme	X/M_{exh}			V_b (BV)			V_{exh} (BV)			Desorption Efficiencies (%)		
	Run Number →	1	2	3	1	2	3	1	2	3	1	2	3
10	0.1 N HCl–0.1 N NaOH	11.5	11.2	NA	210	210	NA	980	960	NA	99	103	NA
10	0.1 N HNO$_3$–0.1 N NaOH	11.5	11.3	NA	225	215	NA	970	970	NA	91	98	NA
50	0.1 N HCl–0.1 N NaOH	16.6	15.8	15.7	84	82	75	240	240	230	100	104	83
50	0.1 N HNO$_3$–0.1 N NaOH	15.6	15.8	NA	84	82	NA	225	240	NA	89	88	NA
50	1 N HCl–1 N NaOH	19.9	18.6	19.5	118	98	100	240	235	245	82	69	74

NA: Not applicable, column runs were not conducted.

243

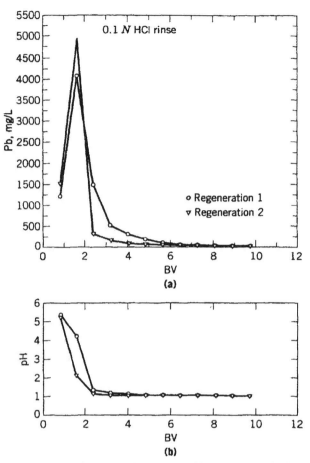

Figure 7.26 Pb concentration and pH of column effluent during acid rinse portion of Norit Americas Hydrodarco 4000 column 0.1 N HCl–0.1 N NaOH regeneration [59].

for the remainder of the base rinse. A reduction in the volume of base used during the reconditioning step appears possible. Similar results were observed for the other column experiments.

The effect of HCl and NaOH concentration on column performance was investigated at 50 mg/L Pb only. Increasing the concentration of the regenerants increased X/M_{exh} and V_b by about 20% and 25%, respectively (see Table 7.10). The retention of OH^- by the carbon in the 1 N regenerant experiments was higher, and correspondingly, the column pH and Pb removal were greater. These results are consistent with earlier work where it was observed that increasing the regenerant concentrations from 0.01 N to 0.1 N improved subsequent column performance [59]. The increase in lead removed at higher regenerant concentrations is most likely due to additional Pb precipitation on the carbon surface and in the carbon pores. It remains to be determined if the improvement in column performance observed at higher regenerant concentrations will offset the increase in chemical cost.

A summary of column performance with and without acid reuse is presented in Table 7.11. In Figure 7.27 [59], Pb breakthrough and effluent pH curves for 10 mg/L Pb experiments with and without the reuse of 0.1 N HCl are presented. The shape of the Pb break-

TABLE 7.11. Summary of Results from the Effect of Acid Reuse Experiments [59].

Pb (mg/L)	Regeneration Scheme Run Number →	Acid Reuse	X/M_{exh} (mg/g)				V_b (BV)				V_{exh} (BV)				Desorption Efficiencies (%)			
			1	2	3	4	1	2	3	4	1	2	3	4	1	2	3	4
10	0.1 N HCl–0.1 N NaOH	Yes	12.0	11.3	11.3	12.1	230	210	210	250	920	880	930	935	80	72	68	85
10	0.1 N HCl–0.1 N NaOH	No	11.5	11.2	NA	NA	210	210	NA	NA	980	960	NA	NA	99	103	NA	NA
50	0.1 N HCl–0.1 N NaOH	Yes	15.8	16.4	15.8	15.9	75	83	78	81	220	220	215	220	98	63	37	75
50	0.1 N HCl–0.1 N NaOH	No	16.6	15.8	15.7	NA	84	82	75	NA	240	240	230	NA	100	104	83	NA
50	1 N HCl–1 N NaOH	Yes	19.5	19.4	19.6	19.3	102	101	102	101	240	235	245	244	75	50	69	70
50	1 N HCl–1 N NaOH	No	19.9	18.6	19.5	NA	118	98	100	NA	240	235	245	NA	82	69	74	NA

NA: Not applicable, column runs were not conducted.

245

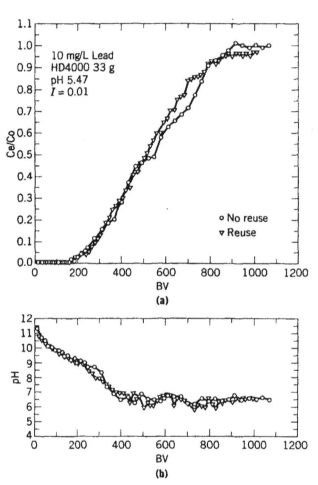

(a)

(b)

Figure 7.27 Pb breakthrough and effluent pH curves for Norit Americas Hydrodarco 4000 with and without the reuse of 0.1 N HCl. Influent Pb concentration and pH were 10 mg/L and 5.47, respectively [59].

through and pH curves were similar, and column performance was not adversely affected by acid reuse. Four cycles were conducted with acid reuse, and column performance was similar from run to run. Desorption efficiencies were lower when acid was reused. It is unclear why this occurred. The influent and effluent pH values of the acid regenerant when acid was reused were similar to those observed in the no-reuse experiments. The presence of Pb in the acid rinse initially (from the previous regeneration) should not have affected desorption because adverse Pb concentration gradients were experienced during normal regeneration (Pb concentrations as high as 6,000 mg/L were observed during some regenerations). Despite the decrease in desorption efficiencies, column performance did not suffer, indicating that the majority of lead removal is via precipitation within the carbon pores.

The formation of $Pb(OH)_4^{2-}$ at high pH values (see Figure 7.1) allows the use of a base-only regeneration. In Table 7.12, column performance parameters for the base-only regeneration experiments as well as for the corresponding HCl-NaOH experiments are presented. In Figure 7.28 [59], Pb breakthrough and effluent pH curves for columns regenerated with 0.1 N NaOH and 0.1 N HCl-NaOH are presented. Three cycles were conducted with base-only regeneration, and the breakthrough curves were similar to the one presented in Figure 7.28. The shape of the breakthrough curves in Figure 7.28 were similar except that the base-only curve was shifted slightly to the left (i.e., less effective treatment), and thus, it was expected that the measures of column performance decreased when base-only regeneration was employed. However, column performance for the 1 N NaOH experiments was better than that observed in the 0.1 N HCl-NaOH regenerations. Thus, the use of a 1 N NaOH regeneration, in place of a 0.1 N HCl-NaOH regeneration, appears to be viable. Elimination of the acid rinse and the subsequent decrease in the volume of regenerant requiring further treatment may make the 1 N NaOH regeneration scheme economically desirable. Desorption efficiencies for the 0.1 N base-only experiments were less than those observed for the HCl-NaOH studies, while the opposite was observed for the 1 N regenerant experiments. In Figure 7.29 [59], the Pb concentrations in the effluent from the 0.1 N and 1 N NaOH base rinses are presented. For 0.1 N NaOH, desorption was still occurring after 10 BVs, while at 1 N NaOH, desorption was essentially completed after about 4 BVs. If the lower concentration of base is used, a larger volume of base regenerant will be required. The effluent pH of the columns after the base rinse were about 12.5 and 13.5 for the 0.1 N and 1 N NaOH, experiments, respectively. The higher Pb removal observed in the 1 N NaOH experiments compared to when 0.1 N NaOH was used was attributed to the higher concentration of OH^- in the carbon pores.

Reed et al. [15] investigated the effect of EBCT and HLR on Pb (1 mg/L) removal by Hydrodarco (HD) 4000. V_b, V_{exh}, X/M_b, X/M_{exh}, CUR, DoCU and desorption efficiencies (DE) were used as measures of column effectiveness. For a HLR of 4.9 m/h (2 gal/min-ft²), EBCTs of 1.85, 3.7, 5.55, 6.75, 12.75 and 22.4 min were investigated. For a HLR of 9.8 m/h (4 gal/min-ft²), EBCTs of 0.93, 1.85 and 6.75 min were used. Breakthrough and exhaustion were defined as Ce/C_o equal to 0.03 and 0.95, respectively. Columns with an inner diameter of 2.5 cm were used for EBCTs of 1.85 min, 3.7 min, 5.55 min and 6.75 min. A 3.8 cm inner diameter column was used for the remainder of the experiments. Columns were prepared and operated in the manner described earlier [59]. The average packed density of the carbon column was 0.45 g/ml. The influent pH, ionic strength and temperature were 5.4, 0.01 (as $NaNO_3$) and ≈22°C, respectively. During the treatment portion of the column run, the effluent pH and Pb concentration were measured periodically.

Average values of column performance parameters and desorption efficiencies for the

TABLE 7.12. Summary of Results from the Effect of Base-Only Regenerant Experiments [59].

Pb (mg/L)	Regeneration Scheme	X/M_{exh} (mg/g)			V_b (BV)			V_{exh} (BV)			Desorption Efficiencies (%)		
	Run Number →	1	2	3	1	2	3	1	2	3	1	2	3
50	0.1 N HCl–0.1 N NaOH	16.6	15.8	15.7	84	82	75	240	240	230	100	104	83
50	0.1 N NaOH	15.8	14.8	14.3	83	70	70	230	215	210	53	40	NA
50	1 N HCl–0.1 N NaOH	19.9	18.6	19.5	118	98	100	240	235	245	82	69	74
50	1 N NaOH	20.5	16.3	16.4	105	83	84	240	225	225	95	89	82

NA: Not applicable, column runs were not conducted.

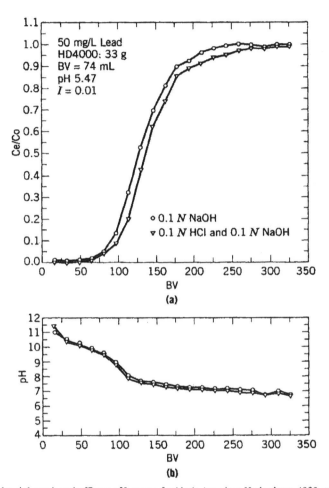

Figure 7.28 Pb breakthrough and effluent pH curves for Norit Americas Hydrodarco 4000 using either 0.1 N NaOH or 0.1 N HCl–0.1 N NaOH regeneration schemes. Influent Pb concentration and pH were 50 mg/L and 5.47, respectively [59].

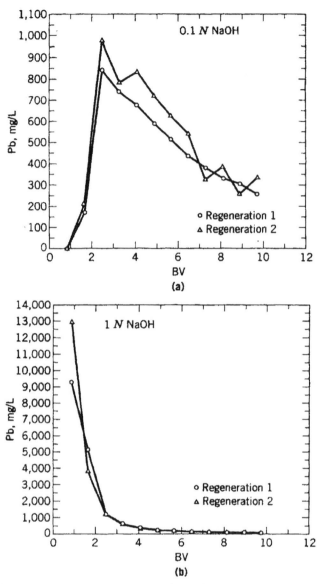

Figure 7.29 Pb concentration in the base rinse using (a) 0.1 N NaOH and (b) 1 N NaOH [59].

EBCTs and HLRs investigated are summarized in Table 7.13. Pb desorption efficiencies for all but one experiment were greater than 80%. For EBCT experiments that used a serial column arrangement, the regeneration efficiencies could only be calculated for largest EBCT (i.e., if three columns were used in series, regeneration efficiencies for the first two columns could not be calculated). An example of typical Pb breakthrough and effluent pH curves are presented in Figure 7.30 for EBCTs ranging from 1.85 and 22.4 minutes [15]. As the EBCT was increased, the breakthrough curves became progressively steeper ($V_b \rightarrow V_{exh}$), indicating that kinetics were less of a factor at higher EBCTs. For a HLR of 4.9 m/h, V_b increased by 63% (520 to 850) as the EBCT was increased from 1.85 to 5.55 minutes,

TABLE 7.13. Summary of Column Performance Parameters [15].*

HLR (m/h)	EBCT (min)	V_b (BV)	V_{exh} (BV)	X/M_b (mg/g)	X/M_{exh} (mg/g)	CUR (g/L)	DoCU (%)	DE (%)
	1.85	520	1,425	1.11	1.97	0.858	55	80
	3.70	755	1,325	1.62	2.02	0.590	80	NA[1]
4.9 m/h	5.55	850	1,275	1.82	2.01	0.525	91	95
	6.75	875	1,160	2.09	2.31	0.524	91	NA[2]
	12.75	945	1,110	2.10	2.31	0.467	92	85
	22.42	910	1,100	2.00	2.17	0.494	92	90
	0.93	275	1,600	0.54	1.81	1.62	30	85
9.8 m/h	1.85	485	1,410	1.07	1.93	0.920	56	82
	6.75	860	1,120	1.99	2.3	0.483	86	86

*Average values for all runs.
[1]Not available, serial column arrangement was used.
[2]Not available, column broke during regeneration.

but it only increased an additional 11% when the EBCT was increased from 5.55 to 12.75 minutes. X/M_{exh} was relatively constant with EBCT, increasing by about 16% as the EBCT was increased from 1.85 to 12.75 minutes. For a HLR of 9.8 m/h, V_b increased by about 77% as the EBCT was increased from 1.85 to 6.75 min. X/M_{exh} was relatively constant with EBCT regardless of the HLR, increasing by about 16% when the EBCT was increased from

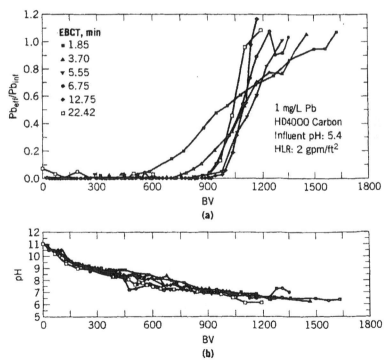

Figure 7.30 Pb breakthrough and effluent pH curves for Norit Americas Hydrodarco 4000 for EBCTs ranging from 1.85 to 22.42 minutes. HLR = 4.9 m/h and influent Pb concentration and pH were 1 mg/L and 5.4, respectively [15].

1.85 to 12.75 minutes. For all EBCTs, desorption/resolubilization of the carbon-bound Pb occurred in the later stages of column operation as the pH decreased. These results were not surprising given the strong relationship between metal removal and pH. Similar results were observed for the 10 mg/L Pb experiments.

The effect of Pb desorption/resolubilization from the first column of a serial column operation on the effectiveness of the overall system was uncertain. Thus, two parallel column experiments (1 mg/L Pb, HLR = 4.9 m/h, three columns in series, EBCTs = 1.85, 3.7 and 5.55 min) were conducted to ascertain if Pb desorption/resolubilization in the first column was detrimental to overall system performance. For one set of columns, the columns were taken off-line after reaching exhaustion, while the second set of columns was operated as described earlier. A summary of column performance parameters for this experiment is presented in Table 7.14. There was little difference in column performance parameters between the two modes of column operation. The desorbed/resolubilized Pb from the first column re-adsorbed/precipitated on the subsequent column without a decrease in process performance.

In Figure 7.31 [15], CUR and DoCU versus EBCT curves for 1 mg/L Pb are presented. For a HLR of 4.9 m/h, DoCU reached a maximum (91%) at EBCTs ≥ 5.55 min. The lowest CUR, 0.467 g/L, occurred at an EBCT = 12.75 minutes, although the difference between the CUR at 5.55 and 12.75 was only 12%. For a HLR of 9.8 m/h, at the highest EBCT tested (6.75 min), the DoCU and CUR were 86% and 0.483 g/L, respectively. The performance of the columns at the different HLRs was similar at EBCTs of 1.85 and 6.75 min (the EBCTs common to both HLRs). Thus, the optimum EBCT for this waste stream is about 6 to 7 minutes regardless of the HLR employed. In Figure 7.32, V_b for 1 mg/L Pb [15], 18 μg/L cis-1,2-dichloroethylene and 100 μg/L 1,1,1-trichloroethane [60] as a function of EBCT are presented. The removal of cis-1,2-dichloroethylene and 1,1,1-trichloroethane is much more dependent on EBCT than is the removal of Pb. Based on the results presented in Figures 7.31 and 7.32, the size of the activated carbon column for a wastewater containing both an organic contaminant and Pb would most likely be based on organic removal, while the regeneration schedule would most likely be controlled by Pb.

In an effort to determine if GAC columns could be used to simultaneously remove inorganic and organic adsorbates, Reed et al. [61] performed column studies on two synthetic wastewaters. Both wastewaters had a Pb concentration of 1 mg/L and an influent pH = 5.4. Phenol (10 mg/L) and TCE (1 mg/L) were chosen as the organic adsorbates. The study carbon was Hydrodarco 4000. The HLR was 4.9 m/h (2 gal/min-ft²), and the EBCT was 12.75 minutes. Columns were prepared and operated as described earlier [58,61,62]. For the TCE experiments, the base rinse was recovered and reused.

In Figure 7.33, Pb and phenol breakthrough and effluent pH curves are presented

TABLE 7.14. Summary of Column Performance Parameters for Effect of Taking Exhausted Column Off-Line [15].

Parameters	Exhausted Column Left On-Line			Exhausted Column Taken Off-Line		
EBCT (min) →	1.85	3.7	5.55	1.85	3.7	5.55
V_b (BV)	510	750	850	530	760	850
V_{exh} (BV)	1,450	1,300	1,250	1,540	1,350	1,300
X/M_{exh} (mg/g)	1.88	2.02	2.02	2.06	2.02	2.00
CUR (g/L)	0.87	0.59	0.52	0.84	0.59	0.52
DoCU (%)	58	80	91	55	80	91

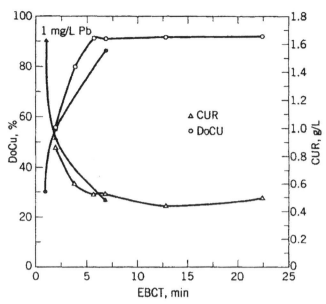

Figure 7.31 Carbon usage rates (CUR) and degree of column utilization (DoCU) versus EBCT for 1 mg/L Pb—Norit Americas Hydrodarco 4000 experiments. Hollow symbols represent HLR = 4.9 m/h (2 gal/min-ft²). Filled symbols represent HLR = 9.8 m/h (4 gal/min-ft²) [15].

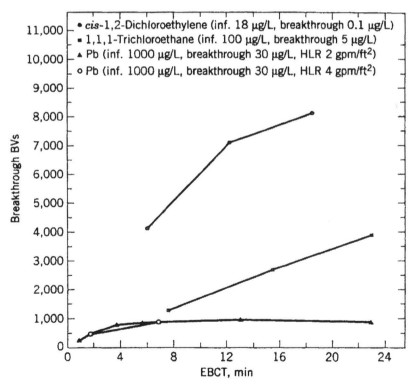

Figure 7.32 Bed volumes treated at breakthrough (V_b) versus EBCT for 1 mg/L Pb [15], 18 µg/L cis-1,2-dichloroethylene and 100 g/L 1,1,1-trichlorethane [60].

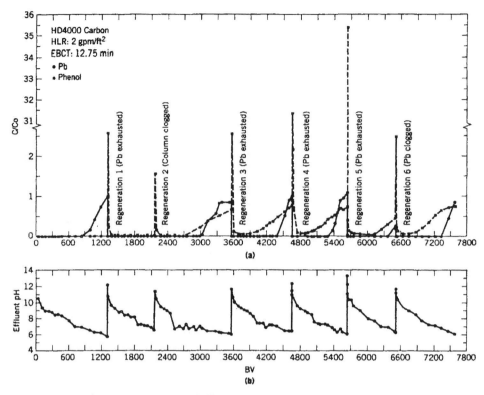

Figure 7.33 Pb and phenol breakthrough and effluent pH curves for Norit Americas Hydrodarco 4000 GAC columns. Influent Pb and phenol concentrations and pH were 1 mg/L, 10 mg/L and 5.47, respectively [61,62].

[61,62]. Seven treatment runs and seven regenerations for Pb were conducted. The column was regenerated using 0.1 N HCl–0.1 N NaOH rinses at Pb exhaustion or when the column clogged (Runs 2 and 6). In Table 7.15, Pb column performance parameters and desorption efficiencies are presented. In Run 1, Pb breakthrough and exhaustion occurred at 870 and 1,270 BVs, respectively. At the time of Pb exhaustion (1,270 BVs), phenol breakthrough had yet to occur. After regeneration, the pH of the column was high ($\approx$ 12), and a portion of the carbon-bound phenol was desorbed (note the spike in phenol concentration immedi-

TABLE 7.15. Summary of Pb Column Performance Parameters
for Pb-Phenol Experiments [61,62].

Run #	V_b (BV)	V_{exh} (BV)	X/M_b (mg Pb/g)	X/M_{exh} (mg Pb/g)	CUR (g/L)	DoCU (%)	RE (%)
1	870	1,270	2.02	2.59	0.49	78	74
2[1]	850	NA	1.92	NA	0.51	NA	90
3	835	1,160	1.92	2.35	0.52	82	77
4	840	1,070	1.91	2.14	0.52	89	89
5	650	870	1.61	1.77	0.67	91	83
6[1]	770	NA	1.81	NA	0.56	NA	97
7	850	1,100	2.07	2.36	0.51	88	92

[1]Column clogged before exhaustion, results represent situation at end of column run. NA: Not applicable.

ately following regeneration). Above pH = 10.5, phenol exists as $C_6H_5O^-$. Obviously, the ionic form of phenol is desorbable at high pH values. As the column pH decreased, the concentration of phenol in the effluent dropped to below the detection limit (0.1 mg/L). In an actual treatment system, the volume of wastewater that had a high phenol concentration would be collected and retreated. If this high pH wastewater was blended with untreated wastewater, column performance could be improved because of the increase in influent pH. In Run 2, Pb breakthrough occurred at 850 BVs. At this point, pressure in the column began to build up and the run was stopped and the column was regenerated. As in Run 1, the pH of the column was high ($\approx$ 12) after regeneration, and a portion of the carbon-bound phenol was desorbed. As the column pH decreased, the concentration of phenol in the effluent dropped to below the detection limit (0.1 mg/L). During Run 3, breakthrough of phenol ($C_e = 0.03C_o = 0.3$ mg/L) occurred after 2,700 BVs. The phenol surface loading at this point was 54.2 mg phenol/g carbon. Pb breakthrough and exhaustion for Run 3 occurred at 835 and 1,160 BVs (relative to the start of Run 3), respectively. The column was not exhausted for phenol at this point, but at the end of Run 3, 3,500 BVs of phenol wastewater had been treated. The effluent phenol concentration and phenol surface loading at this point was 6.60 mg/L and 71.7 mg phenol/g carbon, respectively. Column behavior for Runs 4 through 7 with respect to Pb removal was similar to that observed in earlier runs. In Run 6, pressure buildup occurred, and the column was regenerated prior to Pb exhaustion. Column performance after regeneration of the clogged column (Runs 2 and 6) was not adversely affected. Clogging of GAC columns during full-scale operation with a subsequent buildup of column headloss is common. In full-scale operation, columns will often be taken off-line when a maximum headloss has been reached. Thus, column clogging should not prevent the use of GAC columns for the removal of inorganic and organic contaminants in the field. The carbon usage rate (CUR) for individual runs, based on Pb removal, ranged from 0.500 to 0.668 g /L and averaged 0.541 g/L. The DoCU, based on Pb removal, ranged from 74–91% with a mean value of 84%. For the runs in which the column clogged (Runs 2 and 6), calculation of the DoCU was not possible because the column did not reach exhaustion. The CUR based on phenol can only be calculated for the first phenol breakthrough that occurred at 2,700 BVs. At this point the CUR for phenol was 0.185 g/L. The DoCU with respect to phenol could not be calculated because the column was never exhausted for phenol.

Pb desorption efficiencies ranged from 74–97% with a mean value of 86%. Incomplete lead removal during regeneration did not significantly decrease Pb removal in the subsequent run. The slight decrease in Pb removal that was observed in Runs 5 and 6 was most likely caused by the presence of large amounts of phenol in the effluent. After Regeneration 3, a portion of the phenol adsorption capacity had been reclaimed via the base rinse but the effluent phenol concentration was always greater than the breakthrough concentration (0.03Co). In an actual treatment system, the column at this point would be taken off-line and regenerated for phenol. Regeneration of GAC columns loaded with organics is typically accomplished using thermal methods, however, caustic regeneration is sometimes used. In Regeneration 5, 1 N NaOH was used in place of 0.1 N NaOH to ascertain if better phenol regeneration could be accomplished using a more concentrated caustic solution. Increasing the strength of the base rinse increased phenol desorption (590 mg/L phenol for Regeneration 5 compared to an average of 315 mg/L for the other regenerations) and phenol removal during the subsequent treatment run. Based on these observations, a hot caustic rinse could reclaim a significant amount of the carbon's phenol removal capacity.

In Figure 7.34, Pb and TCE breakthrough and effluent pH curves are presented [61,62].

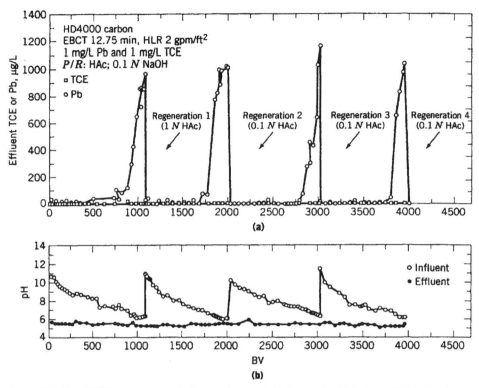

Figure 7.34 Pb and TCE breakthrough and effluent pH curves for Norit Americas Hydrodarco 4000 GAC columns. Influent Pb and TCE concentrations and pH were 1 mg/L, 10 mg/L and 5.47, respectively [61,62].

In Table 7.16, a summary of column performance parameters is presented. Performance decreased slightly from Run 1 to Run 2 but improved in Run 3. Column performance parameters based on Pb removal were similar to those observed in the Pb-phenol experiments. TCE, which does not ionize, did not desorb at the high pH values observed during the base rinse and at the beginning of the treatment run. TCE was not detected in the effluent during any of the four runs. A total of 3,840 BVs (4,530 L) of TCE wastewater were treated for the three runs, resulting in a commutative TCE surface loading of 9.32 mg/g. The Pb CUR averaged 0.51 g/L for an individual run, while the cumulative Pb CUR was 0.13 g/L at the end of Run 4. The Pb DoCU averaged about 90%. Values of CUR and DoCU for TCE could not be determined because TCE did not break through.

Because TCE did not desorb to any great extent in the base rinse, the majority of the base solution was recovered and reused, significantly reducing the volume of residual (V_{res}). For all runs, the amount of base rinse recovered was less than 10 BV ($\approx$8.9 BV), and the normality was less than 0.1 N. Thus, water and NaOH pellets were added to bring the recovered base rinse to 10 BV and 0.1 N NaOH. The cumulative Pb waste concentration factor (WCF $= V_b/V_{res}$) increased from 40 to 57 from Run 1 to 4. The WCF after the third run is similar to WCF's that are typically reported for membrane and ion exchange processes. The cumulative Pb volume reduction factor [VRF $= (V_b - V_{res})/V_b$] averaged about 98%. Thus, for every 1,000 gallons of wastewater treated, about 20 gallons of residual requiring further treatment is produced. The metal concentration of the residual will most likely be high, thus, conventional treatment techniques (e.g., precipitation) can be used.

TABLE 7.16. Summary of Column Performance Parameters for Pb-TCE Experiments [61,62].

Parameter	Run 1	Run 2	Run 3	Run 4
V_b, BV	800	625 (1,425)[1]	800 (2,225)	755 (2,980)
V_{exh}, BV	1,080	850 (1,930)	1,000 (2,930)	910 (3,840)
X/M_b, mg Pb/g	2.04	1.60 (3.64)	2.17 (5.81)	1.94 (7.75)
X/M_{exh}, mg Pb/g	2.36	1.84 (4.20)	2.33 (6.53)	2.05 (8.58)
TCE X/M mg TCE/g	2.66	2.11 (4.77)	2.21 (6.99)	2.32 (9.32)
Pb CUR, g/L	0.47	0.61 (0.27)	0.47 (0.17)	0.50 (0.13)
Pb DoCU, %	86	86	93	94
V_{res},[2] BV	10 + 10 (20)	10 +1.2 (31.2)	10 + 1.1 (41.3)	10 + 1.1 (52.4)
Pb WCF[3]	40	46	54	57
Pb VRF[4]	97.5	97.8	98.1	98.2
Pb Desorption, %	100	93	86	99

[1]Values in ( ) are coumulative.
[2]Volume of residual—first number is acid rinse, second is base rinse.
[3]Cumulative waste concentration factor = V_b/V_{res}.
[4]Cumulative volume reduction factor = $(V_b - V_{res})/V_b$.

Mercury

Pandey and Chaudhuri [46] investigated the removal of Hg(II) using a bituminous coal having a geometric mean size of 0.78 mm. Column diameter and bed depth were 1 cm and 45 cm, respectively, resulting in a bed volume of 35 cm^3. Wastewater was delivered to the column in downflow mode at a HLR of either 2.3 m/h (0.95 gal/min-ft^2), 4.6 m/h (1.9 gal/min-ft^2) or 9.3 m/h (3.8 gal/min-ft^2). Characteristics of the synthetic wastewater are presented in Table 7.17. Influent Hg(II) concentrations of 10, 400 and 2,000 µg/L were investigated. For 400 and 2,000 µg/L Hg, turbidity-free water was used. For one 10 µg/L Hg column experiment (Run 3), kaolinite was added such that the turbidity was initially between 18–22 NTU. This turbid wastewater was then coagulated with 40 mg/L alum, and after settling, the Hg concentration was 8 µg/L, and the turbidity was 1 NTU. In Figure 7.35, Hg breakthrough curves are presented for influent Hg concentrations of 400 and 2,000 µg/L [46]. For the 400 µg/L Hg wastewater, the effluent Hg concentration ranged from 25 to 50 µg/L (88–94% removal efficiency). At the higher influent Hg concentration (2,000 µg/L), the Hg concentration was reduced to about 400 µg/L (80% removal). For both experiments, exhaustion was not reached. In Figure 7.36, Hg breakthrough curves are presented for 10

TABLE 7.17. Characteristics of Synthetic Hg(II) Wastewater
for Study Conducted by Pandey and Chaudhuri [46].

Constituent	Concentration
Hg(II), µg/L	10,400 or 2,000
pH	7.8–8.2
Alkalinity, mg/L as $CaCO_3$	450
Hardness, mg/L as $CaCO_3$	220
Calcium, mg/L as $CaCO_3$	50
Chloride, mg/L	20
Total Dissolved Solids, mg/L	440
Specific Conductance, µS/cm	800–900
Turbidity, NTU	ND to 1

ND: Nondetectable.

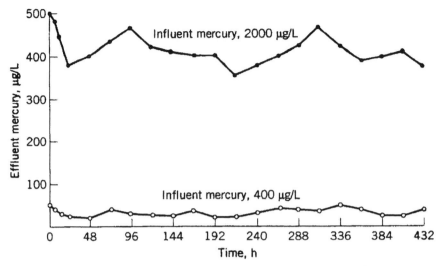

Figure 7.35 Hg breakthrough curves for influent Hg concentrations of 400 and 2,000 µg/L. Flow rate to column = 3.2 ml/min (0.85 gal/min-ft²) and EBCT = 10.9 minutes. Reprinted from *Water Research,* 16(7), M. P. Pandey and M. Chaudhuri, "Removal of Inorganic Mercury from Water by Bituminous Coal," 1113–1118, copyright 1982, with kind permission from Elsevier Science Ltd., The Boulevard, Langford Lane, Kidlington OX5 1GB, UK [46].

µg/L Hg [46]. Runs 1 and 2 were conducted using turbidity-free water, while wastewater for Run 3 was prepared as described above. It should be noted that these experiments were not conducted on the same carbon (i.e., each column experiment employed virgin carbon). For Run 1, Hg concentration was decreased by about 60% using a HLR of 9.3 m/h (3.8 gal/min-ft²). When the HLR was reduced to 4.6 m/h (1.9 gal/min-ft²), the Hg removal efficiency increased to about 80%. The EBCTs for the system at HLRs of 4.6 and 9.3 m/h were 5.46 and 2.73 minutes, respectively. Generally, increasing the EBCT will not affect the effluent concentration but will increase the volume of wastewater treated at breakthrough (longer column runs). The decrease in effluent Hg concentration with increases in EBCT may indicate that other reactions, such as reduction [Hg(II) to Hg⁰] and volatilization, are occurring. For Run 3, influent and effluent Hg concentrations averaged 8 and 1.25 µg/L, respectively (≈85% removal).

Zinc

Chen and Huang [63] investigated the treatment of an actual wastewater containing zinc, toluene, acetic acid and other organic solvents. At the time the study was conducted, the pertinent regulatory agency imposed a zinc effluent limit of 1 mg/L. Precipitation was not able to reach the 1 mg/L Zn limit because of the presence of organic contaminants, especially acetic acid. A combination of chemical precipitation and activated carbon treatment was proposed. To determine the feasibility of GAC columns for Zn removal, the wastewater was diluted, and the pH was adjusted to simulate the effluent from the precipitation process. The characteristics of the undiluted wastewater were as follows: Zn = 2,030 mg/L, TOC = 10,000 mg/L and pH = 4.78. After dilution, the Zn concentration averaged about 43 mg/L, and the pH was 7. The column was packed with 120 g of Filtrasorb 400, and diluted wastewater was introduced to the column at a flow rate of 50 ml/min in the upflow mode.

Following exhaustion for zinc, the column was regenerated using 200 ml of 1 M HCl and then rinsed with 200 ml of NaOH and then put back on-line. Six treatment cycles were performed. For Run 1, about 2,000 ml of wastewater were treated prior to breakthrough, while for Runs 2 through 6, the volume at breakthrough increased to about 4,000 ml. The increase in removal can be attributed to a higher column pH present in Runs 2 through 6 caused by the NaOH portion of the regeneration rinse. Treatment performance decreased slightly with run number for Runs 2 through 6, indicating that a portion of the carbon's metal removal capacity was not restored by the regeneration process. Desorption efficiencies and carbon capacities for Zn were not presented by the authors.

Minicolumn Testing

The use of minicolumn testing with the appropriate scaling equations could greatly reduce the amount of effort required to conduct a GAC column treatability study. Kuennen et al. [28] used minicolumns to investigate Pb removal from DI water and from a municipal raw water as well as the competitive interactions between several heavy metals (Cd, Cu, Pb and Zn). The minicolumn had an inner diameter and bed volume of 0.9 cm and 1.1 cm³, respectively. A flow rate of 8.3 ml/min was used. An H-type carbon was wet sieved to isolate enough GAC having the appropriate diameter. The carbon was acid rinsed prior to its placement in the minicolumn. In Figure 7.37, Pb breakthrough curves for several influent pH values are presented for DI water and Grand Rapids municipal water [28]. The influent Pb concentration was 150 μg/L. Pb removal generally decreased with increasing influent pH and

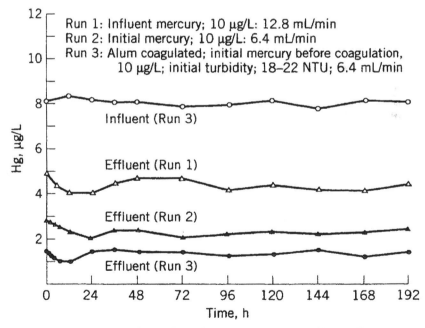

Figure 7.36 Hg breakthrough curves for an influent Hg concentration of 10 μg/L. Reprinted from *Water Research*, 16(7), M. P. Pandey and M. Chaudhuri, "Removal of Inorganic Mercury from Water by Bituminous Coal," 1113–1118, copyright 1982, with kind permission from Elsevier Science Ltd., The Boulevard, Langford Lane, Kidlington OX5 1GB, UK [46].

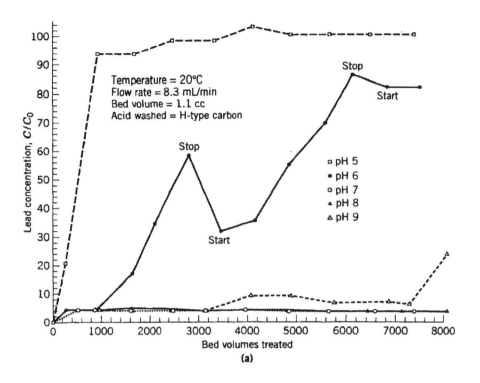

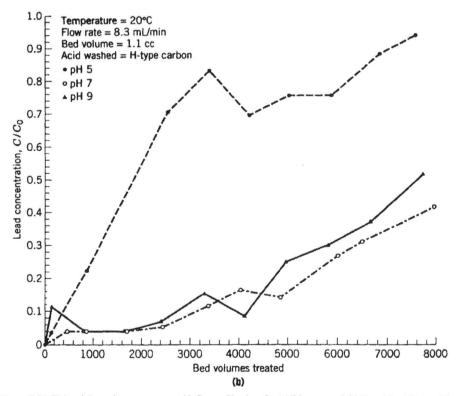

Figure 7.37 Pb breakthrough curves at several influent pH values for (a) DI water and (b) Grand Rapids' municipal water. Influent Pb concentration = 150 µg/L. Reprinted from *Journal American Water Works Association*, Vol. 84, No. 2 (February 1992), by permission. Copyright © 1992, American Water Works Association [28].

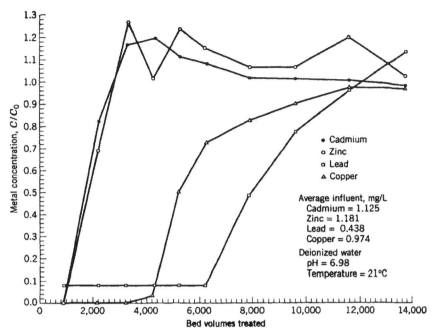

Figure 7.38 Cd, Cu, Pb and Zn breakthrough curves. Reprinted from *Journal American Water Works Association,* Vol. 84, No. 2 (February 1992), by permission. Copyright © 1992, American Water Works Association [28].

was less for the Grand Rapids' water. The formation of $PbCO_{3(aq)}$ in the Grand Rapids, water may have reduced Pb removal over that observed with DI water (alkalinity in the municipal water was 110 to 120 mg/L as $CaCO_3$). In Figure 7.38, breakthrough curves for the metal competition studies are presented. Pb was the easiest to remove followed by Cu. Cd and Zn removal was poor relative to Pb and Cu [28].

REFERENCES

1 Smith, R. A. 1863. "On Adsorption of Gases by Charcoal," *Proc. Roy. Soc. London,* 12(1):424.

2 Rhead, T. F. E. and R. V. Wheeler. 1912. *J. Chem. Soc.,* 101:846.

3 Rhead, T. F. E. and R. V. Wheeler. 1913. *J. Chem. Soc.,* 103:461.

4 Steenberg, B. 1944. *Almquist and Wiksells. Boktryckeri Aktiebolag Uppsala.*

5 Puri, B. R. and R. C. Bansal. 1964. *Carbon,* 1(4):457.

6 Puri, B. R. 1966. *Carbon,* 4(3):391.

7 Garten, V. A. and D. E. Weiss. 1957. *Rev. Pure Appl. Chem.,* 7:69.

8 Boehm, H.P. and M. Voll. 1970. "Basische Oberflaechenoxide auf Kohlenstoff—1. Adsorption von Saeuren," *Carbon,* 8:227.

9 Huang, C. P. 1981. "The Surface Acidity of Hydrous Solids," in *Adsorption of Inorganics at Solid-Liquid Interfaces,* M. A. Anderson and A. J. Rubin ed., Ann Arbor Science Publishers, Inc., Ann Arbor, MI, pp. 183–218.

10 Yates, D. E. and T. R. Healy. 1975. "Mechanisms of Anion Adsorption at the Ferric and Chromic Oxide/Water Interfaces," *J. of Colloid and Interface Science,* 52(2):222.

11 Davis, J. A. and L. O. Leckie. 1978. "Surface Ionization and Complexations at the Oxide/Water Interface: II. Surface Properties of Amorphous Iron Oxyhydroxide and Adsorption of Metal Ions," *J. Colloid and Interface Sci.,* 67(1):90.

12 Corapcioglu, M. O. and C. P. Huang. 1987b. "The Surface Acidity and Characterization of Some Commercial Activated Carbons," *Carbon,* 25(4):569.

13 Reed, B. E. and M. R. Matsumoto. 1991. "Modeling Surface Acidity of Two Powdered Activated Carbons: Comparison of Diprotic and Monoprotic Surface Representations," *Carbon,* 29(8):1191.

14 Sigworth E. A. and S. B. Smith. 1972. "Adsorption of Inorganics by Activated Charcoal," *J AWWA.,* 64:386–391.

15 Reed, B. E., M. Jamil and B. Thomas. 1996. "Effect of pH, Empty Bed Contact Time (EBCT) and Hydraulic Loading (HLR) on Pb Removal by Granular Activated Carbon (GAC) Columns," *Water Environment Research,* 68(5):877.

16 Wu, D. Y., M. H. Hsu and C. P. Hunag. 1978. "Regeneration of Activated Carbon for the Adsorption of Chromium," *Proc. Purdue Ind. Waste Conf.,* Lewis Publishers, Chelsea, MI, pp 409–419.

17 Reed, B., R. Vaughan and L. Jiang. 2000. "Adsorption of Heavy Metals Using Iron Impregnated Activated Carbon," *ASCE J. of Env. Eng.,* 126(9), 869–874.

18 Netzer, A. and D. E. Hughes. 1984. "Adsorption of Copper, Lead, and Cobalt by Activated Carbon," *Water Res.,* 18(8):927.

19 Reed, B. E.. 1995. "Identification of Removal Mechanisms for Lead in Granular Activated Carbon (GAC) Columns," *Separation Science and Technology,* 29(1):1529.

20 Reed, B. E. and M. R. Matsumoto. 1993a. "Modeling Cadmium Adsorption in Single and Binary Adsorbent (Powdered Activated Carbon) Systems," *ASCE J. of Env. Eng.,* 119(2):332.

21 Farley, K. J., D. A. Dzomback and F. M. M. Morel. 1985. "A Surface Precipitation Model for the Sorption of Cations on Metal Oxides," *J. Colloid and Interface Sci.,* 106(1):226.

22 Tan, T. C. and W. K. Teo. 1987 "Combined Effect of Carbon Dosage and Initial Adsorbate Concentration on the Adsorption Isotherm of Heavy Metals on Activated Carbon," *Water Res.,* 21(10):1183–1188.

23 Corapcioglu, M. O. and C. P. Huang. 1987a. "The Adsorption of Heavy Metals onto Hydrous Activated Carbon," *Water Res.,* 9(9):1031.

24 Smith, R. M. and A. E. Martell. 1976. *Critical Stability Constants,* Plenum Press, New York.

25 Huang, C. P. and D. W. Blankenship. 1984. "The Removal of Mercury(II) from Dilute Aqueous Solution by Activated Carbon," *Water Res.,* 18(1):37.

26 Reed, B. E. and M. R. Matsumoto. 1993b. "Modeling Cadmium Adsorption by Activated Carbon Using Langmuir and Freundlich Isotherm Expressions," *Separation Science and Technology,* 28(13,14):97.

27 Reed, B. E. and S. K. Nonavinakere. 1992. "Heavy Metal Adsorption by Activated Carbon: Effect of Complexing Ligands, Competing Adsorbates, and Ionic Strength," *Separation Science and Technology,* 27(14):1985.

28 Kuennen, R. W., R. M Taylor, K. Van Dyke and K. Groenevelt. 1992. "Removing Lead from Drinking Water with Point-of-Use GAC Fixed-Bed Adsorber," *J AWWA,* Feb., pp. 91–101.

29 Boomhower, A. E. 1982. *The Removal of Cd(II) from Water by Activated Carbon Process,* M.S. Thesis, University of Delaware, Newark, Delaware.

30 Huang, C. P. 1984. *Current Removal of Toxic Heavy Metals and Organic Substances by Activated Carbon Process from Contaminated Groundwater,* NTIS PB85-218972.

31 Huang, C. P. and E. H. Smith. 1981. "Removal of Cd(II) from Plating Wastewater by an Activated Carbon Process," in *Chemistry in Water Reuse,* Vol. 2, W.J. Cooper, ed., Ann Arbor Science Publishers, Inc., Ann Arbor, Michigan, pp. 355–400.

32 Rubin, A. J. and D. L. Mercer. 1987. "Effect of Complexation on the Adsorption of Cadmium by Activated Charcoal," *Sep. Sci. and Tech.,* 22(5):1359–1381.

33 Huang, C. P., M. W. Tsanq and Y. S. Hsieh. 1985. "The Removal of Cobalt(II) from Water by Activated Carbon," in *Separation of Heavy Metals and Other Trace Contaminants,* R. W. Peters and B. King, ed., *AIChE* No. 243, 81:85.

34 Chan, D., J. W. Perram, L. R. White and T. W. Healy. 1975. "Regulation of Surface Potential at Amphoteric Surfaces during Particle-Particle Interaction," *J. of the Chemical Society, Faraday Transactions I,* 71:1046.

35 Dzombak D. A. and Morel, F. M. M. 1987. "Adsorption of Inorganic Pollutants in Aquatic Systems," *ASCE J. of Hydraulic Engineering*, 113(4):430.

36 Snoeyink, V. L. and W. J. Weber. 1967. "The Surface Chemistry of Activated Carbon," *Environmental Science and Technology*, 1(3):228.

37 Mattson, J. S. and H. B. Mark. 1971. *Activated Carbon: Surface Chemistry and Adsorption from Solution*, Marcel Dekker, Inc., New York, NY, pp. 129–157.

38 Westall, J. C. and H. Hohl. 1980. "A Comparison of Electrosatic Models for the Oxide/Water Interface," *Advances in Colloid and Interface Sci.*, 12:265.

39 Westall, J. C. 1982. "FITEQL: A Program for the Determination of Chemical Equilibrium Constants from Experimental Data, Version 2.0," United States Department of Energy, Contract # DE-AC06-76Rlo 1830.

40 Reed, B. E., J. J. Jenson and M. R. Matsumoto. 1993. " Acid-Base Characteristics of Powdered Activated Carbon," *ASCE J. of Env. Eng.*, 119(3):585.

41 Stumm, W. and J. Morgan. 1981. *Aquatic Chemistry*, John Wiley and Sons, Inc., New York, NY.

42 Benjamin M. M. and J. O. Leckie. 1981. "Multiple-Site Adsorption of Cd, Cu, Zn, and Pb on Amorphous Iron Oxyhydroxide," *Journal of Colloid and Interface Science*, 79(1):209.

43 Westall, J. C., J. L Zachary and F. M. M. Morel. 1976. "Mineql: A Computer Program for the Calculation of Chemical Equilibrium Composition of Aqueous Systems," Technical Note No. 18, Ralph M. Parsons Laboratory.

44 Bhattacharya, A. K. and C. Venkobachar. 1984. "Removal of Cadmium(II) by Low Cost Adsorbents," *ASCE J. of Env. Eng.*, 110(1):110.

45 Huang, C. P. 1978. "Chemical Interactions between Inorganics and Activated Carbon," in *Carbon Adsorption Handbook*, P. N. Cheremisinoff and F. Ellenbusch, ed., Ann Arbor, Michigan, pp. 281–329.

46 Pandey, M. P. and M. Chaudhuri. 1982. "Removal of Inorganic Mercury from Water by Bituminous Coal," *Water Res.*, 16(7):1113.

47 Rao P. S. 1992. "Kinetic Studies on Adsorption of Chromium by Coconut Shell Carbons from Synthetic Effluents," *J. Environ. Sci. Health*, A27(8):2227–2241.

48 Rajakovic, L. V. 1992. "Sorption of Arsenic onto Activated Carbon Impregnated with Metallic Silver and Copper," *Sep. Sci. and Tech.*, 27(11):1423.

49 Wilczak, A. and T. M. Keinath. 1993. "Kinetics of Sorption and Desorption of Copper(II) and Lead(II) on Activated Carbon," *Water Env. Res.*, 65(3) 238–244.

50 Arulanantham, A., N. Balasubramanian and T. V. Ramakrishna. 1989. "Coconut Shell Carbon for Treatment of Cadmium and Lead Containing Wastewater," *Metal Finishing*, 87:51.

51 Huang, C. P. and P. K. Wirth. 1982. "Activated Carbon for Treatment of Cadmium Wastewater," *ASCE J. of Env. Eng.*, 108(6):1280.

52 Clark, R. M. and B. W. Lykins Jr. 1991. *Granular Activated Carbon: Design, Operation and Cost*, Lewis Publishers, Chelsea, MI.

53 Snoeyink, V. L. 1990. "Adsorption of Organic Compounds," in *Water Quality and Treatment*, F. W. Pontius, ed., McGraw-Hill, Inc., New York, pp. 781–876.

54 Cover, A. E. 1969. *Appraisal of Granular Carbon Contacting. Phase I: Evaluation of the Literature on the Use of Granular Carbon for Tertiary Wastewater Treatment. Phase II: Economic Effect of Design Variables*, U.S. Dept. of Interior, FWPCA, Report No. TWRC-11.

55 Bowers, A. R. and C. P. Huang. 1980. "Activated Carbon Processes for the Treatment of Cr(VI) Containing Industrial Wastewaters," *Prog. Wat. Tech.*, 12:629.

56 Kataoka, T., Y. Ozasa and H. Yoshida. 1977. "Breakthrough Curve in Ion Exchange Column—Particle Diffusion Control," *Nippon Kagaku Kaishi*, No. 3, 387–390.

57 Shay, M. and J. E. Etzel. 1992. "Treatment of Metal-Containing Wastewaters by Carbon Adsorption of Metal-Chelate Complexes," *Proc. Purdue Ind. Waste Conf.*, Lewis Publishers, Chelsea, MI, pp. 563–569.

58 Reed, B. E. and S. Arunachalam. 1994. "Use of Granular Activated Carbon Columns for Lead Removal," *ASCE J. of Env. Eng.*, 120(2):416.

59 Reed, B. E., J. Robertson and M. Jamil. 1995a. "Regeneration of Granular Activated Carbon (GAC) Columns Used for Pb Removal," *ASCE J. of Env. Eng.*, 121(9):653.

60 Love, T. and R. G. Eilers. 1982. "Treatment of Drinking Water Containing Trichloroethylene and Related Industrial Solvents," *J. AWWA*, 74(8):413.

61 Reed, B. E., P. Carriere and B. Thomas. 1995b. "Use of Granular Activated Carbon (GAC) Columns to Remove Heavy Metals and Organics," Proceedings: *SUPERFUND XVI Conference,* Washington, D.C.

62 Vaughan, Jr., R. L., B. E. Reed, R. C. Viadero, M. Jamil and M. Berg. 1999. "Simultaneous Removal of Organic and Heavy Metal Contaminants by Granular Activated Carbon (GAC) Columns," *Advances in Environmental Research,* 3(3):229.

63 Chen, T. N. and C. P. Huang. 1992. "Treatment of Zinc Industrial Wastewater by Combined Chemical Precipitation and Activated Carbon Adsorption," in *24th Mid-Atlantic Industrial Wastewater Conference,* B. E. Reed and W. Sack, ed., Technomic Publishing Co., Inc., Lancaster, PA, pp. 120–134.

CHAPTER 8

Arsenic in Subsurface Water: Its Chemistry and Removal by Engineered Processes

ARUP K. SENGUPTA[1]
JOHN E. GREENLEAF[1]

INTRODUCTION

ARSENIC has been known as a highly toxic element for centuries. Arsenic is ubiquitous in earth's crust in miniscule amounts, e.g., less than 0.01% by mass [1]. Orpiment (As_2S_3), realgar (AsS) and arsenopyrite (FeAsS) are some of the commonly encountered solid phases in which arsenic exists in subsurface soils [2,3]. Note that sulfur and iron are the two common companions of arsenic, and these compounds are formed in subsurface soils under a highly reducing environment. Geochemical or biogeochemical weathering of these solid phases often, if not always, contributes to elevated levels of dissolved arsenic in groundwaters. Historically, anthropogenic sources have also been responsible for discharging arsenic in the environment. Arsenic is commercially used to produce arsenical pesticides, wood preservatives, ceramics, semiconductors and other materials. Mine drainage, smelter wastes and agricultural drainage water have also been found to be responsible for elevated concentrations of As in soil and water [4–6]. Arsenic compounds present in soil and water exist primarily in two oxidation states, namely, (+III) and (+V).

In humans and most experimental mammals, inorganic arsenic is methylated to monomethylarsonic acid (MMA) and dimethylarsinic acid (DMA). Compared with inorganic arsenic, the methylated metabolites are less reactive with tissue constituents, less acutely toxic and more readily excreted in the urine [7,8]. Due primarily to their reduced toxicity when compared with their inorganic counterparts, methylarsenicals, MMA and DMA, are widely used as pesticides, herbicides and defoliants. At low dosages, such as ingestion of arsenic from contaminated drinking water, the chronic toxicity resulting from inorganic As(V) and As(III) is virtually the same because As(V) is rapidly reduced in the body to As(III) [9]. Arsenic toxicity resulting from the ingestion of contaminated water thus remains virtually unaffected by changes in the relative distribution of As(V) and As(III). However, acute toxicity of inorganic As(III) at high dosage is significantly greater than that

[1]Department of Civil and Environmental Engineering, Lehigh University, 13 E. Packer Avenue, Bethlehem, PA 18015, U.S.A., aks0@lehigh.edu

of As(V). Table 8.1 includes the formulas and salient properties of As(V) oxyacid, As(III) oxyacid and their metabolites MMA and DMA.

Two major ongoing events, characteristically unrelated, have sent "arsenic" to front-page headlines in the media several times during the last ten years. First, the maximum contaminant level (MCL) of arsenic in drinking water in the United States has been under careful scrutiny [10–13]. Because it is generally accepted that high levels of arsenic in drinking water have adverse health effects, the United States Environmental Protection Agency (U.S.EPA) gathered and analyzed health effects data to promulgate a new arsenic MCL. In May 2000, U.S.EPA proposed a MCL for arsenic in drinking water of 5 micrograms per liter and requested comments on 3 and 10 micrograms per liter. This debate over a lower arsenic MCL hinges on balancing the costs versus the reduced risks. In 1997, Frey et al. [14] determined that national annualized compliance costs for groundwater utilities (including capital, operating and maintenance costs) regarding potential arsenic MCLs of 2, 5 and 10 micrograms per liter would be 4.4, 2.0 and 0.6 billion dollars, respectively. It is encouraging that due to decreasing costs in arsenic removal technologies and the option of blending, the cost of compliance for potential arsenic MCLs has been found lower than originally thought. In a recent Arsenic Research Partnership study managed by the American Water Works Research Foundation (AWWARF), the revised annualized compliance costs for potential arsenic MCLs of 2, 5 and 10 micrograms per liter were found lower and equal to 3.8, 1.3 and 0.6 billion dollars, respectively, for groundwater systems [15]. U.S.EPA is also proposing to set a public health goal of zero for arsenic. U.S.EPA says it does this for all known carcinogens for which there is no dose considered safe.

The other event that has drawn international attention is the arsenic-inflicted crisis affecting millions of people in the Indian subcontinent. In Bangladesh and the neighboring Indian state of West Bengal, drinking water drawn from underground sources has been responsible for widespread arsenic poisoning affecting over 70 million people [16–22]. According to current estimates, the adverse health effects caused by arsenic poisoning in

TABLE 8.1. Inorganic and Methylated Arsenic Acids.

Name	Formula	Molecular Weight	Negative Logarithm of Acid Dissociation Constant pK_a Values
Arsenic Acid	$\begin{array}{c} O \\ \| \\ As \\ / \mid \backslash \\ OH\ OH\ OH \end{array}$	142	$pK_{a1} = 2.2$ $pK_{a2} = 6.98$ $pK_{a3} = 11.6$
Arsenous Acid	$\begin{array}{c} O \\ \| \\ As \\ \| \\ OH \end{array}$ or $\begin{array}{c} OH \\ \| \\ As-OH \\ \| \\ OH \end{array}$ (hydrated)	124 or 126	$pK_{a1} = 9.2$
Monomethyl Arsonic Acid (MMA)	$\begin{array}{c} O \\ \| \\ OH-As-CH_3 \\ \| \\ OH \end{array}$	138	$pK_{a1} = 4.2$ $pK_{a2} = 8.8$
Dimethyl Arsinic Acid (DMA)	$\begin{array}{c} O \\ \| \\ H_3C-As-CH_3 \\ \| \\ OH \end{array}$	137	$pK_{a1} = 6.27$

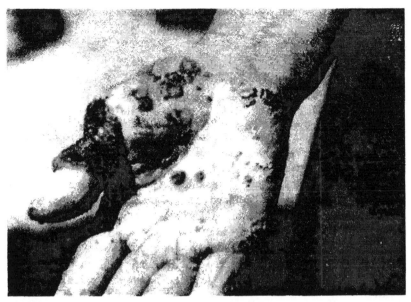

Figure 8.1 Signs of malignant tumors and hyperkeratosis during the advanced stages of arsenicosis. From Lepkowski, *C&EN*, Nov. 16, 1998, 27, with permission from Willard Chappell.

this geographic area are far more catastrophic than any other natural calamity throughout the world in recent times. Figure 8.1 shows the palm of an arsenic-inflicted patient in Bangladesh; signs of malignant tumors and hyperkeratosis are apparent. Although the genesis of arsenic contamination in groundwater for this area is yet to be fully understood, natural geochemical weathering of subsurface soil, and not industrial pollution, is the sole contributor of dissolved arsenic in groundwater. In principle, the arsenic-leaching mechanism in contaminated groundwaters is the same regardless of geographic location. However, the difference in magnitude and severity of arsenic contamination may vary with subsurface soil composition, groundwater withdrawal rate, application of fertilizer and other related human activities.

In all such contaminated groundwaters, dissolved arsenic exists as inorganic compounds in the oxidation states of +III and +V. In this chapter, we will refer to them as As(III) or arsenites, and As(V) or arsenates. Environmental separation of dissolved arsenic from groundwaters essentially involves removal of inorganic arsenic species from contaminated water bodies. Because organoarsenical compounds are less toxic than their inorganic counterparts and result primarily from industrial discharges, the primary focus of this chapter will pertain to the chemistry of inorganic arsenic compounds and engineered processes for their removal from the aqueous phase.

ARSENIC: ITS CHEMISTRY AND NATURAL OCCURRENCE

Arsenic is a borderline element between metals and nonmetals, commonly referred to as "metalloid." It is located in Group VB (or new 15) of the periodic table along with nitrogen, phosphorus, antimony and bismuth. Unlike phosphorus, which is an essential nutrient for all living creatures, arsenic has attained notoriety primarily as a highly toxic and poisonous element. Nevertheless, their chemistries are characteristically similar in several ways. The electronic configurations of As and P are as follows:

$$\text{As:} \quad 1S^2 2S^2 2P^6 3S^2 3P^6 3D^{10} \boxed{4S^2 4P^3} \tag{1}$$

$$\text{P:} \quad 1S^2 2S^2 2P^6 \boxed{3S^2 3P^3} \tag{2}$$

Identical outer electronic configurations (numbers in boxes) result in similar chemical behaviors for many As and P compounds, especially in the +V oxidation states. Table 8.2 summarizes a few such similarities between As(V) and P(V) oxyacids and the ligand properties of their conjugate oxyanions. For comparison, As(III) oxyacid is also included in the same table. Due to their near-identical ligand properties, phosphate solutions are often used to displace arsenates sorbed onto hydrous oxides [23,24]. Phosphate was also found in several arsenic contaminated groundwaters in the Indian subcontinent [25]. It was hypothesized that the leaching of arsenic in groundwater from iron-rich soils had been accelerated due to the extensive application of phosphate-based fertilizer [26]. A recent study has substantiated the possible role of phosphate in increasing arsenic concentration in groundwaters [27]. Between arsenic and phosphorus, there also remain critical dissimilarities. The two-electron reduction of arsenate or As(V) to arsenite or As(III) is favored in acidic solution, whereas the reverse is true in basic solution. In contrast, phosphorus(V) compounds are difficult to reduce [28]. Another major difference between arsenic and phosphorus is that the esters of phosphoric acid are stable, while the esters of As(V) acid are easily hydrolyzed. Thus, free As(V) is more readily available during the metabolic processes, while phosphorus is irreversibly incorporated into esters during ATP (adenosine triphosphate) biosynthesis. Such an availability of free inorganic arsenate and its subsequent reduction to arsenite in the blood supposedly intensifies the severity of arsenic poisoning.

The natural occurrence of arsenic and its fate and mobility in the aqueous and soil environment are seemingly different from the engineered processes often employed for effi-

TABLE 8.2. Oxyacids and Conjugate Anions of As(V) and P(V).

Parent Oxyacid	pK$_a$ Values	Predominant Dissolved Species at pH 5.5	Predominant Dissolved Species at pH 8.5
As(V): H$_3$AsO$_4$	pK$_{a1}$ = 2.2 pK$_{a2}$ = 6.98 pK$_{a3}$ = 11.6	 Monovalent Monodentate Ligand	 Divalent Bidentate Ligand
P(V): H$_3$PO$_4$	pK$_{a1}$ = 2.12 pK$_{a2}$ = 7.21 pK$_{a3}$ = 12.70	 Monovalent Monodentate Ligand	 Divalent Bidentate Ligand
As(III): HAsO$_2$	pK$_{a1}$ = 9.2	O = As — OH Nonionized Monodentate Ligand	O = As — OH Nonionized Monodentate Ligand

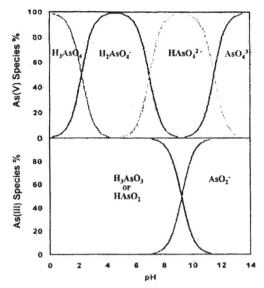

Figure 8.2 Distribution of As(V) and As(III) oxyacids and their conjugate anions as a function of pH.

cient removal of arsenic from contaminated water. However, the underlying chemistry governing these two diverse phenomena is quite similar. For this reason, we will first discuss the chemistry of arsenic in relation to its presence and mobility in the natural environment. Redox conditions, pH and the chemistry of accompanying materials are three primary variables in this regard. Pertinent redox reaction and acid dissociation constants are as follows [29–31]:

$$H_3AsO_4: \quad pK_{a,1} = 2.2; \quad pK_{a,2} = 7.0; \quad pK_{a,3} = 11.5 \tag{3}$$

$$HAsO_2 \text{ or } H_3AsO_3: \quad pK_{a,1} = 9.2 \quad \text{(others not known)} \tag{4}$$

$$H_3AsO_4 + 2H^+ + 2e^- \rightarrow H_3AsO_3 + H_2O; \quad E_h = +0.56V \tag{5}$$

Figure 8.2 shows the distribution of As(V) and As(III) species as a function of pH. Note that at near-neutral pH, monovalent $H_2AsO_4^-$ and divalent $HAsO_4^{2-}$ are the predominant As(V) species. An electrically neutral $HAsO_2$ (or hydrated H_3AsO_3) is the major As(III) species under identical conditions. Figure 8.3 represents the predominance or pe-pH diagram for various As(III) and As(V) species. Also, for meaningful interpretation of arsenic mobility (or immobility) in the environment, some commonly encountered solid-phase arsenic compounds are superimposed on Figure 8.3. Under a highly oxidizing environment and moderate alkaline condition, ferric arsenate precipitate, $FeAsO_{4(s)}$, is the predominant solid phase. On the contrary, under a highly reducing environment, realgar (AsS), orpiment (As_2S_3) and arseno pyrites (FeAsS) are the most thermodynamically favorable solid phases. Around neutral pH under atmospheric conditions (pe values greater than 13.0), As(V) is by far the most predominant species. Oxyanions of As(V) are thus the most commonly encountered arsenic species in surface waters (rivers, lakes, etc.). For groundwaters, prevailing redox and pH conditions as shown by the shaded rectangular box in Figure 8.3 favor the presence of As(III) species along with As(V). Recently gathered data from multiple

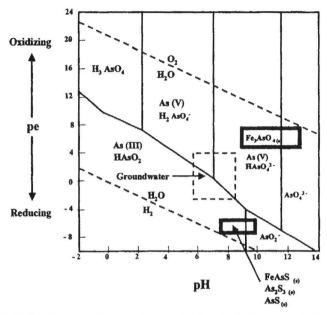

Figure 8.3 Predominance or pe-pH diagram of various As(V) and As(III) species including precipitates.

groundwater wells in Bangladesh and in the city of Hanford, California, USA, show evidence of As(III) constituting well over 50% of total dissolved arsenic in many wells [32,33].

Mobility/Immobility in the Natural Environment

Under highly reducing conditions within the subsurface soil, arsenic exists primarily as $As_2S_{3(s)}$, $AsS_{(s)}$ and $FeAsS_{(s)}$. If these solid phases are exposed to a relatively oxidizing environment, sulfides will tend to be oxidized into more soluble sulfate species, causing geochemical leaching or oxidative dissolution of arsenic in groundwater. In Bangladesh and in the eastern part of India, groundwater has been indiscriminately withdrawn during the last 20 years for agricultural irrigation. As a result, the recharging of aquifers occurred essentially through the percolation of surface water containing dissolved oxygen. Although scientifically unproven, the continuous exposure of arsenic-containing solid phases to the oxidizing conditions generated by aerated water has been postulated as a major pathway for an elevation of arsenic concentration in groundwater through soil leaching [25].

Any solute, dissolved gas or solid phase, that tends to alter the redox condition of the aqueous phase strongly influences the speciation of dissolved arsenic, as illustrated in Figure 8.4. Note that in the presence of oxygen or $MnO_{2(s)}$, i.e., when O_2/H_2O or MnO_2/Mn^{2+} is the prevailing redox pair, As(V) is the predominant species [29]. At near-neutral pH, hydrated Fe(III) oxides (HFO) favor oxidation of As(III) to As(V) thermodynamically. However, oxidation of As(III) by iron solids has not been observed as yet [34,35]. In the presence of dissolved oxygen and/or manganese dioxide, relatively soluble Fe(II) is easily oxidized to insoluble Fe(III) oxides or HFO. Both As(V) oxyanions and As(III) oxyacid are fairly strong ligands that are selectively sorbed onto HFO particles through the formation of inner sphere complexes [23,24] that have been defined as covalent linkages between the adsorbed ion and the sorption site with no water of hydration between the adsorbed ion and the surface functional group. Fate and transport of arsenic in subsurface environment are, there-

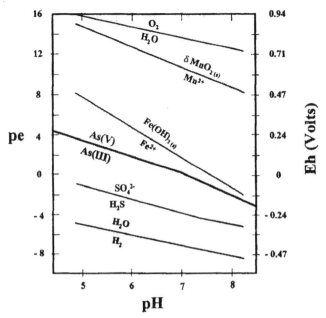

Figure 8.4 Hierarchy and stability of As(III)/As(V) system in comparison with other redox pairs as a function of pH.

fore, controlled not only by pH and redox conditions but also by sorption behaviors of As(V) oxyanions and neutral As(III) molecules. Figure 8.5 illustrates the fate of inorganic arsenic species under varying redox conditions and in the presence of iron.

In many industrial waste sites, the fate and transport of arsenic and other trace metals in a subsurface system are closely linked to biogeochemical reactions. These reactions occur as a result of organic carbon being degraded by different microorganisms using a series of ter-

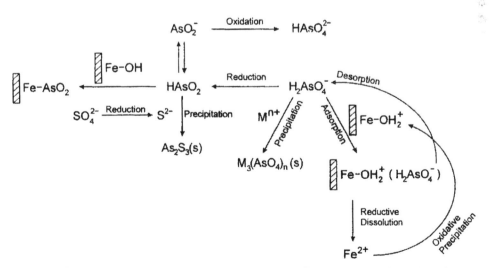

Figure 8.5 A schematic illustrating fates of inorganic arsenic in subsurface environment under varying redox conditions.

minal electron acceptors [36,37]. Arsenic contaminated sites may change over the short term (i.e., decrease of organics due to enhanced bioremediation) or long term (decrease of organic contaminant load due to natural attenuation). Geochemical conditions will continue to change and, consequently, arsenic compounds can be mobilized/immobilized via processes such as oxidation/reduction, sorption/desorption, precipitation/dissolution and/or the formation of complex coordinate compounds. Figure 8.6 illustrates how arsenic speciation and partitioning in groundwater systems change with the changes in electron acceptors responsible for biological reactions. In principle, the integrity of hazardous sites containing biodegradable organic wastes along with inorganic arsenic contaminants is amenable to more reducing environments and thus different from natural subsurface environments free of organic matters.

Mobility of arsenic in the subsurface environment may also be influenced directly by microbially mediated biological reactions. Although the exact role of microbially mediated arsenic release from soil is deemed relatively insignificant, the reduction of As(V) to As(III) is one mechanism by which arsenic may be mobilized. To this effect, the strain MIT-13, a respiratory As(V)-reducing bacterium, has been isolated from the Aberjona Watershed [38]. Because sorption affinity of arsenites onto amorphous iron-oxide materials is less than arsenates, arsenate-reducing bacteria may play a significant role in mobilizing arsenic through the direct reduction of arsenate.

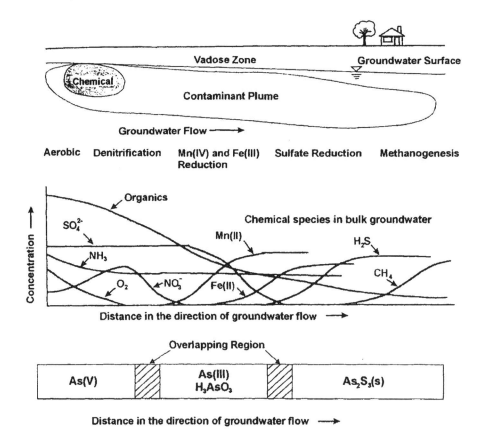

Figure 8.6 An illustration of the change in arsenic speciation with the change in biogeochemistry of the subsurface environment (adapted from Bouwer and Zehnder, 1993, *Trends in Biotechnology*).

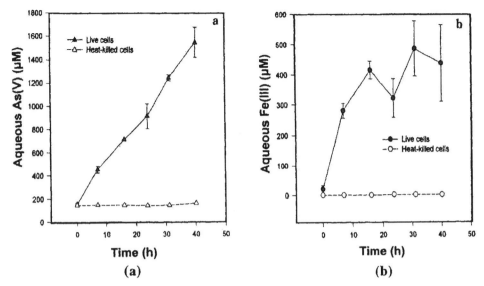

Figure 8.7 (a) Arsenic dissolution from scorodite incubated with S.alga BrY and lactate. As(III) was below the detection limit. (b) Fe(II) produced from scorodite by *S. alga* BrY in the presence of lactate (from Cummings, Caccavo, Fendorf and Rosenweig, *Environ. Sci. Technol.,* 1999, 33, 723–729, with permission from ACS).

The other plausible microbially mediated mechanism is through reductive dissolution of Fe(III) to Fe(II) without any change in the oxidation state of As(V). In a recent study [39], the dissimilatory iron-reducing bacterium, *Shewanella alga* strain promoted arsenic mobilization from a crystalline ferric arsenate as well as from sorption sites within sediments. Spectroscopic analyses confirmed that Fe(III) was reduced to Fe(II), but the oxidation state of As(V) remained unchanged in both liquid and solid phases. Figures 8.7(a) and 8.7(b) illustrate how As(V) and Fe(II) increased in the aqueous phase during reductive dissolution of ferric arsenate (scorodite) in the presence of iron-reducing bacterium.

ARSENIC REMOVAL TECHNOLOGIES: UNDERLYING PRINCIPLES

Removal of dissolved arsenic from natural groundwater and surface water essentially constitutes removals of As(V) oxyanions and neutral As(III) oxyacid. The list of existing and emerging arsenic removal technologies is quite long and is summarized below:

- activated alumina sorption
- polymeric anion exchange
- sorption on iron oxide coated sand (IOCS) particles
- enhanced coagulation with alum or ferric chloride dosage
- ferric chloride coagulation followed by microfiltration
- pressurized granulated iron particles
- polymeric ligand exchange
- iron-doped alginate
- sand with zero-valent iron
- reverse osmosis or nano-filtration

The technical details of the foregoing processes and their relative advantages and disadvantages are recorded in the literature [40–60]. Because arsenic removal processes from contaminated groundwater often pertain to trace concentrations (less than 500 μg/L), precipitation is rarely a major removal mechanism. While the equipment configuration and operational protocol for the above-mentioned arsenic removal processes are often quite different, the underlying chemistry essentially rests on the following two types of interactions. First, As(V) oxyanions (e.g., $H_2AsO_4^-$ and $HAsO_4^{2-}$) possess negative charges and can, therefore, undergo coulombic or ion-exchange (IX)-type interactions. In this instance, conventional anion exchange processes are seemingly quite suitable for removals of As(V) oxyanions. In contrast, As(III) exists as a nonionized oxyacid around neutral pH and is not amenable to removal by ion exchange processes. Second, As(V) and As(III) species are fairly strong ligands or Lewis bases, i.e., they are capable of donating lone pairs of electrons. Thus, they can participate in Lewis acid-base (LAB)-type interactions and often exhibit high sorption affinity toward solid surfaces with Lewis acid characteristic. The following provides a comprehensive discussion on IX- and LAB-type interactions in context to existing arsenic removal processes.

> Note: it is noteworthy that semipermeable membrane processes, namely, reverse osmosis (RO) and nano-filtration (NF) processes also remove arsenates and arsenites [60]. The underlying mechanism of removal is, however, different from all other processes listed before and does not involve IX- or LAB-type interactions. Also, the membrane processes are nonselective, i.e., they remove other dissolved solutes along with arsenic compounds. The process chemistry of arsenic removal by membrane processes is virtually the same as that of any other dissolved species or electrolytes. That is why any further discussion of membrane processes is being avoided here. RO processes may, however, be economically viable in places where simultaneous reduction of arsenic and total dissolved solids (TDS) is warranted.

Ion Exchange (IX)

At above-neutral pH, As(V) exists as a divalent anion and can be removed by an anion exchanger in chloride form as shown below:

$$\overline{2(R^+)Cl^-} + HAsO_{4(aq)}^{2-} \Leftrightarrow \overline{(R^+)_2 HAsO_4^{2-}} + 2Cl_{(aq)}^- \tag{6}$$

The overbar denotes the exchanger phase, and R^+ represents an anion exchanger with a fixed positive charge. Table 8.3 provides the composition and salient properties of a typical anion exchange resin with polystyrene matrix and quaternary ammonium functional group. Note that nonionized As(III) can be oxidized to As(V) oxyanion in order to make it amenable to treatment by ion exchange. In contaminated waters, however, other innocuous anions, namely, chloride, bicarbonate and sulfate, are simultaneously present with dissolved arsenic. In fact, concentrations of these nontoxic competing anions are often several orders of magnitude greater than target arsenates. Thus, the arsenic removal efficiencies by typical ion exchange processes are greatly impaired in the presence of high concentrations of competing ions.

Figure 8.8 shows complete effluent histories during a fixed-bed column run using a strong base anion exchanger in chloride form. Influent composition and pertinent hydrodynamic conditions, namely, empty-bed contact time (EBCT) and superficial liquid-phase velocity (SLV) are also provided in the figure. Two aspects of the effluent history are noteworthy: first, sulfate broke through from the column after As(V) and second, As(V) underwent chromatographic elution, i.e., As(V) concentration after breakthrough exceeded its influent concentration. Similar observations have also been made by other investigators for laboratory and pilot-scale systems [40,41]. The selectivity sequence for various competing anions

TABLE 8.3. Salient Properties of Polymeric Anion Exchangers.

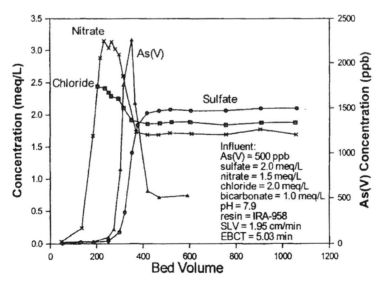

Resin	IRA-900	IRA-958
Structure (Repeating Unit)		
Functional Group	Quaternary ammonium	Quaternary ammonium
Matrix	Polystyrene, macroporous	Polyacrylic, macroporous
Capacity (meq/g air-dried resin)	3.6	3.4
Manufacturer	Rohm and Haas Co., Philadelphia, PA, USA	Rohm and Haas Co., Philadelphia, PA, USA

toward polymeric anion exchangers, based strictly on coulombic interactions, stands as follows:

$$SO_4^{2-} > HAsO_4^{2-} > H_2AsO_4^- > Cl^- > HCO_3^- \tag{7}$$

It is readily recognized that greater sulfate affinity toward polymeric anion exchanger causes reduction in As(V) uptake and chromatographic elution of arsenic. Therefore, for contaminated waters with high sulfate content, ion exchange is not a viable process.

Being a trace species, dissolved arsenic also exhibits an important effluent history characteristic during the fixed-bed ion exchange process. Let us consider the following arsenate-sulfate exchange reaction:

$$\overline{(R^+)_2 SO_4^{2-}} + HAsO_{4(aq)}^{2-} \Leftrightarrow \overline{(R^+)_2 HAsO_4^{2-}} + SO_{4(aq)}^{2-} \tag{8}$$

Figure 8.8 A complete effluent history of an As(V) contaminated water for a fixed-bed column run using a strong-base anion exchanger (IRA-958).

considering ideality, the equilibrium constant for the above exchange reaction is as follows:

$$K_{As/S} = \frac{q_{As} C_S}{q_S C_{As}} \tag{9}$$

where subscripts "As" and "S" represent arsenate and sulfate, respectively, and q_i and Ci denote concentration of species "i" in the exchanger phase (milliequiv/gm) and the aqueous phase (milliequiv/lit), respectively. Again, for an ion exchange reaction,

$$q_{As} + q_S = Q, \text{ Total Exchange Capacity}$$
$$\text{and } C_S + C_{As} = C_T, \text{ Total Aqueous Phase anion concentration}$$

Considering arsenic to be a trace species, i.e., $q_{As} \ll q_S$ and $C_{As} \ll C_S$.

Hence,

$$q_S \simeq Q \tag{10}$$

and,

$$C_S \simeq C_T \tag{11}$$

Applying these equalities in Equation (9) and transposing,

$$\frac{q_{As}}{C_{As}} = K_{As/S} \frac{Q}{C_T} \tag{12}$$

For a particular anion exchanger and a specific body of contaminated water, both Q and C_T are constants. Thus,

$$\frac{q_{As}}{C_{As}} = \text{constant} \tag{13}$$

The number of bed volumes (BVs) treated during a fixed-bed column run prior to breakthrough of a specific contaminant, "i", is proportional to q_i/C_i. For this reason, under trace conditions, BVs do not change with a change in the concentration of the specific contaminant as long as the ion exchanger and background composition of the aqueous phase remain unaltered. Figure 8.9 shows effluent histories of As(V) during three different column runs with varying As(V) concentrations in the influent, all other conditions remaining identical. Note that the arsenic breakthrough bed volume (or time) remained unaltered, while arsenic concentration in the influent changed from 10 μg/L to 2,000 μg/L. This observed phenomenon suggests that the cost-effectiveness of an arsenic removal process by ion exchange is independent of arsenic concentration in groundwater as long as its concentration is very low compared to other competing anions.

Unlike As(V), As(III) oxyacid is nonionized and, thus, is not amenable to removal solely by anion exchange processes. Although thermodynamically favorable, oxidation of As(III) by air or oxygen is kinetically a very slow process. Chlorine or sodium hypochlorite can rapidly oxidize As(III) to As(V) in less than a minute [57]. However, any chlorination or other oxidation step prior to ion exchange makes the removal process operationally complex for relatively small treatment systems serving less than 10,000 people. Use of

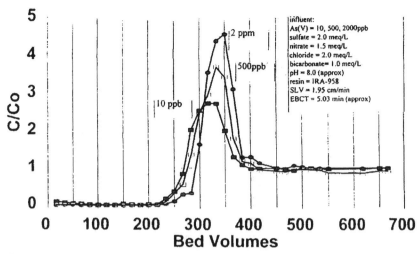

Figure 8.9 Effluent histories of As(V) during three different column runs with varying As(V) concentrations in the influent.

solid-phase oxidizing agents, e.g., $MnO_{2(s)}$ or green sand, is preferred for operational simplicity. Related studies have shown that reduced Mn^{2+} is not released into water [54,58]. Instead, the soluble Mn^{2+} remains bound to the negatively charged surface sorption sites of manganese dioxide.

Lewis Acid-Base (LAB) Interaction

As already shown, arsenates cannot be removed selectively in the presence of sulfate ions solely through ion exchange-type interactions. Also, nonionized As(III) oxyacid is not sorbable onto polymeric anion exchangers. However, both As(V) and As(III) species are Lewis bases or ligands with an ability to donate lone pairs of electrons (see Table 8.2). Note that while $H_2AsO_4^-$ and $HAsO_2$ are monodenate ligands (i.e., one donor atom per molecule), $HAsO_4^{2-}$ is a bidentate ligand with two oxygen donor atoms. Consequently, $HAsO_4^{2-}$ is a stronger ligand and forms stronger inner-sphere complexes with Lewis acids. Compared to As(V) and As(III) species, sulfate and chloride are poor ligands and form only outer-sphere complexes. Thus, Lewis acid-base interaction may be used as an underlying sorption mechanism to separate dissolved As(V) and As(III) species from relatively high background concentrations of sulfate, chloride or bicarbonate.

Several hydrous inorganic oxides (both naturally occurring and processed) and tailored polymeric sorbents contain Lewis acid-type functional groups and, hence, are effective for arsenic removal through LAB interaction. Among the inorganic materials used in engineered processes, Al(III) and Fe(III) oxides are the most common because they are relatively inexpensive and environmentally benign. Lewis acid properties of the surface functional groups of these inorganic metal oxides are pH dependent and can be presented as follows:

$$\equiv \overline{SOH_2^+} \Leftrightarrow \overline{SOH} + H^+ \tag{14}$$

$$\equiv \overline{SOH} \Leftrightarrow \overline{SO^-} + H^+ \tag{15}$$

where ≡SOH represents electrically neutral surface functional groups of hydrated metal oxides. Acid-base properties of hydrated metal oxides have been studied extensively and the findings are summarized in the literature [61,62]. In the presence of noncomplexing electrolytes (e.g., sodium perchlorate or sodium nitrate), pH corresponding to point of zero charge (PZC) represents conditions with no residual surface charges, i.e., negative surface charges essentially equal the positive surface charges. At pH below PZC, hydrous metal oxides possess fixed positive charges and can thus act as anion exchangers. For hydrous aluminum oxide and iron oxide, pHs corresponding to points of zero charge are approximately 7.8 and 8.3, respectively [43,61–63]. As(V) oxyanions (say $H_2AsO_4^-$) can be sorbed onto such sites through an exchange reaction (say with chloride) as shown below:

$$\equiv \overline{SOH_2^+(Cl^-)} + H_2AsO_4^- \Leftrightarrow \overline{SOH_2^+(H_2AsO_4^-)} + Cl^- \tag{16}$$

Note that electrostatically, both Cl^- and $H_2AsO_4^-$ are identical, i.e., both have one negative charge. Strictly based on ion exchange or coulombic interaction, monovalent arsenate cannot exhibit much greater affinity than competing chloride anion. In reality, however, $H_2AsO_4^-$ is highly preferred over Cl^- by hydrated aluminum or iron oxides. Such a high sorption affinity results from the formation of inner-sphere complexes through LAB interaction between arsenate (Lewis base) and terminal Fe or Al atom in the solid oxide. Thermodynamically, the overall free energy change for the reaction in Equation (16) includes contributions from both ion-exchange (IX) and Lewis acid-base (LAB)-type interactions, i.e.,

$$\Delta G_{overall}^0 = \Delta G_{IX}^0 + \Delta G_{LAB}^0 \tag{17}$$

$$\text{or} \quad -RT \ln K_{overall} = -RT \ln K_{IX} - RT \ln K_{LAB} \tag{18}$$

$$\text{or} \quad K_{overall} = K_{IX} K_{LAB} \tag{19}$$

where ΔG^0 denotes free energy changes at the standard state, R is the universal gas constant, T is the temperature in Kelvin and K is the equilibrium constant. Note that for $H_2AsO_4^-$ and Cl^- exchange, K_{LAB} is significantly greater than unity, and consequently, $H_2AsO_4^-$ uptake is preferred over Cl^-.

For As(V) compounds, namely, $H_2AsO_4^-$ and $HAsO_4^{2-}$, both coulombic and LAB interactions are operative. However, the sorption of monodentate nonionized arsenite (i.e., $HAsO_2$) takes place solely through LAB interactions. Various surface complexation reactions leading to the generation of new surface charges and consequent release of H^+ or OH^- in the aqueous phase are also possible as shown below for $H_2AsO_4^-$ and $HAsO_2$:

$$\equiv \overline{SOH} + H_2AsO_{4(aq)}^- + H_2O \Leftrightarrow \overline{SOH_2^+(H_2AsO_4^-)} + OH_{(aq)}^- \tag{20}$$

$$\equiv \overline{SO^-} + H_2AsO_{4(aq)}^- + 2H_2O \Leftrightarrow \overline{SOH_2^+(H_2AsO_4^-)} + 2OH_{(aq)}^- \tag{21}$$

$$\equiv \overline{SOH_2^+} + HAsO_{2(aq)} \Leftrightarrow \overline{SOH(HAsO_2)} + H_{(aq)}^+ \tag{22}$$

Other possible stoichiometries pertaining to arsenate/arsenite sorption have been provided elsewhere and are not duplicated here [63]. Ion exchange-type sorption reactions that do not introduce new surface charges or add aqueous-phase acidity or basicity are thermodynamically most favorable. Figure 8.10 depicts the sorption of several solutes onto hydrated iron oxide surfaces through formation of inner-sphere and outer-sphere complexes. Note that sulfate and chloride anions form only outer-sphere complexes and thus have much lower sorption affinity compared to arsenates and arsenites. Phosphate, however, forms bidentate inner-sphere complexes like divalent arsenate. Application of phosphate-based fertilizers has recently been cited as a major factor responsible for desorption of arsenate from iron-rich soil in the subsurface environment of the Indian subcontinent and consequent elevation of arsenic concentration in groundwater [27].

Points of zero charge (PZC) or isoelectric points of hydrated metal oxides, as already indicated, are pH values where the net surface charge is zero. Isoelectric pH or pH_{pzc}, however, is not constant for a specific oxide material and depends on the solution composition. For solutes or electrolytes forming only outer-sphere complexes, pH_{pzc} is essentially independent of the solute concentrations or ionic strengths. Figure 8.11 shows that pH_{pzc} of ferrihydrite is equal to 8.3 and remains unchanged at three different sodium chloride concentrations. However, pH_{pzc} is reduced in the presence of a ligand due to enhanced surface protonation caused by the ligand uptake. Figure 8.12 shows how pH_{pzc} of ferrihydrite is reduced significantly when arsenites and arsenates are present in the solution even at fairly low concentrations compared to sodium chloride [63]. The observation in Figure 8.12 that the reduction in pH_{pzc} is greater with As(V) than with As(III) also suggests that As(V) forms stronger inner-sphere complexes with the surface functional groups of ferrihydrite than As(III).

Due to their weak acid-base properties, hydrated metal oxides have high affinities to-

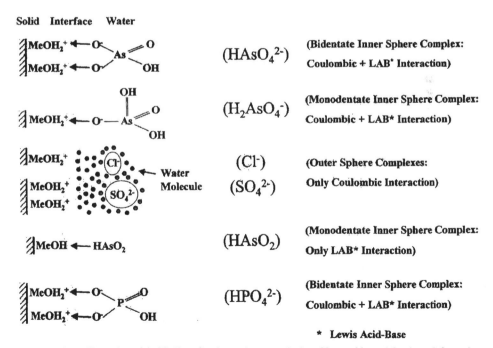

Figure 8.10 An illustration of the binding of various solutes onto hydrated iron oxide particles through formation of inner- and outer-sphere complexes.

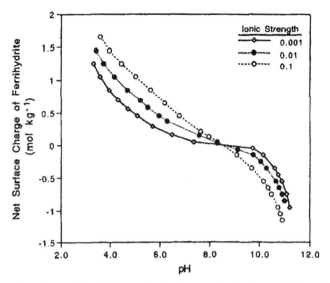

Figure 8.11 Net surface charge of ferrihydrite at three ionic strengths of sodium chloride and determination of pH$_{pzc}$ (from Jain, Raven, and Loeppert, *Environ. Sci. Technol.*, 1999, 33, 1179–1184, with permission from ACS).

ward hydroxyl ions (OH⁻). Thus, ligand sorption is greatly reduced at alkaline pH. The predominant reactions at alkaline pH can be presented as follows:

$$\equiv \overline{(SOH_2^+)_2 HAsO_4^{2-}} + 5OH^- \rightarrow 2 \equiv \overline{SO^-} + AsO_4^{3-} + 5H_2O \tag{23a}$$

$$\overline{(\equiv SOH)HAsO_2} + 2OH^- \rightarrow 2 \equiv \overline{SO^-} + AsO_2^- + 2H_2O \tag{23b}$$

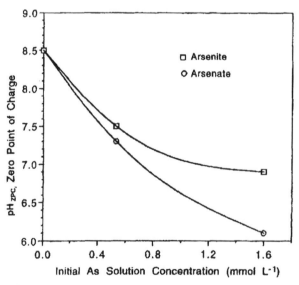

Figure 8.12 Relationship between the initial arsenite and arsenate concentration and pH of zero point charge of ferrihydrite (from Jain, Raven, and Loeppert, *Environ. Sci. Technol.*, 1999, 33, 1179–1184, with permission from ACS).

Equations (23a) and (23b) essentially represent the desorption of anionic ligands from hydrated metal oxides in alkaline medium. Note that under highly alkaline conditions, hydrated metal oxides are negatively charged, as are As(V) and As(III) oxyanions. Due to Donnan co-ion exclusion effects, both As(V) and As(III) are, therefore, rejected, and regeneration is favorable thermodynamically. In engineered processes, these reactions form the basis of the regeneration step, allowing reuse of hydrated metal oxides (namely, activated alumina, specialty iron oxide particles) for multiple cycles.

Polymeric Ligand Exchange

It is known that both As(III) and As(V) compounds form complexes with transition metal cations, such as copper(II). Thus, conceptually, Cu(II) ions, if immobilized onto a polymer phase at high concentrations, may act as sorption sites with relatively high affinities toward ligands like arsenates and arsenites. Such a tailored sorbent, referred to as polymeric ligand exchanger (PLE), is expected to remove trace amounts of ligands from the background of relatively high concentrations of sulfate and chloride. Equation (24) shows the exchange reaction between sulfate and arsenate for such a hypothetical PLE, where the shaded rectangle represents the polymeric substrate anchoring copper(II) ions.

$$-\!\!\!\boxed{/\!/\!/\!/\!/\!/}\!\!\!- \; + HAsO_{4(aq)}^{2-} \rightarrow -\!\!\!\boxed{/\!/\!/\!/\!/\!/}\!\!\!- \; + SO_{4(aq)}^{2-} + (n-m)H_2O$$

$$[Cu(H_2O)_n]^{2-}(SO_4^{2-}) \qquad\qquad [Cu(H_2O)_m]^{2+}(HAsO_4^{2-})$$

(24)

The species in parentheses are exchangeable anions (counterions), and they maintain the solid-phase electroneutrality. Between arsenate and sulfate, both have two negative charges, but arsenate is a stronger ligand (bidentate with two oxygen donor atoms) and, therefore, can satisfy the coordination requirements of Cu(II).

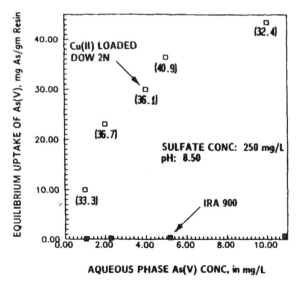

Figure 8.13 Comparison of arsenate uptakes between the polymeric ligand exchanger (DOW 2N-Cu) and strong base anion exchanger (IRA-900). Numbers in parentheses indicate arsenate/sulfate separation factors at pH = 8.5 (from Ramana and SenGupta, *Env. Eng. Div. J.*, 1992, 118, 5, 755–775, with permission from ASCE).

The concept was validated by loading Cu(II) onto a specialty chelating exchanger containing only nitrogen donor atoms (bispicolylamine functional group) [50,64]. The chelating exchanger (M 4195) is commercially available from Dow Chemical Co., Michigan, USA. Figure 8.13 shows arsenate-sulfate isotherms for the PLE and a strong-base anion exchanger (IRA-900) under identical experimental conditions. Much higher As(V) loading for PLE can be readily noted. PLEs are mechanically strong, chemically stable and amenable to efficient regeneration. Other details about the properties of PLE and their abilities to remove inorganic and organic ligands are available in the open literature [64–67].

SORPTION BEHAVIORS OF As(III)

Dissolved arsenic materials, both As(III) and As(V), are removable by a variety of unit processes that include coagulation/flocculation with alum and Fe(III) salts, fixed-bed activated alumina or ion exchange columns, coagulation followed by microfiltration, etc. While the subject of As(V) sorption has received significant attention, an in-depth discussion of the mechanisms of As(III) removal is lacking in the literature. An attempt is made in this section to provide a meaningful, scientific discussion of some conspicuous sorption behaviors of dissolved As(III) species.

Effect of pH on As(III) vs. As(V) Removal

Removals of As(III) and As(V) by activated alumina, coagulant alum or hydrated Fe oxides are influenced by pH. However, the nature of pH effect is not exactly the same. While the maximum As(V) removal capacity is recorded at slightly acidic pH (between 4 and 6), the same for As(III) is consistently observed at slightly alkaline pH, e.g., around 8.0. Figures 8.14 and 8.15 show As(III) and As(V) uptake data for alumina and HFO particles [24,43].

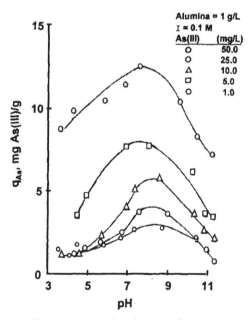

Figure 8.14 Effect of pH on As(III) sorption onto activated alumina (from Ghosh and Yuan, *Envir. Prog.*, 1987, 6, 3, 150–157, with permission from AIChE).

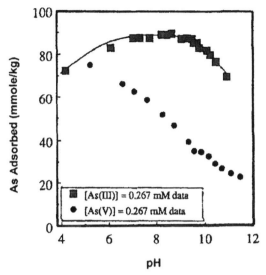

Figure 8.15 Effect of pH on As(III) and As(V) adsorption envelopes on hydrated iron oxides (from Manning, Fendorf, and Goldberg, *Environ. Sci. Technol.*, 1996, 32, 2383–2388, with permission from ACS.)

As(V) compounds or arsenates exist as anions, and they are best removed by hydrated metal oxides in protonated forms, i.e., $\equiv SOH_2^+$. Selective uptake is favored by the concurrent presence of IX and LAB interactions. On the contrary, IX-type interaction is altogether absent for As(III) or arsenite sorption. Between $\equiv SOH_2^+$, $\equiv SOH$ and $\equiv SO^-$, $HAsO_2$ is most favorably sorbed onto nonionized surface functional groups, i.e., $\equiv SOH$. Figure 8.16 shows the surface speciation of HFO (e.g., hematite) as a function of pH [68]. Note that nonionized surface functional groups $\equiv SOH$ predominate at around pH ≈ 8.0. At this pH, $HAsO_2$ also exists primarily as a nonionized species. Thus, from a law of mass action consideration, $HAsO_2$ sorption is most favored at slightly greater than neutral pH. Because the acid dissociation constants of hydrated aluminum oxide particles are in the same vicinity, the same

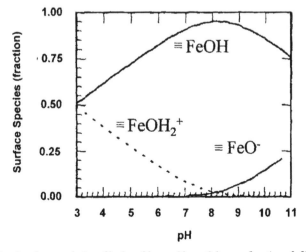

Figure 8.16 Simulated surface speciation of hydrated iron oxide particles as a function of pH (adapted from Au, Penisson, Yang, and O'Melia, *Geochemica et cosmochimica Acta.*, 1999, 63, 19/20, 2903–2917).

observation [i.e., maximum As(III) adsorption at slightly alkaline pH] is also valid for activated alumina particles. However, As(III) removal by oxides of aluminum is always lower than oxides of iron for reasons explained in the next section. From an application viewpoint, the foregoing theoretical analysis has a major ramification. For contaminated waters containing primarily As(III) species at neutral to slightly alkaline pH, no pH reduction is required or advisable to enhance arsenic removal capacity. Effects of pH may best be summarized as follows;

Aqueous pH	Sorption/Desorption Behavior
5.0–6.0	Hydrated metal oxides positively charged due to protonation; As(V) sorption maximum
7.0–8.5	Neutral surface sites predominate; As(III) sorption maximum
Above 11.0	Surface sites negatively charged; both As(III) and As(V) rejected due to Donnan co-ion exclusion effect

Removal of As(III): Alum vs. Ferric Coagulants

Both alum and ferric salts are very effective in removing dissolved As(V) compounds during coagulation/flocculation processes. Also, freshly precipitated hydrated aluminum oxides (HAO) and hydrated Fe(III) oxides (HFO) could almost completely remove low concentrations (100 µg/L or lower) of As(V) through selective sorption. In contrast, the performances of alum and Fe(III) coagulants for As(III) removal are drastically different. While ferric chloride coagulant or HFO microparticles significantly reduce dissolved As(III), alum coagulant and HAO particles are practically ineffective in removing arsenites. Figure 8.17 shows comparison of alum and ferric chloride coagulants for As(V) and As(III) removals under otherwise identical conditions [69]. Note that while As(V) removal was very high for alum and Fe(III) coagulants, As(III) removal was considerably lower for alum compared to Fe(III) coagulant. Sorption results with HAO and HFO particles (Figure 8.18) also yielded the same trend, i.e., As(III) removal was poor with HAO particles. Similar observations, i.e., poor As(III) removal by Al(III) coagulant, compared to Fe(III) coagulant, were also recorded by Sorg and Logsdon [47] and Edwards [42]. From a strictly engineering viewpoint, this observation clearly suggests that alum is an inappropriate coagulant if arsenic occurs in the +III oxidation state in contaminated source water. In such instances, the oxidation of As(III) to As(V) is a required pretreatment for efficient arsenic removal.

The obvious scientific question is as follows: what is the mechanistic explanation for such a striking difference in the effectiveness of alum and ferric chloride coagulants (or HFO and HAO particles) for As(III) removal? The redox chemistry of Fe(III) is quite different from that of Al(III). It has been discussed that Fe(III) solids first oxidize As(III) to As(V), which is subsequently sorbed strongly onto HFO particles. Dissolved Fe(II), thus formed, is oxidized back to Fe(III) by dissolved oxygen as shown below [42]:

$$HAsO_2(aq) + 2Fe(OH)_3(s) + 3H^+ \rightarrow H_2AsO_4^- + 2Fe^{2+} + 4H_2O \tag{25}$$

Dissolved Oxygen

However, such a pathway for enhanced As(III) removal by HFO particles can be ruled out because As(III) removal has also been observed in our laboratory under nitrogen pres-

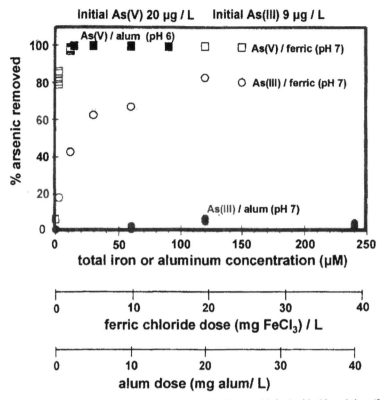

Figure 8.17 Comparison of As(III) and As(V) removals by coagulation with ferric chloride and alum (from Hering and Elimelech, AWWARF Report "Arsenic removal by enhanced coagulation and membrane processes," 1996, with permission from AWWA).

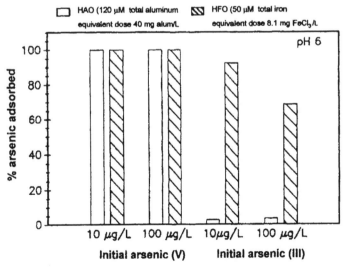

Figure 8.18 Comparison of As(III) and As(V) removals by sorption onto HAO and HFO particles (from Hering and Elimelech, AWWARF Report "Arsenic removal by enhanced coagulation and membrane processes," 1996, with permission from AWWA).

285

sure. Spectroscopic studies confirmed the presence of As(III) inside HFO particles through formation of inner-sphere complexes [23,24] and oxidation of As(III) by HFO particles was not observed by previous researchers [34,35].

The difference in Lewis acid properties of Al(III) and Fe(III) is hypothesized as the primary reason for such a drastic difference in the performance of HAO and HFO toward As(III) removal. Electronic configurations of Al^{3+} and Fe^{3+} are provided below:

$$Al^{3+}: \ 1S^2 2S^2 2P^6 \text{ (same as argon)} \tag{26}$$

$$Fe^{3+}: \ 1S^2 2S^2 2P^6 3S^2 3P^6 3d^5 \tag{27}$$

Al^{3+} is a very hard cation with an electronic configuration similar to that of inert argon. Conversely, Fe^{3+} is a transition metal cation with an incomplete 3d orbital. Although classified as a hard cation, Fe(III) is a relatively soft Lewis acid compared to Al^{3+}. $HAsO_2$ or hydrated $HAsO_2$ (H_3AsO_3) is the predominant As(III) compound where arsenic is the donor atom, as shown:

$$O = \ddot{A}s - OH \tag{28}$$

Compared to As(V) oxyanions where oxygens are the primary donor atoms, $HAsO_2$ is a softer Lewis base. The formation of an inner-sphere complex (i.e., Lewis acid-base interaction) is thus favored between $HAsO_2$ and a relatively soft Fe(III) Lewis acid. Such an interaction is practically absent between $HAsO_2$ and extremely hard Al(III). Higher sorption affinity of dissolved As(III) toward HFO particles stems from stronger Lewis acid-base interaction and is the underlying reason for superior performance of ferric chloride as a coagulant when compared with alum. Similarly, fluoride (F^-) is an extremely hard ligand with an electronic configuration similar to Ne, which is why fluoride removal by sorption is extremely effective with HAO particles or activated alumina [41] due to favorable hard acid-hard base interaction.

EXPERIENCE AT ALBUQUERQUE

Three treatment technologies for removing arsenic from groundwater in Albuquerque, NM, USA, were tried at a pilot-scale level: ion exchange (IX), activated alumina (AA) sorption and ferric chloride coagulation followed by microfiltration (C/MF). The composition of average feedwater to Albuquerque's proposed arsenic treatment facility is provided in Table 8.4. The details of the pilot-plant studies are available elsewhere [70]. The purpose of this section is to provide a brief review of these three processes and then discuss how findings from the pilot-scale study and, especially, their relative economic viability can be extended to other locations with contaminated groundwaters.

Ion Exchange (IX)

Figure 8.19 shows the schematic of the anion exchange process carried out at Albuquerque. The selectivity sequence for various anions of interest during the ion exchange process are as follows:

$$SO_4^{2-} > HAsO_4^{2-} > NO_3^- > H_2AsO_4^- > Cl^- > HCO_3^- > OH^- \tag{29}$$

TABLE 8.4. Composition of Average Feedwater to Albuquerque's
Proposed Arsenic Treatment Facility.

Constituent or Parameter	Average Concentration, mg/L
Alkalinity	164
As (total)	0.052
As(III)	0.00035
Ca	3.9
Cl	5.7
F	1.47
Fe	0.017
K	1.21
Mg	0.38
Na	100.5
NO_3—mg N/L	2.9
Se	<0.002
Si—mg SiO_2/L	28.7
SO_4	54.6
Total dissolved solids	320
pH	8.54

Once the ion exchange bed is exhausted, brine regeneration is carried out to desorb As(V) species. The operating cost of the process is greatly reduced by precipitating desorbed As(V) as ferric arsenate precipitates and recycling and reusing the brine as regenerant. The key steps are as follows:

Arsenate Sorption:

$$\overline{2R^+Cl^-} + HAsO_4^{2-} \rightleftharpoons \overline{(R^+)_2 HAsO_4^{2-}} + 2Cl^- \tag{30}$$

where $\overline{R^+Cl^-}$ denotes an anion exchanger pre-saturated with exchangeable chloride ion.

Arsenate Desorption with Brine:

$$\overline{(R^+)HAsO_4^{2-}} + 2Cl^- \rightleftharpoons \overline{2R^+Cl^-} + HAsO_4^{2-} \tag{31}$$

Arsenate Precipitation with Fe(III) and Brine Recycle:

$$2FeCl_3 + 3Na_2HAsO_4 + 3H_2O \rightarrow Fe(H_2AsO_4)_{3(s)} + Fe(OH)_{3(s)} + 6NaCl \tag{32}$$

Further details regarding the effects of pH and ferric chloride dosage on removal and fixation of arsenic as ferric arsenate are available in the literature [71]. Experimental results in Albuquerque showed that the regenerant could be recycled ten times without any loss of effectiveness in regeneration. The principal advantage of this process is that it reduces the volume of wastewater produced by the IX process and, at the same time, reduces the salt consumption.

Activated Alumina (AA) Sorption

Figure 8.20 shows the process schematic for AA treatment using granular-activated alumina particles as the sorbent medium. The chemistry of arsenic removal by granular AA is essentially the same as that of hydrated aluminum oxide (HAO), discussed earlier. Like ion exchange, AA treatment is a fixed-bed columnar operation, but it requires sodium hydroxide as the primary regenerant. Also, a lowering of pH to 6.0 increases As(V) removal capacity through enhanced protonation of surface functional groups. The selectivity sequence for AA sorption is as follows:

$$OH^- > HAsO_4^{2-} > H_2AsO_4^- > H_3SiO_4^- > SO_4^{2-} > HCO_3^- > Cl^- \tag{33}$$

Spent alkaline regenerant from the AA process contains a high concentration of dissolved arsenic and aluminum. Following pH adjustment, As(V) is finally transformed into a solid phase through precipitation/adsorption.

Ferric Chloride Coagulation/Microfiltration (C/MF)

In principle, this process is similar to sorption of dissolved arsenic onto HFO flocs that are formed due to the hydrolysis of ferric chloride solution added to the raw water in a rapid-mix vessel. Figure 8.21 shows that the process schematic and the hydraulic detention time in the rapid-mix unit is about 20 seconds. The sorption process is essentially instantaneous, and the microfilter removes HFO particles almost completely. The microfiltration units require backwashing every 20 minutes; suspended iron oxide particles are then col-

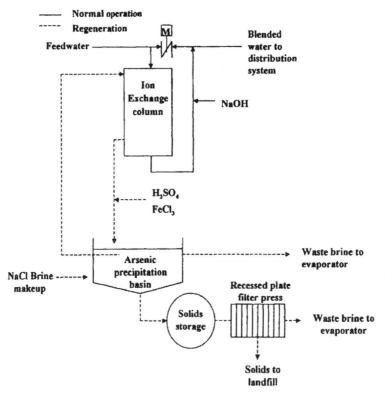

Figure 8.19 Schematic of the ion exchange process carried out at Albuquerque, NM, USA (from Chwirka, Thomson, and Stomp III, *AWWA Jour.*, 2000, 92, 3, 79–88, with permission from AWWA).

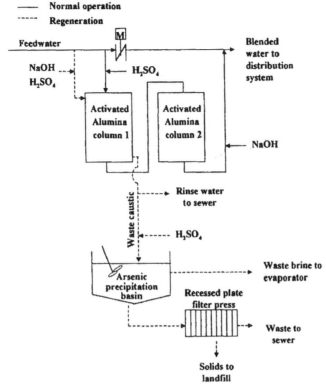

Figure 8.20 Schematic of the activated alumina adsorption process carried out at Albuquerque, NM, USA (from Chwirka, Thomson, and Stomp III, *AWWA Jour.*, 2000, 92, 3, 79–88, with permission from AWWA).

lected and dewatered for disposal. Figure 8.22 shows residual arsenic and filtrate pH with varying ferric chloride dosages. Note that less than 5 μg/L arsenic concentration can be easily achieved through the C/MF process.

Cost Comparison and Important Process Parameters

Based on the results of pilot-scale studies, fixed and operating costs of the three technologies were determined for the proposed 8,700 m³/d (2.3 mgd) groundwater treatment plant to be installed in Albuquerque [70]. The estimated costs are provided below:

Ion Exchange (IX):
  Capital cost: $5.2 million
  Annual operating and maintenance cost: $447,000

Activated Alumina (AA):
  Capital cost: $4.6 million
  Annual operating and maintenance cost: $444,000

Ferric Chloride Coagulation/Microfiltration (C/MF):
  Capital cost: $4.1 million
  Annual operating and maintenance cost: $273,000

Primarily due to its lower operating cost, the city of Albuquerque selected the C/MF process to use at its 8,700 m³/day (2.3 mgd) well. This facility is expected to be operational late in the year 2001 and will be the largest drinking water treatment facility in the United States designed specifically to remove arsenic.

It is noteworthy that the relative cost effectiveness of these three processes may not be valid in other locations and is greatly influenced by plant-specific process parameters as discussed below:

- Should As(V) concentration in groundwater increase from 52 µg/L to a higher value (say 100 µg/L), the cost of the IX process essentially remains unchanged because resin and regenerant requirements are independent of the trace species concentration, as discussed earlier. However, a higher ferric chloride dosage and handling of a larger amount of sludge would be required for the C/MF process.
- If As(III) constitutes a major portion of the total dissolved arsenic, neither IX nor AA treatment will be effective without preoxidation of As(III) to As(V). The performance of C/MF will, however, remain practically unchanged due to its ability to remove As(III).
- For groundwater with higher alkalinity, both AA and C/MF treatment processes would need an enhanced level of acid dosage for the same degree of arsenic removal. The IX process will not need any additional treatment for contaminated

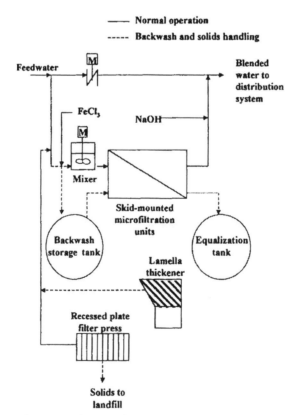

Figure 8.21 Schematic of the coagulation/microfiltration process carried out at Albuquerque, NM, USA (from Chwirka, Thomson, and Stomp III, *AWWA Jour.*, 2000, 92, 3, 79–88, with permission from AWWA).

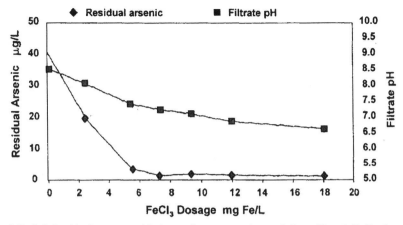

Figure 8.22 Relationship between residual arsenic concentration, solution pH and $FeCl_3$ for coagulation/microfiltration process.

water with a higher alkalinity. In the same vein, a higher sulfate concentration in source water will make the IX process more expensive by reducing its arsenic removal capacity.

- The presence of dissolved silica or natural organic matter will alter arsenic removal efficiencies of all three processes [72,73] but not to the same degree and, hence, will influence their relative economic viability.
- Disposal of spent regenerant or arsenic-laden solids poses a significant portion of the operating expenses. Geographic location and prevailing environmental regulations would greatly influence relative cost effectiveness of various treatment processes. For example, spent liquid-phase regenerants from IX and AA treatment in the Albuquerque study were classified as hazardous wastes. Hence, an elaborate treatment of residuals was warranted within the premise of the drinking water treatment plant. In a location where such small-volume liquid-phase regenerants can be directly transported to a neighboring hazardous waste treatment facility, the operating costs of IX and AA processes will be considerably reduced.

ARSENIC REMOVAL: EXPERIENCE IN INDIAN SUBCONTINENT

Millions of people in Bangladesh and West Bengal, India, are threatened with arsenic poisoning caused by drinking contaminated groundwater [16–18]. In many affected areas, villagers do not have access to any secondary source of arsenic-free water. Surface water treatment and subsequent distribution of treated water in remote areas will be extremely complex and cost prohibitive. Nickson et al. [18] earlier reported that a simple aeration of anoxic groundwater will be able to scavenge dissolved As by adsorption onto precipitated iron oxyhydroxide. Such an arsenic removal methodology is strongly dependent on the composition of groundwater and, more specifically, on its arsenic, iron and bicarbonate content, and pH. A number of tests confirmed the inability of this approach to guarantee reduction of dissolved arsenic concentration below 50 μgL^{-1}, the maximum limit set for drinking water in Bangladesh and India. Here we report the proven performance of a simple-to-operate treatment unit in remote villages and elucidate underlying mechanisms responsible for attaining excellent arsenic removal.

Under the auspices of Bengal Engineering College in Howrah, India, 12 well-head arsenic removal units were installed since 1997 in remote areas bordering West Bengal and
Bangladesh. The project was funded through a grant from Water For People (WFP), an af-
filiation of American Water Works Association (AWWA), located in Denver, Colorado,
USA, and private donations [74]. Each well-head unit is attached with a hand pump and
serves approximately 200 households. Figure 8.23 shows the schematic of a manually operated well-head unit for arsenic removal. The entire unit was manufactured locally using indigenous materials. For months, most of these units have been consistently producing water
with arsenic concentration well below 50 μgL^{-1}. The units are operationally simple; one
needs to operate the hand pump manually for a few minutes and then open the outlet valve
and collect arsenic-free water in a container. Locally manufactured granular-activated alumina (Oxide India Ltd., Durgapur, West Bengal, India) is the primary adsorbent used in the
treatment units. Figure 8.24 presents total arsenic concentrations in influent and treated waters plotted against the number of bed volumes of water treated. Regardless of wide fluctuations in influent arsenic concentrations, arsenic has remained well below 50 μgL^{-1} in the
treated water for over a year.

Both As(III) and As(V) species are significantly present in every contaminated groundwater location; As(III) constitutes about 20–40% of total dissolved arsenic [33]. At
circum-neutral pH of groundwater, As(V) exists as $H_2AsO_4^-$ and $HAsO_4^{2-}$, while As(III)
exists as nonionized H_3AsO_3. Because activated alumina is ineffective in removing As(III)
species, such excellent removal of total arsenic for a prolonged time period (i.e., months) is
counterintuitive.

The top part of the alumina column is designed with a large void space and a vent open to
the atmosphere. As the hand pump is operated manually, the water entering the top part of
the column is partially oxygenated, triggering oxidation of dissolved Fe(II) and subsequent
precipitation of Fe(III) hydroxides and minimal conversion of As(III) to As(V) oxyanion at
near-neutral pH.

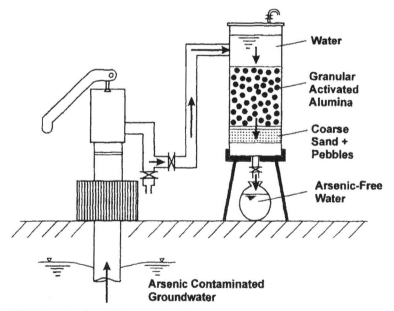

Figure 8.23 Schematic of manually operated well-head arsenic removal unit using indigenous materials.

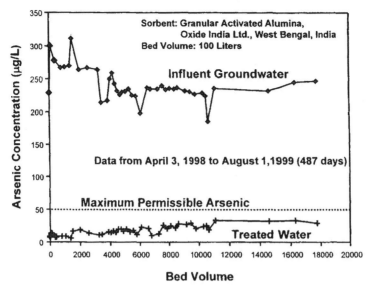

Figure 8.24 Influent and effluent arsenic concentration histories for a well-head column in a remote village for a prolonged time period (courtesy of Anirban Gupta, B.E. College, India).

$$4Fe^{2+} + O_2 + 10H_2O \rightarrow 4Fe(OH)_{3(s)} + 8H^+ \tag{34}$$

$$2HAsO_2 + O_2 + 2H_2O = 2H_2AsO_4^- + 2H^+ \tag{35}$$

The first reaction essentially goes to completion, and dissolved iron is practically absent in the treated water. Both reactions are accompanied by the generation of hydrogen ions (i.e., acid), which are neutralized by relatively high concentrations of alkalinity (over 150 mgL^{-1} of HCO_3^-) present in contaminated groundwater. Consequently, treated water pH always remains slightly alkaline. At neutral to above-neutral pH, freshly precipitated hydrated iron oxides have relatively high sorption affinities toward As(III) oxyacid through formation of inner-sphere complexes aided by structural Fe atoms. Thus, sorption onto iron precipitates is the sole removal mechanism of the unoxidized As(III). The bulk of the dissolved As(V) is subsequently removed by the activated alumina as the water percolates through the column. For groundwaters with relatively low As(III) content, Fe(III) hydroxides greatly supplement As(V) removal capacity of activated alumina. Neither precipitation of Fe(II) from groundwater by aeration nor use of activated alumina without hydrated iron oxides can separately guarantee such near-complete removal of total dissolved arsenic. The concentrations of dissolved As(III), As(V), Fe(II) and HCO_3^- in contaminated groundwaters vary from one location to another but remain within an envelope where the foregoing mechanism is operative. Figure 8.25 depicts the proposed reactions responsible for near-complete removal of total dissolved arsenic for a prolonged period of time.

The units are ordinarily backwashed everyday for 10–15 minutes with the hand pump to remove iron hydroxide precipitates that are subsequently collected at the top of a coarse sand filter located in the same premise. These treatment units in remote villages do not require any electricity, pH adjustment or external addition of chemicals. As the arsenic concentration in treated water approaches 50 µg/L, each unit is regenerated using caustic soda

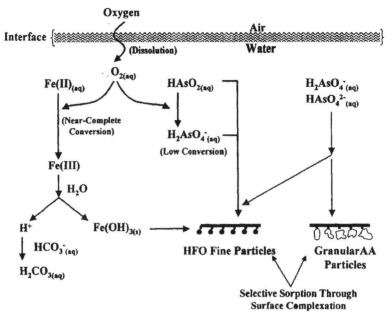

Figure 8.25 An illustration of the underlying reactions for near-complete removal of dissolved arsenic in well-head units.

followed by sulfuric acid. In every location, the well-head units are operated and maintained by a citizens' committee formed by villagers.

DEVELOPMENT OF A POLYMERIC/INORGANIC HYBRID SORBENT

It is universally recognized that a fixed-bed sorption process is operationally simple, requires virtually no start-up time and is forgiving toward fluctuations in feed compositions. However, in order for the fixed-bed process to be viable and economically competitive, the sorbent must exhibit high selectivity toward the target contaminant, be amenable to efficient regeneration and be durable. Amorphous and crystalline hydrated Fe oxide (HFO) precipitates show strong affinity toward both As(III) and As(V) oxyanions through ligand exchange in the coordination spheres of structural Fe atoms. Recent investigations using extended X-ray absorption fine structure (EXAFS) spectroscopy confirmed that As(III) and As(V) species are selectively bound to the oxide surface through formation of inner-sphere complexes [23,24]. Note that strong base polymeric anion exchangers or activated alumina do not have the desired property to remove As(III) species at near-neutral pH. Understandably, hydrated iron oxides including ferrihydrites, hematites and goethites possess the requisite attributes to be effective arsenic-selective sorbents. However, the current process of syntheses, although straightforward, produces only very fine submicron particles that are unusable in fixed beds because of excessive pressure drops and poor mechanical strength of the particles.

During the last two years, researchers at Lehigh University have perfected a simple chemical-thermal technique to produce a hybrid (polymeric/inorganic) sorbent that is selective toward arsenic(III) and (V) compounds and is, at the same time, compatible with fixed-bed column operation [75]. The sorbent is referred to as Hybrid Ion Exchanger (HIX).

It consists of spherical macroporous polymeric cation exchanger beads within which submicron hydrated Fe oxide (HFO) particles have been uniformly and irreversibly dispersed.

The preparation of HIX consists of the following three steps:

- *Step 1:* loading of Fe(III) onto the sulfonic acid sites of the porous cation exchanger at an acidic pH
- *Step 2:* desorption of Fe(III) and simultaneous precipitation of Fe(III) hydroxide within the pores of the cation exchangers
- *Step 3:* ethanol wash followed by a mild thermal treatment to partially convert amorphous iron hydroxides into crystalline goethite and hematite

Figure 8.26 provides an illustration of the entire procedure. At the conclusion of Step 3, submicron HFO particles form agglomerates and are irreversibly encapsulated within the spherical exchanger beads. Turbulence and mechanical stirring did not result in any loss of HFO particles. A high concentration of sulfonic acid functional groups in the polymeric exchanger allows high and uniform loading of HFO particles (approximately 12% Fe by weight) within the polymeric material. Purolite C-145, a commercially available cation exchanger with polystyrene matrix and sulfonic acid functional group, was used as the parent ion exchanger. Figure 8.27(a) shows hybrid spherical beads; Figure 8.27(b) shows the

Step 1. Loading with FeCl$_3$ Solution at pH < 2.0

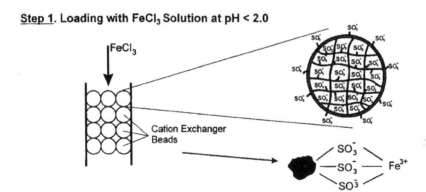

Step 2. Desorption and simultaneously precipitation in the gel phase and pores

$$Fe^{3+}(aq) + OH^- \xrightarrow{\text{Precipitation}} Fe(OH)_3 (S)$$

Step 3. Alcohol wash and mils thermal treatment

$$Fe(OH)_3 (S) \underset{}{\overset{60\,^{\circ}C}{\rightleftharpoons}} FeOOH (S)$$
(Crystalline)

Figure 8.26 Step-wise procedure for the preparation of arsenic selective hybrid ion exchanger (HIX).

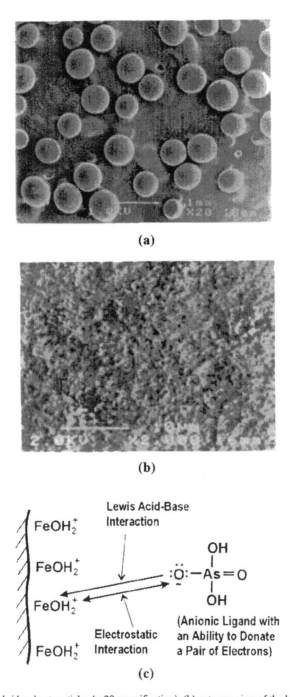

(a)

(b)

Lewis Acid-Base
Interaction

FeOH$_2^+$

FeOH$_2^+$

$\ddot{\text{O}}:-\text{As}=\text{O}$ with OH above and OH below

FeOH$_2^+$

FeOH$_2^+$

Electrostatic
Interaction

(Anionic Ligand with
an Ability to Donate
a Pair of Electrons)

(c)

Figure 8.27 (a) Hybrid sorbent particles ($\times$ 20 magnification), (b) cutaway view of the hybrid sorbent with uniformly dispersed HFO particles ($\times$ 2,000 magnification) and (c) the presence of electrostatic and Lewis acid-base interactions for selective sorption of $H_2AsO_4^-$ onto HFO particles (for H_3AsO_3, only Lewis acid-base interaction is present).

scanning electron microphotograph (SEM) of a sliced sorbent particle and Figure 8.27(c) provides the underlying interactions resulting in selective sorption of $H_2AsO_4^-$ onto dispersed HFO particles. The hybrid ion exchanger (HIX) essentially combines excellent hydraulic properties of spherical cation exchanger beads and high arsenic sorption affinities of HFO particles. It is important to note that Jekel and coworkers in Germany [45] have recently developed a pressurized granular ferric hydroxide material and tested it for high As(V) removal capacity. German materials are, however, granular and are disposed of after one run without regeneration and reuse.

Results with Hybrid Ion Exchanger (HIX)

FIXED-BED COLUMN RUNS

Figure 8.28 shows complete effluent histories during a fixed-bed column run using HIX. The influent composition and the hydrodynamic conditions during the column run, i.e., EBCT (empty-bed contact time) and SLV (superficial liquid-phase velocity) are also provided in Figure 8.28. The following information is noteworthy:

- Other anions, namely, sulfate and chloride, broke through almost immediately after the start of the column run, while As(V) breakthrough was not complete even after 8,000 bed volumes. Mass-balance calculations confirmed that in the absence of any kinetic limitations, HIX may treat well over 10,000 bed volumes of contaminated water prior to arsenic breakthrough.
- The effluent pH during the entire column run was slightly above neutral. Thus, a very high arsenic removal capacity can be obtained with HIX without necessitating pH adjustment.
- Iron content of HIX is approximately 12% by mass, the balance being polymeric

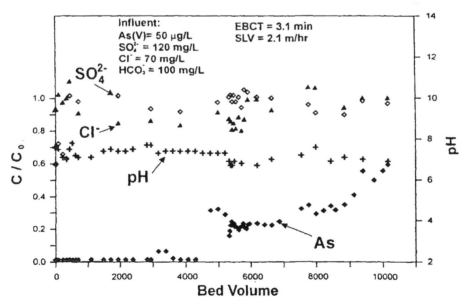

Figure 8.28 Complete effluent histories during a fixed-base column run using HIX.

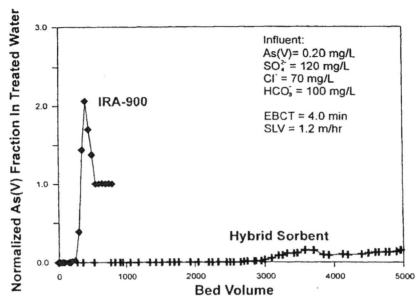

Figure 8.29 Comparison of effluent histories of As(V) using a strong-base polymeric anion exchanger and the HIX.

material. Once the iron content within HIX is normalized for total iron in line with pure iron oxide material, HIX can treat 30,000–40,000 bed volumes prior to arsenic breakthrough.

Figure 8.29 shows a comparison of effluent histories of As(V) during two separate column runs with two different sorbents, namely, a strong-base polymeric anion exchanger (IRA-900 from Rohm and Haas Co., Philadelphia, PA, USA) and the HIX [75]. While complete arsenic breakthrough for IRA-900 started at less than 300 bed volumes, HIX was removing arsenic well after 5,000 bed volumes. Arsenic breakthrough from IRA-900 underwent chromatographic elution, i.e., concentration of arsenic at the exit of the column became significantly greater than its influent concentration. This phenomenon is attributed to the anion exchanger's greater sulfate selectivity over As(V) oxyanions. There is also enough experimental evidence to suggest that As(III) is completely oxidized to As(V) during the fixed-bed column run using HIX. Details to this effect are being deliberately avoided here for the sake of brevity.

As(III) vs. As(V) Removal Capacities

As(III) species or arsenites are nonionized at pH 7.0. Thus, the electrostatic interaction is essentially absent at near-neutral pH. Any sorption of As(III) species onto the surface sites of HIX particles has to take place through formation of inner-sphere complexes. Figure 8.30 shows comparable As(III) and As(V) removal capacities with HIX in the pH range of 5.5–8.0. Note that As(V) removal capacity showed a downward trend with an increase in pH from 6.0 to 8.0. Contrary to that, As(III) removal remained fairly unaffected with an increase in pH. This observation is consistent with experimental findings of other researchers who studied As(III) sorption on freshly precipitated ferrihydrite or goethite [24,35].

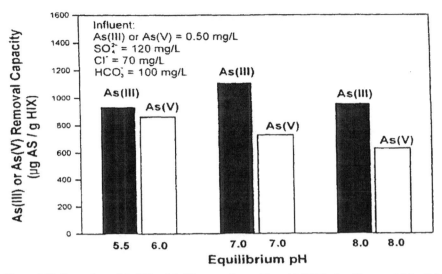

Figure 8.30 Comparison of As(III) and As(V) removal capacities with HIX in the pH range of 5.5 to 8.0.

REGENERATION, RINSING AND REUSE

In order for the sorption process to be viable, the hybrid exchanger (HIX) has to be amenable to efficient regeneration and reuse. To this end, a portion of the exhausted HIX after the lengthy run in Figure 8.29 was regenerated using 10% NaOH. Concentration profiles of arsenic during the desorption process and other salient hydrodynamic parameters are provided in Figure 8.31. Note that in less than eight bed volumes, almost the entire amount of arsenic(V) was completely desorbed from the bed. Carbon dioxide-sparged influent was

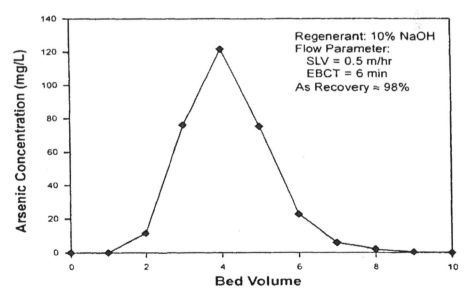

Figure 8.31 As(V) concentration profile during regeneration with sodium hydroxide.

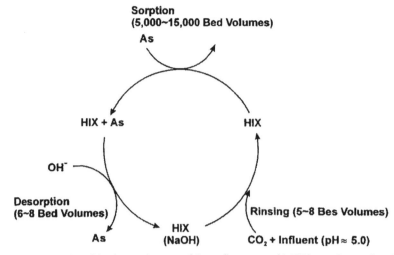

Figure 8.32 An illustration of the three major steps of the cyclic process with HIX, manely, sorption, desorption and rinsing.

used for rinsing and neutralizing the bed following caustic soda regeneration. The modified rinsing procedure avoided use of strong hydrochloric acid or sulfuric acid in the process. Figure 8.32 shows a schematic of the cyclic arsenic removal process with HIX. In order to substantiate the reusability of HIX, two consecutive cycles (i.e., sorption, regeneration and rinsing) were carried out in minicolumns for As(V) removal. As may be seen from Figure 8.33, no significant trend or difference in As(V) removal capacity occurred between Cycle 1 and Cycle 2. Laboratory trial runs carried out at the Bengal Engineering College, Howrah, India, using contaminated groundwater collected from an affected village showed that HIX containing approximately 10% Fe performed better than activated alumina for nearly 10,000 bed volumes.

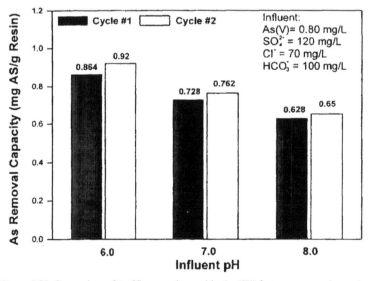

Figure 8.33 Comparison of As(V) removal capacities by HIX for two consecutive cycles.

CONCLUSIONS

The United States is in the process of setting a new standard for arsenic Maximum Contaminant Level (MCL) in drinking water. In January 2001, under a federal rule signed by the United States Environmental Protection Agency (USEPA) and the former US president, a revised arsenic Maximum Contaminant Level (MCL) to 10 µg/L was made effective five years from its promulgation; however, this ruling was later reversed in March 2001, by the current USEPA administration. Currently, the 60-year-old arsenic MCL of 50 µg/L is the highest among all the developed countries in the world, but a revised standard is scheduled to come into effect by January 2002. According to an earlier estimate of USEPA, a 10 µg/L arsenic MCL would require corrective action for over four thousand (4,000) water supply systems serving an approximate population of twenty million.

Over 50 million people in Bangladesh are routinely exposed to arsenic poisoning through drinking water. Arsenic-inflicted health crises, namely, bladder cancer, hyperkeratosis, cardiovascular diseases, lung cancer and frequent human deaths resulting from them abound in the region. Natural geochemical contamination through soil leaching is the sole contributor of arsenic in groundwater. Rainfall in this region is quite significant (over 1,500 mm/year) and much higher than that in North America. Ironically, however, arsenic-free surface water is practically unusable due to poor sanitation and consequent contamination with the potential for an outbreak of waterborne diseases. In many remote villages, arsenic-contaminated groundwater is the only viable source of drinking water, and cost-effective arsenic removal technology is a bare necessity to alleviate people's sufferings. We close this chapter by providing a concise summary of key aspects pertaining to arsenic contamination of groundwater and its removal by engineered processes.

- Barring a few exceptions, arsenic contamination of groundwaters around the world is essentially a natural phenomenon where geochemical soil leaching is the sole contributor of dissolved arsenic in groundwater. Nevertheless, there are suggestive evidences that human activities, such as indiscriminate withdrawal of groundwater, application of phosphate-based fertilizer, etc., tend to accelerate such leaching processes.

- At low dosages (i.e., less than 500 µg/L of water), As(V) compounds are readily reduced to As(III) inside human bodies. Contrary to popular belief, chronic toxicity caused by ingestion of arsenic-contaminated water is unaffected by the relative distribution of As(III) and As(V) compounds.

- Contaminated groundwaters normally contain arsenites and arsenates. Between them, arsenates are more amenable to removal by a majority of treatment processes. Pre-oxidation of arsenites to arsenates followed by pH lowering through acid dosage is a commonly accepted proposition in many quarters for treating arsenite. For small treatment systems serving less than 5,000 people, pre-oxidation of arsenite is more operationally complex than currently recognized. Alternately, adopting a treatment process capable of removing both arsenites and arsenates is more desirable.

- Both alum and ferric chloride coagulants are equally effective in removing dissolved As(V) compounds. Selective sorption onto the surfaces of microparticles of hydrated aluminum oxide (HAO) and hydrated Fe oxide (HFO) is the underlying mechanism for excellent As(V) removal. However, the effectiveness of HAO and HFO particles varies markedly for As(III) removal. While HFO particles can achieve good As(III) removal, HAO particles are essentially ineffective. It is hypothesized that the difference in Lewis acid property between Fe(III) and Al(III) is primarily responsible for the observed disparity in their As(III) removal efficiencies.

- Many arsenic removal technologies are currently in existence. As diverse as they may appear in their physical configurations, ion exchange and/or Lewis acid base interaction are the primary arsenic removal mechanisms in each instance.

- In Albuquerque, three different processes were tried at pilot-scale level, namely, ion exchange, activated alumina treatment and coagulation/microfiltration. Of them, coagulation/microfiltration was found to be the most cost-effective treatment process. It is, however, to be noted that the cost-effectiveness may vary from one location to the other and is greatly influenced by the concentration of arsenic, alkalinity and sulfate in source water and the cost of residuals treatment and disposal. Besides anion exchanger and activated alumina, several new sorbent materials seem to hold promise based on laboratory investigations. Field trials are, however, necessary to validate their performance and cost-effectiveness.

- Reverse osmosis and nano-filtration membrane processes may also be viable options in places where a reduction in total dissolved solids (TDS) or nitrate is desired along with arsenic removal.

- Last but not least, authors of this chapter earnestly believe that U.S.EPA's decision to slash arsenic MCL to any concentration below 10 μg/L is unwarranted and, based on the currently available epidemiological data in the literature, too radical to offer any true health benefits to U.S. citizens. The European Union (EU) set the arsenic MCL standard at 10 μg/L. The difference in the annual cost of compliance between 10 μg/L and 5 μg/L may run in the vicinity of one billion dollars nationally. The money thus saved by setting the new arsenic MCL at 10 μg/L can be better spent in protecting our surface and groundwater sources.

ACKNOWLEDGEMENTS

Authors are thankful to Matthew DeMarco for his work on the development of HIX material. Lori Stewart carried out the fixed-bed column runs in Figure 8.8. Active cooperation from Anirban Gupta of Bengal Engineering College, West Bengal, India, and Arun Deb of Roy F. Weston Inc., One Weston Way, West Chester, Pennsylvania, USA, regarding arsenic-related work in the Indian subcontinent is gratefully acknowledged. We are also thankful to the United States Environmental Protection Agency (U.S.EPA) and the Pennsylvania Infrastructure Technology Alliance (PITA) for their partial financial support.

REFERENCES

1 Bowen, H. J. M. (1979). *Elemental Chemistry of the Elements,* Academic Press, London and New York.

2 Moore, J. N., Ficklin, W. H. and Johns, C. (1988). Partitioning of arsenic and metals in reducing sulfide sediments. *Environ. Sci. Technol.,* 22, 432–437.

3 Brannon, J. M. and Patrick, W. H. (1987). Fixation, transformation and mobilization of arsenic in sediments. *Environ. Sci. Technol.,* 21, 450–459.

4 Mok, W. M. and Tai, C. M. (1988). Mobilization of arsenic in contaminated river waters. In *Arsenic in the Environment*, Nriagu, J. O., Ed., Wiley-Interscience, New York.

5 Johnson, C. A. and Thornton, I. (1987). Hydrological and chemical factors controlling the concentrations of Fe, Cu, Zn, and As in a river system contaminated by acid mine drainage. *Water Res.*, 21, 359.

6 Davis, A., Ruby, M. V., Bloom, M., Schoof, R., Freeman, G. and Bergstrom, P. D. (1996). *Environ. Sci. Technol.*, 30, 392–399.

7 Buchert, J. P., Lauwerys, R. and Roels, H. (1981). Comparison of the urinary excretion of arsenic metabolites after a single dose of sodium arsenite, monomethylarsonate or dimethylarsinate in man. *Int. Arch. Occup. Envir. Health*, 48, 71–79.

8 Vahter, M. and Marafante, E. (1983). Intercellular interaction and metabolic fate of arsenate and arsenite in mice and rabbits. *Chem. Biol. Interact.*, 47, 29–44.

9 Smith, A. H. (1999). Personal Communication.

10 Pontius, F. W., Brown, K. G. and Chen, C. J. (1994). Health implications of arsenic in drinking water. *Jour. AWWA*, 86(9), 52–63.

11 Smith, A. H., Hoperhayn-Rich, C., Bates, M. N., Goeden, H. M. et al. (1992). Cancer risks from As in drinking water. *Envir. Health Persp.*, 97, 259.

12 U.S.EPA (1998). Research plan for arsenic in drinking water. EPA/600/R-98/042, Office of Research and Development, Cincinnati, OH.

13 Guhamajumder, D. N., Haque, N., Ghosh, B. K., De, A., Santos, A., Chakraloti, D. and Smith, A. H. (1998). Arsenic levels in drinking water and the prevalence of skin lesions in West Bengal, India. *Int. J. Epidemiol.*, 27, 871–877.

14 Frey, M. M., Owen, D. M., Chowdhury, Z. K., Raucher, R. S. and Edwards, M. A. (1998). Cost to utilities of a lower MCL for arsenic. *Jour. AWWA*, 90(3), 89–102.

15 AWWA Research Foundation (2000). *Drinking Water Research*, 10, 3, p. 6.

16 Lepkowski, W. (1998). Arsenic crisis in Bangladesh. *C&EN News*, Nov. 16, 27–29.

17 Bearak, D. (1998). "New Bangladesh disaster: Wells that pump poison." *The New York Times*, Nov. 10.

18 Nickson, R., McArthur, J., Burges, W., Ahmedy, K. M., Ravenscroft, P. and Rahman, M. (1998). Arsenic poisoning of Bangladesh groundwater. *Nature*, 395, 338.

19 Dhar, R. K., Biswas, B. Kr., Samanta, G., Mandal, B. Kr., Chakraborti, D., Roy, S., Jafar, A., Islam, A., Ara, G., Kabir, S., Khan, A. W., Ahmed, S. A. and Hadi, S. A. (1997). Groundwater arsenic calamity in Bangladesh. *Current Science*, 73: 1, 48–59.

20 Proceedings on International Conference on Arsenic in Groundwater: Cause, Effect and Remedy. 6–8 February, 1995. Jadavpur University, Calcutta, India, 79 pages.

21 Proceedings on International Conference on Arsenic Pollution of Groundwater in Bangladesh. 8–12 February, 1998. Dhaka, Bangladesh, 179 pages.

22 Subramanian, K.S. and Kosnett, M.J. (1998). Human exposures to arsenic from consumption of well water in West Bengal, India. *Int. J. Occup. Environ. Health*, 4, 217–230.

23 Manning, B. and Goldberg, S. (1997). Adsorption and stability of arsenic(III) at the clay mineral-water interface. *Environ. Sci. Technol.*, 31, 2005–2011.

24 Manning, B. A., Fendorf, S. E. and Goldberg, S. (1998). Surface structures and stability of As(III) on goethite: spectroscopic evidence for inner-sphere complexes. *Envir. Sci. Technol.*, 32, 16, 2383–2388.

25 Das, D. (1995). Ph.D. Dissertation, Jadavpur University, India.

26 Bagla, P. and Kaiser, J. (1996). India's spreading health crisis draws global arsenic experts. *Science*, 274, 174–175.

27 Acharyya, S. K., Chakrobarty, P., Lahiri, S., Raymahaskay, B. C., Guha, S. and Bhowmik, A. (1999). Arsenic poisoning in the Ganges Delta. *Nature*, 401, October 7, p. 545.

28 Latimer, W. M. and Hildebrand, J. H. (1951). *Reference Book of Inorganic Chemistry*. 3rd Ed., MacMillan, New York.

29 Drever, J. I. (1988). *The Geochemistry of Natural Water;* Prentice-Hall: Englewood Cliffs, New Jersey.

30 Ferguson, J. F. and Gavis, J. (1972). A review of arsenic cycle in natural waters. *Wat. Res.,* 6, 1259–1274.

31 Morel, F. M. F. and Hering, J. G. (1993). *Principles and Applications of Aquatic Chemistry.* Wiley-Interscience, New York.

32 Hering, J. G. and Chiu, Van. Q. (2000). Arsenic occurrence and speciation in municipal ground-water-based supply system. *Jour. Envir. Engr., ASCE,* 126, 5, 471–474.

33 Safiullah, S., Sarker, A. K., Zahid, A., Islam, M. Z. and Halder, S. Z. (1998). Proc. Int. Conf. on Arsenic in Bangladesh (Dhaka). 8–12 February.

34 Oscarson, D. W., Huang, P. M., Defosse, C. and Herbillon, A. (1981). Oxidative power of Mn(IV) and Fe(III) oxides with respect to As(III) in terrestrial and aquatic environments. *Nature,* 291, 50.

35 Pierce, M. L. and Moore, C. B. (1982). Adsorption of As(III) and As(V) on amorphous iron hydroxide. *Wat. Res.,* (6) 1247.

36 Smith, S. L. and Jaffe, P. R. (1998). Modeling the transport and reaction of trace metals in water saturated soils and sediments. *Wat. Res.,* 34, 3135–3147.

37 Masscheleyn, P. H, Delaune, R. D. and Patrick, W. H. (1991). Effect of redox potential and pH on arsenic speciation and solubility in a contaminated soil. *Environ. Sci. Technol.,* 25, 1414–1419.

38 Ahmann, D., Roberts, A. L., Krumholz, L. R. and Morel, F. M. M. (1994). Microbe grows by reducing arsenic. *Nature,* 371, 750.

39 Cummings, D. E., Caccavo, F., Fendorf, S., and Rosenzweig, R. F. (1999). Arsenic mobilization by the dissimilatory Fe(III)-reducing bacterium *Shewanella alga* Bry. *Envir. Sci. Technol.,* 33, 723–729.

40 Horng, L. L. and Clifford, D. (1997). The behavior of polyprotic anions in ion-exchange resins. *Reactive and Functional Polymers,* 35, 41–54.

41 Clifford, D. (1999). Ion exchange and inorganic adsorption. *Water Quality and Treatment,* R. D. Letterman, Ed., 5th Edition, McGraw-Hill Inc., New York, NY, Chapter 9.

42 Edwards, M. (1994). Chemistry of arsenic removal during coagulation and Fe-Mn oxidation. *Jour. AWWA,* 76, 64–78.

43 Ghosh, M. M. and Yuan, J. R. (1987). Adsorption of inorganic arsenic and organicoarsenicals on hydrous oxides. *Envir. Progress,* 3(3), 150–157.

44 Cheng, R. C., Liang, S., Wang, H. C. and Beuhler, M. D. (1994). Enhanced coagulation for arsenic removal. *Jour. AWWA,* 86, 9, 79–90.

45 Driehaus, W., Jekel, M. and Hildebrandt, U. (1998). Granular ferric hydroxide—a new adsorbent for the removal of arsenic from natural water. *J. Water SRT Aqua,* 47, 1, 30–35.

46 McNeil, L. S. and Edwards, M. (1995). Soluble arsenic removal at water treatment plants. *Jour. AWWA,* 87, 4, 105–114.

47 Sorg, T. J. and Logsdon, G. S. (1978). Treatment technology to meet the interim primary drinking water regulations for inorganics: Part 2. *Jour. AWWA,* 70, 7, 379.

48 Hering, J., Chen, P-Y., Wilkie, J. A. and Elimelech, M. (1997). Arsenic removal from drinking water during coagulation. *Jour. Envir. Engr. ASCE,* 123, 8, 801–807.

49 Min, J. M. and Hering, J. (1998). Arsenate sorption by Fe(III)-doped alginate gels. *Wat. Res.,* 32, 5, 1544–1552.

50 Ramana, A. and SenGupta, A. K. (1992). Removing selenium (IV) and arsenic(V) oxyanions with tailored chelating polymers. *Jour. Env. Eng. Div., ASCE,* 118, 5, 755–775.

51 Vagliagandi, F. G. A. and Benjamin, M. M. (1995). Adsorption of arsenic by ion exchange, activated alumina and iron-oxide coated sand (IOCS). Water Quality Technology Conference, *AWWA,* Nov. 12–16, New Orleans, LA.

52 Shen, Y. S. (1973). Study of arsenic removal from drinking water. *J. AWWA,* 65, 8, 543.

53 Lackovic, J. A., Nikolaidis and Dobbs, G. M. (2000). Inorganic arsenic removal by zero-valent iron. *Environmental Engineering and Science,* 17, 1, 29–39.

54 Bajpai, S. and Chaudhury, M. (1999). Removal of arsenic from manganese dioxide coated sand. *Jour. Env. Engr., ASCE,* 125, 8, 782–784.

55 Torrens, K. D. (1999). Evaluating arsenic removal technologies. *Pollution Engineering,* July, 25–28.

56 Kartinen, E. O. and Martin, C. J. (1995). An overview of arsenic removal processes. *Desalination,* 103, 79–88.

57 Frank, P. and Clifford, D. (1986). As(III) oxidation and removal from drinking water. EPA Project Summary, Report No. EPA/600/S2-86/021. Water Engrg. Res. Lab., Envir. Protection Agency, Office of Research and Development, Cincinnati, OH.

58 Driefhaus, W., Seith, R. and Jekel, M. (1995). Oxidation of arsenic(III) with manganese oxides in water treatment. *Wat. Res.,* 29, 1, 297–305.

59 Clifford, D., Subramonian, S. and Sorg, T. (1986). Removing dissolved inorganic contaminants from water. *Environ. Sci. Technol.,* 20, 1072–1080.

60 Waypa, J. J., Elimelech, M., and Hering, J. (1997). Arsenic removal by RO and NF membranes. *Jour. AWWA,* 89, 10, 102–114.

61 Stumm, W. and Morgan, J. J. (1996). *Aquatic Chemistry.* 3rd Ed. John Wiley & Sons Inc., New York.

62 Dzomabak, D. A. and Morel, F. M. M. (1990) Surface Complexation Modeling: Hydrous Ferric Oxide. John Wiley & Sons, Inc., New York.

63 Jain, A., Raven, K. P. and Loeppert, R. H. (1999). Arsenite and arsenate adsorption on ferrihydrite: surface charge reduction and net OH⁻ release stoichiometry. *Envir. Sci. Technol.,* 33, 1179–1184.

64 Zhu, Y. (1992). Ph. D. Dissertation. Chelating Polymers with Nitrogen Donor Atoms. Department of Civil Engineering. Lehigh University, Bethlehem, PA, USA.

65 Zhu, Y. and SenGupta, A. K. (1992). Sorption enhancement of some hydrophilic organic solutes through polymeric ligand exchange. *Environ. Sci. Technol.,* 26, 10, 1990–1998.

66 Zhao, D. and SenGupta, A. K. (1998). Ultimate removal of phosphate using a new class of anion exchanger. *Wat. Res.,* 32, 5, 1613–25.

67 Zhao, D. (1997). Ph.D. Dissertation. Polymeric Ligand Exchange: A new approach toward enhanced separation of environmental contaminants. Department of Civil and Environmental Engineering, Lehigh University, Bethlehem, PA, USA.

68 Au, K.-K., Penisson, A. C., Yang, S., and O'Melia, C. R. (1999). Natural organic matter at oxide/water interface: complexation and conformation. *Geochimica et Cosmochimica Acta,* 63, 19/20, 2903–2917.

69 Hering, J. G. and Elimelech, M. (1996). Arsenic removal by enhanced coagulation and membrane processes. AWWARF Report. Denver, CO.

70 Chwirka, J. D., Thomson, B. M. and Stomp, J. M. (2000). Removing arsenic from groundwater. *Jour. American WaterWorks Assoc.,* 92, 3, 79–88.

71 Papassiopi, N., Vircikova, E., Valendin, N., Kontopoulos, A. and Molnar, L. (1996). Removal and fixation of arsenic in the form of ferric arsenates. *Hydrometallurgy,* 41, 243–253.

72 Vagliasandi, F. G. A. and Benjamin, M. M. (1998). Arsenic removal in fresh and NOM-preloaded ion exchange packed bed adsorption reactors. *Water Sci. Technol.,* 38, 6, 337–343.

73 Meng, X., Bang, S. and Korfiatis, G. P. (2000). Effects of silicate, sulfate and carbonate on arsenic removal by ferric chloride. *Wat. Res.,* 34, 4, 1255–1261.

74 Murphy, M. (2000). Helping people help themselves. *AWWA Jour.,* 92, 4, 139–48.

75 SenGupta, A. K., DeMarco, M. and Greenleaf, J. (2000). A new polymeric/inorganic hybrid sorbent for selective arsenic removal, pp. 142–149. Proceedings of IEX 2000: Ion Exchange at the Millennium (Ed. J. A. Greig); Churchill College, Cambridge University, England; July 16–21, 2000. Imperial College Press, London.

CHAPTER 9

Cr(III) Separation and Recovery from Tannery Wastes: Research, Pilot and Demonstration Scale Investigation

DOMENICO PETRUZZELLI[1]
GIOVANNI TIRAVANTI[2]
ROBERTO PASSINO[3]

INTRODUCTION

THE use of leather goes back to prehistoric times when the principal raw material was hides or skins of mammals including, to a small extent, reptiles, fish and birds.

A tanning operation consists of converting the raw skin, a highly putrescible material, into leather, a stable and mechanically resistant matter, which can be used in the manufacture of a wide range of products (shoes, clothing, bags, furniture and many other items of daily use). The whole operation involves a sequence of complex chemical processes and mechanical treatments, of which the fundamental stage is tanning. When coupled to pre- and posttreatments, this confers to the final product specific properties as stability, appearance, water resistance, temperature resistance, elasticity, permeability to perspiration and air, etc. [1].

The valorization of hides and skins also generates by-products that find outlets in several industry sectors such as pet and animal food, fine chemicals including cosmetics, photography intermediates, soil conditioners and fertilizers.

The process of making leather, however, has always been associated with odors and pollution, as they seem to be inevitable consequences of these productive activities. Environmental impacts of tanning and related activities are essentially associated with the chemicals applied and raw materials used, thus inducing air, soil, surface and groundwater pollution. Therefore, provisions for pollution control, waste generation and disposal, chemical safety, raw materials control, wastewater treatment as well as energy consumption minimization are essential in tanning industry management [1,2].

The present chapter deals with the application of conservative technologies to the solution of environmental problems related to the tannery industry. Most of them are based on

[1]Department of Civil and Environmental Engineering, The Polytechnic University of Bari, 4, Via Orabona, 70126 Bari, Italy, d.petruzzelli@poliba.it
[2]Istituto di Ricerca sulle Acque, National Research Council, 5, Via De Blasio, 70123 Bari, Italy, petruzzelli@irsa.ba.cnr.it
[3]Istituto di Ricerca sulle Acque, National Research Council, 1, Via Reno, 00198 Rome, Italy.

the use of reactive polymers (ion exchange resins), but all rely on a better organization of the production lines.

Generally speaking, technologies for metals control in industry wastewaters are based on precipitation and co-precipitation methods [3] coupled with pre- or post-oxidation [4], reduction [5] or concentration [6]. Reference technologies, although efficient in terms of pollutant control, defer the environmental problem from a diluted liquid-phase (wastewater) to a concentrated inert solid phase (sludge). Potentially hazardous, this sludge must be disposed of in environmentally safe, controlled landfills, according to the enforced legislation [7]. As opposed to precipitation methods, which are "destructive" toward pollutants, "conservative" technologies allow for removal, recovery and recycling of persistent pollutants, thus taking advantage of the economic revenue associated with the commercial value of the recycled products [4].

The Istituto di Ricerca sulle Acque (Water Research Institute), National Research Council, Bari, Italy, in cooperation with national and foreign institutions, has long been involved in the development of conservative technologies for environmental control [8–14]. Initial investigation focused on determining the best sorbents for metals removal and recovery from solutions simulating the average composition of spent tanning baths or segregated wastewaters from tannery operations or the acidic extracts from sludge. Later investigations developed through different stages, including laboratory-scale batch and dynamic (column) experiments, process optimization and up-scaling to pilot-plant prototypes, process development to demonstration scale units.

To test technical-economical feasibility of the innovation, a 10 m^3/h mobile installation was designed and assembled [10,11]. The IERECHROM® process (Ion Exchange REcovery of CHROMium) successfully passed several demonstration campaigns in north and south Italy (Milan and Naples) as well as in France (Graulhet). The patented process is currently being examined by the International Tanning Authorities for possible full-scale applications [14].

The production of rawhides is strictly related to cattle availability. On a worldwide basis, important areas are localized in the United States, Argentina, Russia, the European Union, New Zealand and Australia. Typically, hides and skins are traded in the salt state or in the form of partially tanned or "wet blue" conditions. Globally, approximately 6 Mt/y of rawhides (cattle, goats and sheep) on a wet blue or salted basis are processed to yield about 460,000 t/y of heavy leather and about 1,300 Mm^2/y of light leather. Europe produced 74,000 t/y of heavy leather and about 240 Mm^2/y of light leather [1].

With a positive trade balance exceeding 750 MEuro in finished products, the EU tanning industry resists the gradual increase in imports from extra-community countries and is still the world's largest supplier to the international market. Exports account for about 40–90% of the tanning turnover in different EU member states. Asia, and in particular the Far East, with its fast-growing economy, has taken an increasingly important position in the global market.

Italy, by far the most important location in terms of establishment, production and turnover, accounts for 65% of the EU production, as a result of more than 2,400 installations spread throughout the country. Spain ranks second (255 installations), followed by Portugal (110) and Germany (40) (Figure 9.1) [1]. Other regions within EU member states present highly concentrated tannery pools with neighboring municipalities strongly dependent on this economic activity. This is frequent in Southern Europe, specifically Italy, with Tuscany (Florence), Veneto (Venice), Piedmont (Turin) and Campania (Naples) regions representing typical examples. Incidentally, the above areas appear particularly vulnerable from the environmental, archeological and artistic point of view.

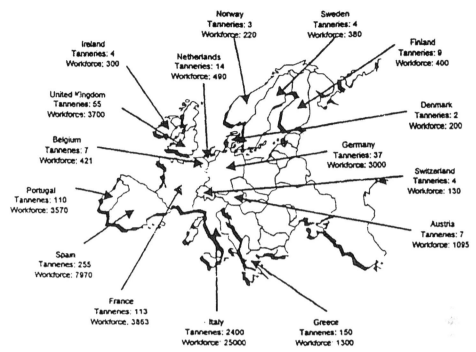

Figure 9.1 Distribution of European tanning industry and workforce.

Tanneries in Spain are located in the Catalunia (Barcelona), Valencia, Murcia and Madrid regions. Minor industrial aggregations are located in the central region of France and in Portugal, Porto areas. Germany and the United Kingdom suffered a sharp decline in the number of tanneries (Figure 9.1).

The tanning industry is very conservative with labor-intensive, low-technology procedures involving a series of very specific treatments and manipulations, mostly based on craftsman activities, whose "recipes" are protected by secrecy from small-medium enterprises. In this way, they fulfill very strict qualitative requirements from fashion ateliers, the elective buyers of leather.

Main innovations are related to the environmental performance of production lines after stricter EU regulations. These refer essentially to the use of alternative tanning agents with respect to chromium rather than on machinery and procedures. However, the increasingly high costs of raw materials make it risky to experiment with new and more environmentally friendly procedures. This is a barrier to new investment on one side, and a particularly difficult obstacle to managing the environmental impacts on the other side. Another barrier to the introduction of innovations is that new technologies often have to be managed with production needs. This is particularly true for those tanners who compete in the very high-quality niche markets and traditional family-run small enterprises.

THE ENVIRONMENTAL IMPACT OF TRIVALENT CHROMIUM
ON WATER-SOIL COMPARTMENTS

Chromium(III) is widely used as tanning agent and is an important source of contamination due to the large volumes of exhaust liquid and solid effluents produced. Although chromium is an essential component for human nutrition at trace levels [15,16], there is no doubt that its compounds at higher concentrations are acutely and chronically toxic [16,17]. The dose threshold effects for this element have not yet been determined with accuracy to allow for specific regulations, although it appears to be definitely less toxic to mamalian and aquatic organisms than other heavy metal species such as Hg, Cd, Pb and Ni, probably due to the low solubility of this element in the reduced form [18]. Cr(III) compounds, indeed, have a very low solubility in soils and are thus relatively unavailable to plants [19]. This consideration encourages application of tannery sludge, rich in proteic matter, to land as soil conditioners [20,21]. The reducing characteristics of reference sludge, owing to the presence of Fe(II) and organic matter, stabilize Cr(III) species, thus reducing their environmental mobility [19]. Field investigations on Cr(III) migration in soils treated with tannery sludge [22] confirmed the low mobility of the metal, but reliability of results was affected by indeterminations concerning the monitoring methods of contaminant concentrations and estimating migration fluxes through water-saturated soils. The lack of threshold values and long-term toxicity effects strongly deters land application of tannery sludge. Moreover, the possibility that organic ligands and/or acidic environments favor Cr mobility in soils [23] and that MnO_2 allows for its oxidation to the more toxic and mobile Cr(VI) species [24,25] is further evidence that the use of tannery sludge in agriculture should be drastically restricted.

In conclusion, the environmental impact of the tannery industry is highly blamed for the presence of hazardous pollutants such as Cr, Al and Fe in the wastes. At least 50% of dosed tanning agents to the productive lines [Cr(III) derivatives] are wasted. About $100 \, Mm^3/y$ of liquid-effluents ($\approx 500 \, mgCr/L$) are disposed of in Europe, and this leads to the formation of 300,000 t/y dry sludge (1–5% Cr content) as a result of the precipitation technologies currently applied for metals control [5,21,26].

THE TANNING PROCESS

Tanning operations include a series of chemical and mechanical treatments, converting a putrescible organic material into a biochemically and mechanically stable product. After treatment, the leather appears soft, elastic and resistant to mechanical stress after cross-linking of the long-chain proteic structure of the collagen, the main constituent of the rawhides and skins.

Two main processes are applied for this purpose: chrome tanning and vegetal tanning. The first process is applied in the majority of cases (70%), while vegetal tanning is applied for specific preparations, e.g., shoe soles manufacturing [2]. Chrome tanning operations include the following main steps (Figure 9.2):

a. reviving

b. lime treatment (de-hairing)

c. skudding, cleaning and neutralization

d. pickling

e. chrome tanning

f. pressing and washing

g. shaving

h. coloring

i. finishing

j. polishing

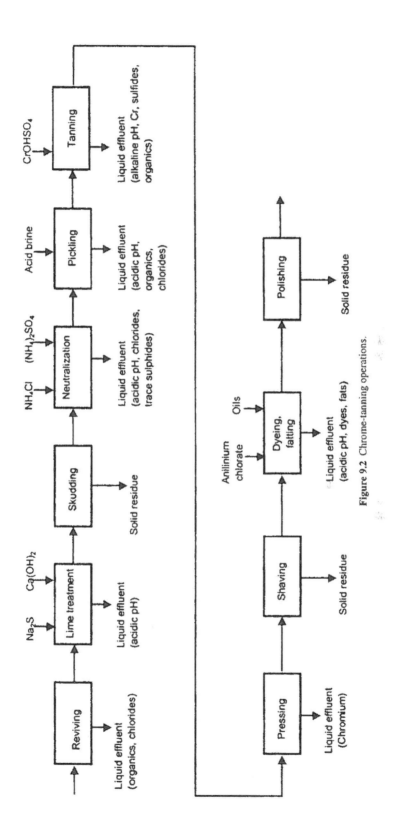

Figure 9.2 Chrome-tanning operations.

The *reviving* step consists of soaking dry (salted) hides in water to remove excess salinity and to soften after rehydration of the natural fibers. As a result of the operation, saline waters are formed with a slight organic content after the use of preservants. This is followed by a *de-hairing* operation that is carried out by treating raw materials with excess lime or caustic solutions to which is added sodium sulfide (0.2–3%). The operation is facilitated by the addition of specific enzymes to open up skin follicles thus releasing hairs and making the material more permeable to the tanning agents. *Skudding* operations are aimed at the removal of residual fats from pores and superficial dermis. This is followed by neutralization and cleaning for residual sulfide and elimination of excess alkalinity. The *pickling* operation is based on the treatment of the material with acidic brines (NaCl 10–15%; H_2SO_4 1–2%) for better control of the chrome tanning operation. In this step, the structure of collagen is modified for better permeation of chromium into the pores of the leather.

The *tanning* operation per se consists of contact with concentrated solutions of basic chromium sulfate ($CrOHSO_4$ 10%) in large rotating drums for a contact time of several hours (max. 24 h). The aim of this operation is the preservation and consolidation of raw materials through cross-linking of the structure of natural collagen. This is followed by the *washing* and *pressing* of the swollen leather from the excess chromium. Final operations include *shaving* from residual hairs, *coloring*, *polishing* and *finishing* according to market demand.

All the mentioned operations lead to the formation of liquid and/or solid wastes, some streams include the presence of chromium, while others do not, however, all must be disposed of properly for environmental protection requirements (Figure 9.2).

ENVIRONMENTAL TECHNOLOGIES FOR Cr(III) CONTROL

Current physicochemical methods for metals control in the environment involve either dispersion and/or conversion into a different physical form, generally as sludge. Most technologies are based on "end-of-pipe" treatments that are considered quite satisfying in the context of purging the wastewaters from metals prior to effluent discharge, but they generate a treatment residue, the sludge, which is difficult to dispose of in conditions of environmental safety. This is recognized by the industrial generators who pay the expensive price of the clean-up and sludge disposal operations which are carried out in controlled landfills.

Based on the current EU legislation, the limit for Cr(III) discharge in confined water bodies (rivers, lakes) and coastal seawaters is set at 2 mgCr/L [7]. More restrictive limits are imposed for solid-waste disposal which are generally classified as hazardous for Cr(III) contents ranging from 100 mgCr/kg. Tannery solid wastes must be landfilled in conditions of environmental safety [7]. In addition to the overall content of pollutants in the solid phase, specific reference is also made to their leachability under different conditions. On these premises, treatment of tannery effluents (liquid and solid) is imposed for environmental protection in most industrialized countries.

Problems in the management of tannery wastes are associated with variability of their composition depending on type and quality of leather products; geographic location of the installations, local market demand and niche production; and type of applied tanning process. Moreover, in considering that the amounts of chromium chemicals dosed change sensibly with the type and quality of hides and that metals retention onto the finished product is limited (30–50% of the amounts dosed), it is realized that there is large variability in the quantity and quality of the tannery wastes formed. Table 9.1 shows the average composition

TABLE 9.1. Average Composition of Tannery
Wastewaters (mg/L).

Parameter	Spent Tanning Bath	Segregated Wastewaters*
Cr(III)	3,500–4,000	100–500
Al(III)	80–150	2–60
Fe(III)	40–100	2–10
Ca^{2+}	1,000–5,000	100–300
Mg^{2+}	500–1,500	50–100
SO_4^{2-}	10,000–12,000	300–600
NaCl	50,000–60,000	1,000–2,000
TOC (as CH_3COOH)	1,200–1,800	80–200
pH	2.5–3.5	3.5–3.8

*Spent tanning bath diluted (≈1:15) with leather washing waters.

of a typical spent tanning bath together with composition of "segregated" wastewaters from tanning operations. Segregated wastewaters refer to that fraction of liquid wastes effectively containing Cr(III), which, on average, amount to only 10% of the overall effluent volume resulting from operations. These wastes include spent tanning baths and leather washing waters.

Conventional wastewater treatments are aimed at the abatement of metals and organic matter through reduction of BOD (Biochemical Oxygen Demand) and COD (Chemical Oxygen Demand), ammonia, salinity and suspended solids. Bio-persistent organics such as surfactants, biocides, preservants such as naphthalin derivatives and chlorinated pesticides are removed by specific treatments.

Figure 9.3 shows conventional treatment schemes for tannery wastewaters that include separate lines for physicochemical and biological treatments of liquid streams, followed by sludge management of mixed (physicochemical + biological) sludge (discussed later). Physicochemical treatments include straining, equalization and primary settling for the abatement of suspended solids, followed by oxidation and chemical precipitation of metals through lime and polyelectrolyte addition and final settling of metal-containing sludge [Figure 9.3(a)]. Metals precipitation may be associated with oxidation or reduction operations. The effluent from physicochemical treatment is conveyed to biological treatment for the oxidation of biodegradable organic matter and nitrification of ammonia, followed by secondary settling and final disinfection [Figure 3(b)]. Advanced treatments for the abatement of specific bio-persistent organic substrates may also be included through chemical oxidation and/or sorption processes.

Cr(III) abatement is needed before biological treatments to minimize metals toxicity on biomasses. Both physicochemical and biological treatments end up with the formation of sludge that is usually mixed and disposed of in controlled landfills via thickening, conditioning, digesting and filterpressing operations [Figure 3(c)].

Chemical precipitation of metals defers the environmental problem from a diluted liquid phase to a concentrated solid phase (hazardous sludge). Moreover, the potential resource associated with the economic value of pollutants recovered is "destroyed" through costly confinement in artificial repositories (landfills).

As a result, the most important environmental problem in the tannery industry is related to sludge management which includes aspects still waiting for reliable and economically feasible solutions.

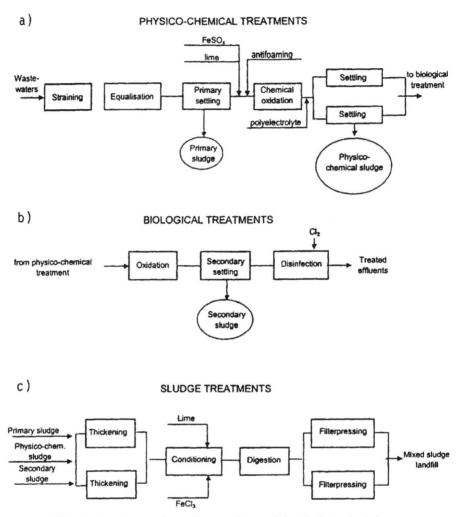

Figure 9.3 Environmental protection operations carried out in the tanning industry.

Two different types of sludge are produced: hazardous metals-laden mineral sludge from physicochemical operations and nonhazardous biological sludge. The overall volume of mixed sludge (physicochemical and biological) is quantified in about 10–15% of the incoming volume of wastewaters treated, with an average concentration (after thickening) in the range of 2–5% dry solids and max 20% dry solids after filterpressing operations [Figure 3(c)]. Mixed sludge includes metal hydroxides [1–5% Cr(III) content] and organic residues (hairs, fats, proteic matter) from pre- and post-tanning operations and dead biomass from the biological treatment of wastewaters.

CONSERVATIVE TECHNOLOGIES FOR POLLUTANT CONTROL

The general opinion in this context involves several basic strategies to eliminate or reduce the masses of metal sludge now disposed of into the environment: source avoidance and/or waste minimization; direct metals concentration after selective separation of liquid waste streams; waste exchange; and selective extraction of metals for recovery, recycle and reuse.

The greatest potential for applications is represented by the last option (recycle, recovery and reuse), which is mainly limited to precious metals. Among the available technologies, it is worth mentioning advanced precipitation through fractional separation of solid phases, selective ion exchange, membrane processes (reverse osmosis, electrodialysis), evaporation and liquid extraction [3].

The rising costs of raw materials together with the stringent need for environmental control justifies the growing interest toward conservative technologies for removal, recovery and recycling of pollutants to the industry production lines of origin.

Developed countries have enforced strict regulations for environmental control, especially on persistent pollutants such as heavy metals and recalcitrant organics [7]. To keep with these constraints, industry has developed and applied pollution control facilities that, in the case of metals, are based on precipitation and coprecipitation methods [5,26] coupled with pre- or post-oxidation or reduction [3–5], and, in the case of persistent organics, on advanced oxidation processes [26].

Reference technologies, although efficient in terms of pollutant control, defer the problem from a diluted liquid phase (wastewater) to a concentrated inert solid phase (sludge), potentially hazardous, which must be disposed of in conditions of environmental safety according to the enforced legislation. The environment has no assimilative and/or degradation capacity for metals, accordingly, the ultimate repository for metals-laden hazardous wastes is their destructive confinement in controlled landfilling.

As opposed to precipitation-oxidation methods, "conservative" technologies allow for removal of pollutants to the strictest limits imposed by the enforced legislation, and recovery and recycling of pollutants, thus taking advantage of the economic revenue associated with the commercial value of recovered products. The concept is better substantiated as follows [4,27]:

Pollutants (wastes) + Knowledge (technology) = Potential Resources

In the following is illustrated research and development of new conservative technologies, mostly based on the use of the concerted action of simpler chemical procedures and/or on the use of reactive polymers (ion exchange resins). The final objective in all cases was aimed at determining feasible and economically reliable solutions to the environmental problems related to the tannery industry. In a preliminary step of the investigation, specific reference was made to the Cr(III)-containing sludge that was leached in acidic media and the supernatant solution submitted to oxidation of Cr(III) to Cr(VI) or eluted onto selective ion exchange resins for metals separation and recovery.

Lately, having realized that most chromium was confined in a minority fraction (10%) of the tannery effluents (the segregated wastewaters, Table 9.1), the object of the investigation was focused only on these latter wastes. The resulting process, namely, IERECHROM® (Ion Exchange Recovery of CHROMium) [14], developed to demonstration scale, allowed for quantitative removal and recovery of Cr(III) from segregated effluents. The innovation is under the attention of tannery authorities for possible full-scale applications.

METALS SPECIATION IN TANNERY WASTES

Special attention was focused on the chemical forms of metals (species) present in the liquid phase as well as at the liquid-solid interphase in the case of separative processes. This is for better control of the mentioned variations in the composition of tannery wastes in order to gain a deeper insight into the fundamentals of different phenomena, and to better control principles of the proposed innovations.

Another important aspect of kinetic order was related to the presence of specific polyvalent metals complexes contributing to the chemical speciation in systems, which were not in true thermodynamic equilibrium and were leading to a steady, slow reconversion of species toward equilibrium. Reference phenomena were completed in a matter of weeks or months, in most cases. The mentioned time span largely exceeded completion of laboratory tests. Polyvalent metals solutions (Cr, Al, Fe) appeared as "living" in terms of composition stability that was continuously evolving toward equilibrium conditions.

As a consequence, aged and freshly prepared solutions had different behaviors with respect to retention properties of reactive polymers (constant at all the controlling parameters such as pH, ionic strength and temperature). Timing of the operations and a knowledge of metals speciation in the complex background of tannery solutions (i.e., the presence of organic and inorganic ligands, the relative concentrations of components, the solution pH, etc.) was of paramount importance. Metal species distribution curves (Figure 9.4) were used as orientation maps for insights into reaction mechanisms. Speciation data, however, did not include the mentioned kinetic effects related to metal complexes reconversion phenomena, which is a problem still open to solutions.

Figure 9.4 shows typical distribution of species for the Cr/Al/Fe system in the segregated tannery wastewaters. Reference curves were obtained by simulations of the thermodynamic equilibrium conditions by the use of the computer code MINEQL+ from J.C. Westall at Massachussets Institute of Technology, USA [28].

Reference computer code, among the most credited in the literature, includes different subprograms for variable parameters such as solution pH (titrations), ionic strength or incipient precipitation phenomena. The program computes the effective distribution of species at equilibrium through evaluation of mass and charge balance that are matched with thermodynamic equilibrium constants of different chemical forms. The solution to different problems corresponds to the minimum value of the overall free energy of the system under consideration.

Other more sophisticated simulation programs, including the influence of the metal activities on thermodynamic equilibrium constants and the presence of precipitates and/or adsorbing solid phases, are available in the literature. Among others, it is worth mentioning the HYDRAQL [29] and MINTEQA2 [30] codes. All the mentioned codes are based strictly on thermodynamic backgrounds and do not include transient pseudoequilibrium conditions and/or kinetic effects.

PRETREATMENT OF TANNERY SLUDGE

Three hundred thousand t/y mixed dry sludge are formed in the EU as a result of the precipitation methods adopted for metals control in the tannery industry. With an average chromium content of 1–5%, reference sludge is classified as hazardous and must be disposed of in controlled landfills. The landfilling operation becomes increasingly more expensive with the saturation of landfilling sites in the vicinity of tannery installations, with increased disposal costs for the incidence of longer-distance transport.

To find a solution to the problem, the first approach focused on mixed sludge (Table 9.2)

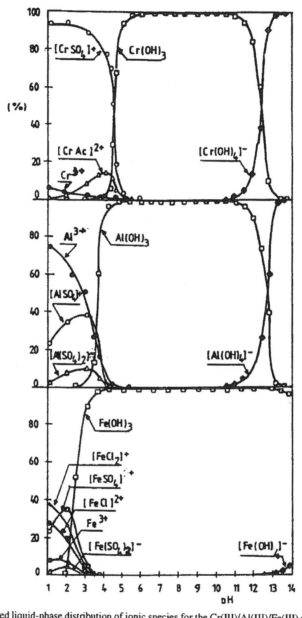

Figure 9.4 Computed liquid-phase distribution of ionic species for the Cr(III)/Al(III)/Fe(III) system in the segregated tannery effluents (MINEQL computer code, Reference [28]).

TABLE 9.2. Average Composition of Tannery Sludge from Italian Installations and the Solid Residue after the Acidic Leaching Operation (% Dry Sludge) [9].

Parameter	Sample A	Sample B	Sample C	Solid Residue after Leaching
Water content	81.8	79	76	20
Organic fraction	39.1	53	32	80–90
Cd(III)	0.4	4.2	4.6	0.1–0.2
Fe(II)	0.85	2.4	0.9	0.01–0.05
Al(III)	1.1	0.15	2.6	0.01–0.05
Ca^{2+}	19.8	3.4	1	5–15
Mg^{2+}	2.3	0.3	0.05	1–2
Alkalinity (meq/g dry sludge)	10.3	11.5	10.9	—

A: Soges sludge, Florence; B: Arzignano mixed sludge, Vicenza; C: Turbigo mixed sludge, Milan.

that was submitted to acidic leaching by the use of mineral acids (H_2SO_4, HCl, HNO_3) with the aim of their detoxification through selective extraction of the metals. Figure 9.5 shows chromium extraction vs. pH for three typical sludges contacted with 98% w/v sulfuric acid. After washing and filtration, relatively low residual quantities of metals (e.g., 0.1–0.2% Cr, Table 9.2) were detected in the leached residue. Detoxified (metals-free) solid residue was safely applied to land as soil amendant and/or used as animal fodder.

The sludge leachate was submitted to direct re-precipitation of metals and separation of hazardous solid wastes which were disposed of in controlled landfills [9] and oxidation of Cr(III) in alkaline media, followed by metals separation and concentration through ion exchange, for recovery and recycling purposes [8].

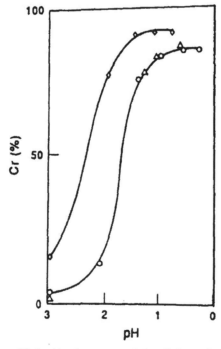

Figure 9.5 Chromium acidic leaching from tannery sludge: ○ Soges; △ Arzignano (Table 9.2).

TABLE 9.3. Composition of the Sulfuric Acid Extracts (pH 1) of Representative Tannery Sludge from Different Tanneries (mg/L)

Parameter	Soges	Sludge* Arzignano	Turbigo
Cr(III)	109	1,160	1,260
Fe(II, III)	271	740	210
Al(III)	340	79	1,060
Ca^{2+}	535	740	530
Mg^{2+}	500	116	60
TOC**	1,300	2,000	680

*From indicated Italian tanneries.
**Total organic carbon.

Sludge samples reported in Table 9.3 cover a wide range of representative compositions from Italian installations.

Metals leaching tests, carried out by the use of different acids (H_2SO_4, HCl, HNO_3), did not show significant differences. Concentrated sulfuric acid (98% w/v) was preferred due to the lower cost and lower extraction capacity toward calcium. The acid was dosed directly to the sludge suspension (5% solids) under vigorous stirring until pH 1 was reached, in excess of 1.6 times with respect to the total alkalinity. Higher dosages did not significantly increase chromium extraction and proved to be uneconomical in consideration of the subsequent chromium oxidation reaction that was carried out in alkaline media (which will be discussed later). In these conditions, >90% of the Cr(III) in tannery mixed sludge was leached (Figure 9.5).

PROCESS 1: RE-PRECIPITATION OF METALS FROM MIXED SLUDGE ACIDIC LEACHATE

Figure 9.6 shows the conceptual scheme of the process based on the following main steps [9]: acidic leaching of mixed sludge, filtration and lime neutralization of the detoxified solid residues and final washing before disposal by land application as soil amendant.

As illustrated, the leaching operation is carried out by the addition of 98% w/v H_2SO_4 to pH 1.5 (1.6 eq H_2SO_4/eq alkalinity). Figure 9.5 shows chromium extraction curves for two real sludges from Italian installations.

The acidic leachate is neutralized by the addition of stoichiometric amounts of caustic, thus inducing re-precipitation of the metals (Cr, Al, Fe) in the form of hydroxides. The resulting hazardous sludge is safely disposed of in controlled landfills.

The procedure is essentially aimed at the separation of the nonhazardous organic fraction (70%) of mixed sludge from the hazardous mineral fraction, including most of the metals. From the economic point of view, the above operations lead to remarkable economic advantages with respect to deliberate disposal of the whole mixed sludge in controlled landfills. The process is, therefore, justified on a strictly economic basis.

PROCESS 2: OXIDATION OF Cr(III) TO Cr(VI), RECOVERY AND SEPARATION OF METALS BY ION EXCHANGE

Figure 9.7 shows the conceptual scheme of the process [8].

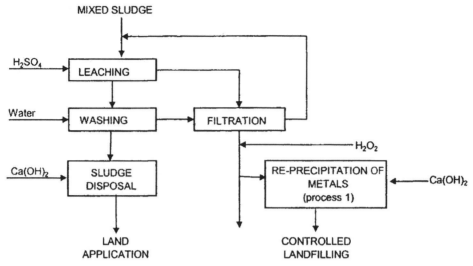

Figure 9.6 Conceptual scheme of Process 1.

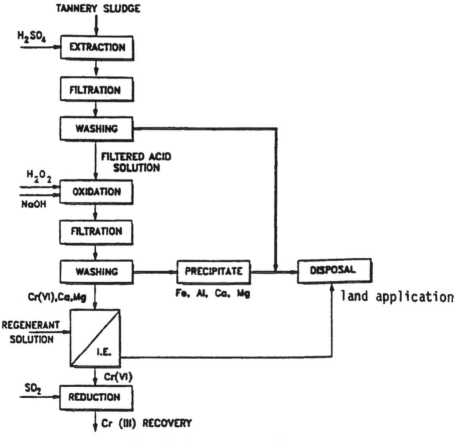

Figure 9.7 Conceptual scheme of Process 2.

After acidic extraction to pH 1.5, the detoxified solid residue is applied to land or recycled for multipurpose reuse. The supernatant solution is sent to Cr(III) oxidation by the use of hydrogen peroxide in alkaline media (H_2O_2/NaOH). Different oxidant-to-chromium ratios ($X = H_2O_2$/Cr) in the range of 1 to 10 were tested. $X = 1$ corresponds to the stoichiometric molar ratio according to the following reaction:

$$2[Cr(OH)]^{2+} + 3H_2O_2 + 8OH^- \rightarrow 2CrO_4^{2-} + 8H_2O \qquad (1)$$

The reaction solution is filtered for Fe, Al, Mg and Ca hydroxide separation, whereas the soluble chromate ion is purified and concentrated by ion exchange [6]. Chromates in the spent resin regeneration eluates are reduced back to Cr(III), by sulfite addition, for reuse in the tannery industry, or are recycled as Cr(VI) in other productive activities (e.g., the plating industry).

The resulting nonhazardous solid residue (Fe, Al and Mg hydroxide) may be reused as coagulants in tannery wastewater treatment operations.

Cr(III) Oxidation

Figure 9.8 shows kinetic curves for the Cr(III) oxidation reaction from real sludge leachate. Experiments, carried out at different pH (5–11), temperature (20, 60°C); H_2O_2/Cr ratios, X (1, 1.7, 2), indicated 30 min as the best reaction time. Longer contact time impaired process economy with a slight increase in the chromium oxidation yield [Figure 9.8(a)].

The dependence of Cr(VI) formation on H_2O_2/Cr ratios at different pH, constant temperature (60°C) and ionic strength ($I = 0.5M$, $NaNO_3$), is shown in Figure 9.8(b). An increase in the oxidation yield with pH and X ratios is observed within 5 min after the addition of the reagents. This is consistent with first-order kinetic dependence of the Cr(III) species over the OH^- and H_2O_2 concentrations. The influence of temperature on chromium oxidation yield is shown in Figure 9.8(c); temperature in the range of 60°C appears a reasonable compromise between oxidation yield and operational costs. Lower temperatures led to longer reaction time for acceptable oxidation yield.

The presence of anions such as NO_3^-, Cl^- and SO_4^{2-} in the concentration range 0.5–1.5M did not influence the oxidation reaction.

Generally speaking, chromium(III) oxidation by hydrogen peroxide in alkaline media appears to be a kinetically controlled process, with $[Cr(OH)]^{2+}$ acting as the rate-determining species for the overall reaction. Longer reaction times at 20°C, as compared to 60°C, appear inconsistent with the calculated activation energy values for the oxidation reaction [Equation (1)] in the mentioned temperature range. This confirms that the process is probably affected by the simultaneous formation of less reactive solid phases, as the solubility product for chromium hydroxide is exceeded in most cases, and/or to the formation of stable polynuclear Cr(III) complexes (see "Pretreatment of Tannery Sludge"). The formation of stable adducts would make the overall oxidation process slower depending on the redissolution kinetics of solids and/or decomposition kinetics of complexes. Other authors found a strong effect of solution aging on the oxidation kinetics, attributing this phenomenon to the formation of slow-forming polynuclear Cr(III) complexes, refractory to the oxidation reaction with hydrogen peroxide [31,32].

Quantitative oxidation was obtained for the oxidant-to-chromium ratio $X = 1$, pH 10, $t = 60°C$ and reaction time exceeding 30 min. The above conditions were adopted throughout process development steps.

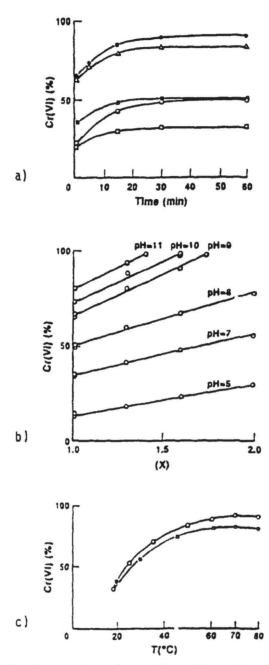

Figure 9.8 Chromium(III) oxidation reaction as a function of time at different pHs (a), H_2O_2/Cr ratios, X (b), and temperature (c): (a) $\triangle$: $X=1$, pH 10, $t=60°C$; $\square$: $X=1$, pH 10, $t=20°C$; $\bigcirc$: $X=1.7$, pH 3.5, $t=60°C$; $\bullet$: $X=5$, pH 10, $t=60°C$; $\blacksquare$: $X=1$, pH 10, $t=60°C$; c) $\blacksquare$: $X=1$, pH 10; $\bigcirc$: $X=5$, pH 10.

Chromium Separation and Recovery

pH control of the oxidation solution to 8 allowed for Fe and Al precipitation in the form of hydroxides and quantitative separation from chromate ion. Beyond pH 10, the formation of tetrahydroxoaluminate species $[Al(OH)_4]^-$ (Figure 9.4), increased Al solubility, thus impairing metals separation. Digestion of the suspension at 60°C for 30 min allowed for better separation of the solid phases. Solubility of Ca and Mg ions at pH 8 is still high for quantitative separation from Cr(VI) solutions.

With the aim of improving the quality of the Cr(VI) solution recovered, these were further processed through cation exchange for calcium and magnesium separation on commercial strong acid resins, e.g., Amberlite IR200, and anion exchange for selective recovery and concentration of chromate ion [6].

Once purified, Cr(VI) solutions may be reduced back to Cr(III) in the form of basic sulfate for recycling to tannery operations. To this aim, many reducing agents are considered in technological applications [33]. Among others, SO_2 is preferred for direct recovery of basic chromium sulfate:

$$Cr_2O_7^{2-} + 3SO_2 + 11H_2O \rightarrow 2Cr(OH)(H_2O)_5SO_4 + SO_4^{2-} \qquad (2)$$

The reaction is easy and fast at pH 2 and room temperature by dosing minimum excess of reactant over the stoichiometric value.

Process Development

Preliminary economic considerations showed that the operative costs of the process (chemicals and energy consumption) are double with respect to landfilling disposal of mixed tannery sludge. The commercial value of recovered chromium derivatives is, on the other hand, too low to warrant economic revenue from its recovery. However, progressive saturation of landfilling sites, especially in densely populated areas such as Europe, will lead to an inevitable rise in landfilling costs. This latter consideration and the continuous steady increase of the costs of raw materials will enhance interest in the proposed innovation.

Although the process never passed pilot-scale laboratory investigation, it is now under consideration by the International Tannery Authority for future developments.

PROCESS 3: LEACHING OF SLUDGE AND METALS RECOVERY BY SELECTIVE ION EXCHANGE

Figure 9.9 shows the conceptual scheme of the process [9]. The detoxified solid residue after leaching is safely applied to land. Sludge acidic leachate (Table 9.3) is filtered and submitted to oxidation of Fe(II) by the addition of hydrogen peroxide. The resulting solution is processed by ion exchange. Figure 9.10 shows details of the ion exchange section, which is based on the use of an aminophenol sulfate-form weak electrolyte anion resin (Duolite A7, from Rohm and Haas Co., Philadelphia, Pa, USA), for selective chelation of ferric species and an H-form strong cation resin (Purolite C 160, from Purolite Co., UK) for separation of the Al(III) from Cr(III) species. Trivalent chromium breaks through both ion exchange sections, as a nonexchangeable polynuclear sulfato-complex, whereas aluminum species bind to the strong cation resin.

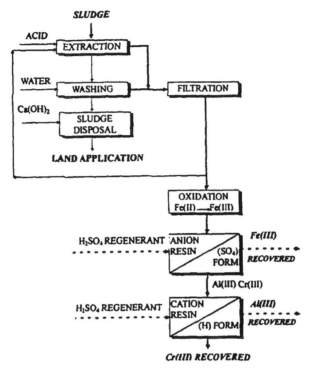

Figure 9.9 Conceptual scheme of Process 3.

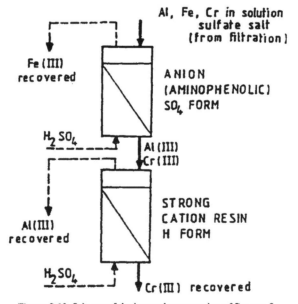

Figure 9.10 Scheme of the ion exchange section of Process 3.

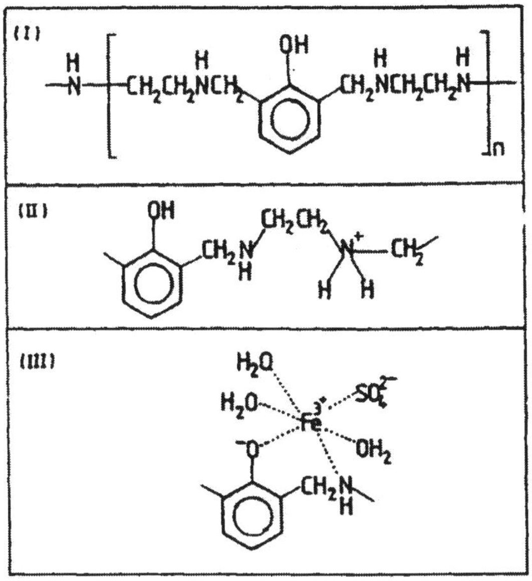

Figure 9.11 Aminophenol resin Duolite A7 (I) Structure of the aminophenol matrix, (II) secondary amine functional groups, (III) ferric complex.

Selection of the Ion Exchange Resins

Based on literature data and direct experience, the weak electrolyte anion resin Duolite A7 was determined to be the best sorbent for ferric species retention from the acidic sludge leachate. These ferric species could specifically bind by chelation onto the sulfate-form amino-functional groups at pH 1.5–2 (Figure 9.11) [34].

A systematic investigation was carried out on a set of cation exchangers with different porosity, cross-linking degree and exchange loading to find out the best sorbent for separation of chromium from aluminum species. Table 9.4 shows the main physicochemical properties of the resins investigated.

Before testing, cation and anion resins were sieved in the range 20–30 US mesh and conditioned by three consecutive acid-base reconversions (H_2SO_4/NaOH 1 M, 5 Bed Volumes, BV). Conditioned resin samples, in the regenerated SO_4^{2-}- and H^+-form, respectively, were loaded into separate laboratory columns and submitted to exhaustion-regeneration cycles. Preliminary experiments showed that the exchange kinetics were relatively slow, therefore,

TABLE 9.4. Main Physicochemical Characteristics of the Resin Investigated.

Resin	Functional Group	Matrix	Cross-linking (%DVB)	Porosity	Physical Form	Exchange Loading (eq/lr)
Amberlite IR120	$-SO_3H$	Styrene	8	gel	beads	1.9
Amberlite 252	$-SO_3H$	Styrene	15	macro	beads	1.8
Amberlite 200	$-SO_3H$	Styrene	20	macro	beads	1.7
Purolite C100	$-SO_3H$	Styrene	10	gel	beads	2.2
Purolite C160	$-SO_3H$	Styrene	20	macro	berads	2.3
Duolite C3	$-SO_3H$	Styrene	—	macro	granular	1.2
Duolite A7	$-NH_2R$; $-NHR_2$	Phenol	—	macro	granular	2.5

the exhaustion step was carried out at low flow rates not exceeding 5 BV/h. These cycles were performed by eluting the sludge acidic leachate (Table 9.3) on the anion resin and the effluent from this on the cation resin columns.

Resin regeneration was carried out at lower flow rates (3 BV/h) by using 1 M sulfuric acid for both the anion and cation sections. After regeneration, the Duolite A7 resin was washed to pH 2 for best performance.

The influence of critical variables such as solution pH, ionic strength, and anionic background were tested to evaluate the behavior of different resins as a function of metals speciation at the liquid-solid interface.

Selective Removal of Fe(III) on the Aminophenol Resin

As mentioned, the weak electrolyte anion resin Duolite A7 in sulfate form shows good chelating performance toward ferric species in acidic media [34]. The aminophenol matrix of the resin contributes to the chelation reaction of the mentioned species (Figure 9.11). Figure 9.12 shows the exhaustion breakthrough curves of the sulfate-form resin at pH 1 and 2, respectively. The pH of the influent solution has a critical effect on the resin behavior with best performance in the range of 1.5–2. At pH lower than 1.5 (Figure 9.13), the predominant ferric species in the acidic leachate is the cationic complex $[FeSO_4]^+$ that is excluded by the protonated functional groups of the anion resin by electrostatic repulsion (Donnan effect) [35]. Protonated quaternary ammonium groups, on the other hand, have no donor electron pairs available for metals chelation. Both factors contribute to the poor resin performance toward ferric species at pH lower than 2. At pH 2, together with a steady reduction of the cationic complex $[FeSO_4]^+$, a peak is observed with respect to the anion complex $[Fe(SO_4)_2]^-$ (Figure 9.13) which freely diffuses in the resin phase for best retention performance of the resin. Beyond pH 2.5, precipitation of ferric species occurs in the form of hydroxide.

Cr(III) and Al(III) species are present, respectively, as cationic species $[CrSO_4]^+$ (97% and $[AlSO_4]^+ + Al^{3+} \geq 80\%$ (Figure 9.13), which break through (together with Ca^{2+}) the anion resin column. Figure 9.14 shows operative exhaustion-regeneration performance of the sulfate-form resin Duolite A7 with real tannery sludge leachate.

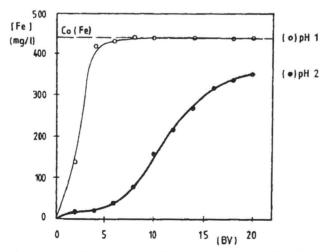

Figure 9.12 Exhaustion breakthrough curves for ferric species on resin Duolite A7 at different pHs.

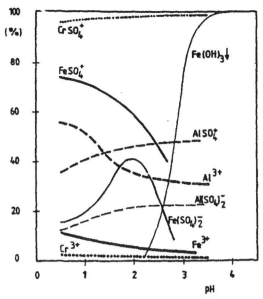

Figure 9.13 Computed liquid-phase distribution of ionic species for the Cr(III)/Al(III)/Fe(III) system in the tannery sludge leachate (MINEQL computer code, Reference [28]).

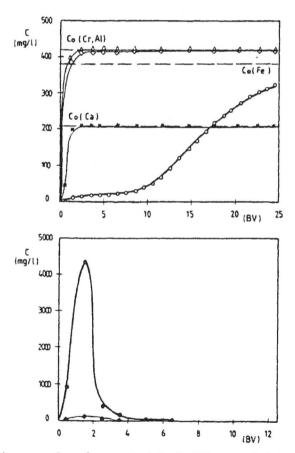

Figure 9.14 Exhaustion-regeneration performance of resin Duolite A7 for tannery sludge leachate (exhaustion pH: 1.5, flow rate: 4 BV/h; regenerant solution: 1 M H_2SO_4, flow rate: 2 BV/h).

327

Separation of Cr(III) from Al(III) by the Strong Cation Resins

To separate Cr(III) from Al(III) present in the effluent solutions from the Duolite A7 resin, experiments were carried out at different solution pH, metals concentration, and ionic strength on the set of cation resins reported in Table 9.4. Exhaustion performance of all resins investigated was essentially equivalent, thus confirming a slight (if any) influence of resin characteristics on metals separation behavior. Figure 9.15 shows Cr and Al breakthrough curves for the representative resin Purolite C160, for experiments at different solution pH.

The influence of ionic speciation on resin performances is clearly evidenced by the conformation of the curves. At pH lower than 1.0 (Figure 9.13), the predominant cationic species in solution are in the order: $[CrSO_4]^+$ ($\approx$98%); $[AlSO_4]^+$ ($\approx$40%); Al^{3+} ($\approx$50%). The high charge-density free trivalent aluminum ion is preferentially bound to the resin functional groups due to the strong electroselectivity effect of this latter species [36]. Trivalent aluminium ions are selectively retained with respect to the mentioned more abundant chromium cationic complexes. At pH higher than 1.0, a significant reduction of the Al^{3+} species in favor of $[AlSO_4]^+$ is observed, thus inducing relaxation of the electroselectivity effect favoring aluminium retention. As a consequence, an increased selectivity for chromium species over aluminium is evidenced by the marked inflexions of the Cr breakthrough and the corresponding regression of the Al breakthrough curves at pH 2 and 2.5 (Figure 9.15).

To better substantiate the influence of the accompanying anions on resin performance, experiments at different sulfate ion concentrations were carried out, while maintaining constant solution pH and metals concentrations. At high sulfate ion concentration, i.e., 1 N (Figure 9.16), better separation of chromium from aluminum species is observed, although

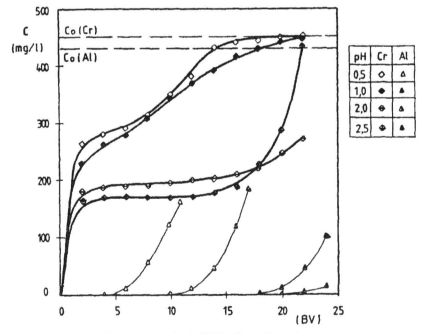

Figure 9.15 Exhaustion breakthrough curves for Cr(III)/Al(III) species in the tannery leachate on strong cation resin Purolite C160 at different pHs (flow rate: 4 BV/h).

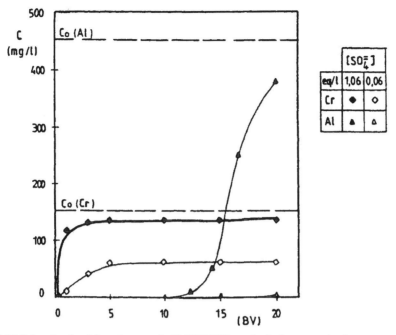

Figure 9.16 Exhaustion breakthrough curves for Cr(III)/Al(III) species in the tannery leachate on strong cation resin Purolite C160 at different sulfate ion concentrations (flow rate: 4 BV/h).

with a lower loading capacity for these latter species. Once again, the above findings are confirmed by metals speciation: at higher sulfate ion concentration, the fraction of free Cr^{3+} species is definitely low (5% vs. 40%) for retention, whereas free aluminum ions are still substantially present and ready for preferential retention by the resin functional groups.

Regeneration of anion and cation resins was carried out by the use of 1 M sulfuric acid for quantitative recovery of almost pure ferric and aluminum sulfate from spent regeneration eluates, ready for recycling as coagulants in the tannery wastewaters treatment.

Process Development

The process is under development on a larger scale through laboratory pilot-plant investigation. In this context, particular attention will be devoted to the influence of the presence of organic matter in the acid extract potentially fouling ion exchangers, specifically, the anion section; kinetic aspects of metals complex formation in the liquid phase; and mass transfer phenomena and ion exchange kinetics of metals retention at resin functional groups. Better insight into the above aspects might improve or simplify the basic concept of the process.

PROCESS 4: CHROMIUM REMOVAL AND RECOVERY FROM SEGREGATED TANNERY WASTEWATERS: THE IERECHROM® PROCESS

The IERECHROM® process (Ion Exchange REcovery of CHROMium) was specifically developed for removal and recovery of Cr(III) from "segregated" streams of tannery operations (see "Conservative Technologies for Pollutant Control" and Table 9.1) [10,11,14].

The process is based on the use of a macroporous weak electrolyte cation resin with carboxylate functional groups (i.e., Purolite C106, from Purolite Co., UK). The resin binds all polyvalent (Cr, Al, Fe) species. Separation and recovery of metals is carried out by a two-step resin regeneration including alkaline hydrogen peroxide for quantitative recovery of chromate and aluminate ions and 1 M sulfuric acid for ferric species recovery. Aluminate ion is later separated from chromate by pH control of the spent regeneration eluates. Chromium is recycled for tannery operations or for related productive activities, while ferric and aluminum sulfate are recycled as coagulants for wastewater treatments.

Figure 9.17 shows a simplified conceptual scheme of the process.

Process Optimization

EXHAUSTION BEHAVIOR OF THE RESIN

Strong electrolyte cation resins manifested poor performance toward chromium removal from segregated streams of tanning operations. Accordingly, their use was dismissed in favor of carboxylate resins. Better performances of these latter resins are attributed to the complexing ability of the carboxylate groups toward polyvalent metal species which is added to the normal ion exchange properties of the resins.

Table 9.5 shows the main physicochemical properties of the resin investigated.

As a reactive ion exchange process, i.e., ion exchange accompanied by chemical reactions [37,41], preliminary experiments showed that metals retention kinetics were relatively slow [12]. For this reason, the resin exhaustion step was carried out at flow rates not exceeding 5 BV/h on partially hydrolyzed (H/Na-form) resin in strictly controlled pH conditions (8.5–9) [13]. Reference pH was sufficiently high to avoid hydrolysis and permanent

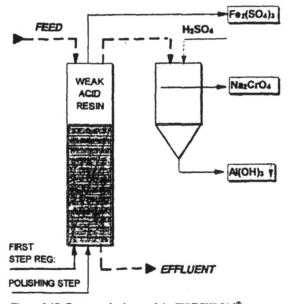

Figure 9.17 Conceptual scheme of the IERECHROM® process.

TABLE 9.5. Main Physicochemical Properties of the Commercial Resin Purolite C106.*

Resin	Functional Group	Matrix	Porosity	Exchange Loading (eq/lr)
Purolite C106	carboxylate	acrylic	macroporous	3.0

*From Purolite Company International, Pontyclyn, Wales, UK.

inactivation of the resin carboxylate groups, but sufficiently low to minimize in situ (column) precipitation of metals in the form of hydroxides. Partial hydrolysis was carried out by controlled washing of the regenerated Na-form resin to pH 9 with soft (cation free) water (30 BV).

Figure 9.18 shows the exhaustion breakthrough curves for (partially hydrolyzed) Na/H-form resin Purolite C106, after elution of the tannery segregated streams (Table 9.1). Polyvalent metal species (Cr, Al, Fe) are nonspecifically removed from the influent solution (pH 3.5–4), thus confirming the good affinity of the resin toward all metal species. The shape of breakthrough curves confirms the selectivity sequence of the carboxylate groups toward all polyvalent free metal species in the order: $Fe^{3+} > Al^{3+} > Cr^{3+}$, thus in good agreement with hydrated ionic radii and the related charge density scale [38,42].

Analysis of the speciation equilibria in the segregated wastewaters suggests that in spite of the relevant fractions of other cation species (i.e., $[CrSO_4]^+$, $[AlSO_4]^+$, $[FeCl_2]^+$, $[FeCl]^{2+}$) (Figure 9.4), the free metal ions are preferentially retained by the resin carboxylate groups, given the strong electroselectivity effect [36]. As a consequence of metals depletion from the liquid phase, speciation equilibria of the mentioned cationic complexes are shifted toward the formation of free trivalent metal ions, replenishing the equivalent amounts retained by the resin. Retention of ferric species is also attributed to the strong interactions of carboxylate groups, via mechanisms other than ion exchange, i.e., covalent

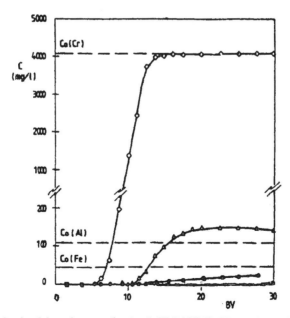

Figure 9.18 Exhaustion breakthrough curves for the Cr(III)/Al(III)/Fe(III) species in the tannery segregated wastewaters on the weak electrolyte cation resin Purolite C106 (◆ Cr, △ Al, ● Fe; downflow rate 5 BV/h, pH 3.0).

bonding through formation of stable carboxy-metal complexes, as reported in the literature [39,40].

It follows that the kinetically retained aluminum species are "rolled up" over the influent concentration (Figure 9.18) by the more thermodynamically stable ferric-carboxylate adduct (notice the peak of the Al breakthrough curve at the incipient ferric ions breakthrough).

REGENERATION BEHAVIOR OF THE RESIN

It is not an easy task to recover polyvalent metal ions from carboxylate resin due to the strong affinity of the functional groups for the reference ions. To this aim, five regeneration protocols for the metals-exhausted resin were tested in order to optimize metals regeneration from resin [13]:

a. Elution with 1 M NaOH

b. Elution with 1 M H_2SO_4

c. Elution with 1 M NaOH followed by 1 M H_2SO_4 or vice versa

d. Elution of 1 M NaOH followed by 1 M H_2SO_4/Na_2SO_4 solution

e. Elution of alkaline 0.15 M $H_2O_2/NaCl/NaOH$ solution followed by resin polishing with 1 M H_2SO_4

Protocol c was run by eluting sequentially 1 M NaOH followed by 1M H_2SO_4; the reverse procedure (e.g., H_2SO_4 first) was also tested. Protocol d was run by eluting 1 M NaOH first, followed by 1 M H_2SO_4/Na_2SO_4 mixture (50% v/v). Protocol e was run by eluting a freshly prepared mixture of 0.15 M $H_2O_2/1$ M NaCl/NaOH pH 12, followed by 1 M H_2SO_4. Up-flow or down-flow elution was run to minimize dilation phenomena (swelling-shrinking) of the resin matrix passing from metals to other ionic forms. Table 9.6 shows the regeneration fractions for Cr, Al and Fe of resin Purolite C106 after application of the above regeneration protocols [13].

Regeneration with conventional chemicals (mineral acids, bases, brine solutions) (protocols a–d) gave only partial elution of metals with maximum figures less than 50% in most cases (Table 9.6).

Regeneration of metals-converted carboxylate resin in acidic media proceeds according to the following reaction:

$$(RCOO^-)_n Me^{n+} + nH^+ \rightarrow nRCOOH + Me^{n+} \quad (R = \text{resin matrix}) \tag{3}$$

TABLE 9.6. Metals Regeneration Fractions after Adoption of the Five Regeneration Ptotocols on Resin Purolite C106.

Regeneration Protocol	Chemicals	BV (L/Ir)	Regeneration Fraction (%)		
			Cr	Al	Fe
a	H_2SO_4 1 M	10	5	20	53
b	NaOH 1 M	10	6	25	36
c	NaOH 1 M/H_2SO_4 1 M	10/5	13	45	100
d	Na_2SO_4 1 M + H_2SO_4 1M	10	54	70	100
e	H_2O_2 0.15 M + NaCl 1.0 M + NaOH	20	95	100	2
	H_2SO_4 1 M*	5	5	—	98

*Polishing operation after protocol e.
BV = bed volumes of regenerant used.

The reaction occurs with sensible shrinking of the polymer matrix due to the hydrophobic nature of the undissociated H-form resin matrix [43]. Partial regeneration, by using acidic media, is ascribed to interdiffusion hindrances of the hydrogen ions in and of the metal ions out of the resin phase (protocol *a*). On the other hand, sodium ion alone is not able to displace polyvalent metal ions from the resin-phase (protocol *b*) by mass action. By carrying on with consecutive acid-base elutions, the resin matrix is sequentially reconverted from the stiff H-form to the more hydrophilic (swollen) Na-form and vice versa, thus opening periods of ionic permeation in between the acidic and alkaline steps. Quantitative release of the metals is obtained only after several uneconomical acid-base reconversions of the resin in the H- and Na-forms (protocol *c*).

In the experiments carried out using 1:1 mixtures of 1 M H_2SO_4/Na_2SO_4 (protocol *d*), sodium sulfate was added as an osmotic adjuvant, thus favoring more hydrophilic conditions for relaxation (swelling) of the polymer matrix. In these latter conditions, a better, but still partial, Cr recovery was obtained (Table 9.6).

In situ (column) oxidation of Cr(III) to Cr(VI) as anionic chromate led to the most favorable solution. By using 0.15 M $H_2O_2/1$ M NaCl brines in alkaline media (NaOH pH 12) (protocol *e*), the anionic chromates formed were suddenly rejected by the resin phase by electrostatic repulsion (Donnan effect) [35] for very efficient regeneration of chromium from the resin carboxylate groups (Figure 9.19).

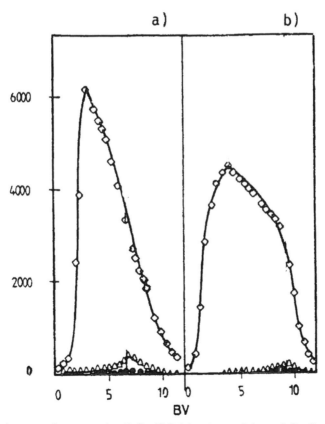

Figure 9.19 Metals regeneration curves using alkaline H_2O_2 brines (protocol *e*) on resin Purolite C106, at different pHs. (◆ Cr, △ Al, ● Fe; upflow rate 5 BV/h): a) H_2O_2 0.15 M, NaCl 1 M, NaOH, pH 12; b)H_2O_2 0.15 M, NaCl 1 M, NaOH, pH 10.

Quantitative elution of chromates was obtained according to the reactive ion exchange process [37,41]:

$$2(RCOO)_3Cr + 3H_2O_2 + 10NaOH \Leftrightarrow 2Na_2CrO_4 + 8H_2O + 6(RCOO^-Na^+) \quad (4)$$
$$(R = \text{resin matrix})$$

It is likely that the overall mechanism proceeds initially via the thermodynamically unfavored Cr^{3+}/Na^+ ion exchange reaction:

$$(RCOO)_3Cr + 3Na^+ \Leftrightarrow 3RCOO^-Na^+ + Cr^{3+} \quad (5)$$

followed by hydrolysis of trivalent chromium in alkaline media:

$$Cr^{3+} + 4OH^- \Leftrightarrow [Cr(OH)_4]^- \quad (6)$$

and final oxidation of tetrahydroxochromate ion:

$$2[Cr(OH)_4]^- + 3H_2O_2 + 2OH^- \Leftrightarrow 2CrO_4^{2-} + 8H_2O \quad (7)$$

As the hydrolysis [Equation (6)] and oxidation [Equation (7)] reactions proceed, the unfavored ion exchange reaction [Equation (5)] is shifted to the right-hand side, thus allowing for easy quantitative elution of chromium species from the resin phase.

From a kinetic point of view, a "shell progressive" conversion mechanism of the resin beads may be assumed for the case at hand [41,44]. An outer shell of sodium-form carboxylate resin is formed around a bead core still in the metal form, thus hindering further migration of hydroxide ions that are excluded from the solid phase after the Donnan effect. As a consequence, the overall oxidation reaction [Equation (4)] proceeds at decreasing rates due to the increasing lack of hydroxide ions needed for the hydrolysis [Equation (6)] and oxidation [Equation (7)] reactions.

Sodium hydroxide plays multipurpose roles in the resin regeneration process: it acts as a counter-ion source for the ion exchange reaction [Equation (5)] and it acts as alkylating agent for the chromium hydrolysis [Equation (6)] and oxidation reactions [Equation (7)].

Another redox reaction has to be taken into account during resin regeneration according to protocol e: the disproportion reaction of the hydrogen peroxide in alkaline media:

$$H_2O_2 \rightarrow H_2O + 1/2O_2 \quad (8)$$

On a theoretical basis, the chromium oxidation reaction [Equation (7)] is thermodynamically favored over the competitive side reaction [Equation (8)] which occurs later and only in case of significant excess of H_2O_2. Resin regeneration levels were carefully optimized to control the effect of potential side reactions.

Hydroxide ion plays a determinant role in the separation of metals (Cr, Al, Fe) from spent regeneration eluates. In moderately alkaline media (i.e., $10 < pH < 12$), ferric species are not hydrolyzed (Figure 9.4, bottom), and, accordingly, they stay in the resin phase, whereas aluminum species are quantitatively co-eluted with chromate after the hydrolysis reaction to aluminate ion:

$$Al^{3+} + 4OH^- \Leftrightarrow [Al(OH)_4]^- \Leftrightarrow AlO_2^- + 2H_2O \quad (9)$$

Recovery of ferric species is finally obtained after elution of 4 BV of 1 M H_2SO_4 (Figure 9.20).

The IERECHROM® Process

Based on laboratory experience, a 10 m³/d mobile plant was designed and assembled to demonstrate process reliability at a larger scale. The project was funded by the Italian Ministry of the Environment and coordinated by the "Unione Nazionale Industria Conciaria" (National Tannery Industry Association) with the technical support of the National Research Council. The plant was assembled on a trailer and trucked for demonstration campaigns at tannery installations in Naples and Milan (Italy) and in Graulhet (France) [10].

Figure 9.21 shows the scheme of the plant that included two ion exchange sections, each filled with 100 L of resin (columns C1, C2), two precipitators for metal hydroxides separation and recovery (SP1, SP2), two filters (F1, F2), one de-gasser (SI5) and several service vessels for feed, product and regenerant stock solutions. Figure 9.22 shows a general view of the mobile plant.

More than 30 exhaustion-regeneration cycles were run with satisfactory performance. Figure 9.23 summarizes the average performances of the plant in terms of fractional chromium removal, which was monitored systematically 98% [Figure 9.23(a)]; influent and effluent chromium concentrations, the latter systematically below the Maximum Acceptable Concentration (MAC) for discharge in internal and coastal seawaters according to the European regulations (2 mgCr/L) [Figure 9.23(b)]; and resin operative loading for chromium, Q_{Cr} [Figure 9.23(c)]. The coincidence of the two superimposed lines in Figure 9.23(c), refer

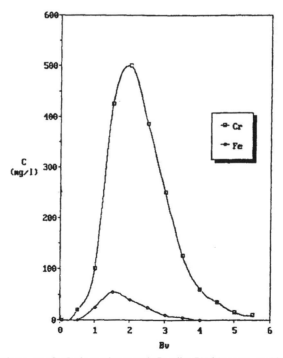

Figure 9.20 Regeneration curves for ferric species on resin Purolite C106 (regenerant: 1 M H_2SO_4, upflow rate 5 BV/h).

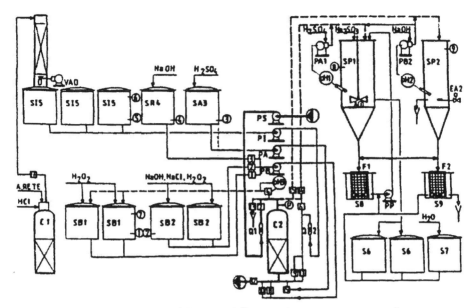

Figure 9.21 Scheme of the IERECHROM® process demonstration plant (10 m³/h).

to the resin chromium loading during the exhaustion and regeneration steps, respectively, confirming the absence of metal accumulation in the resin phase for steady and reproducible operation of the plant in the long run. Mass balance extended to other components confirmed that no accumulation of other metals (Fe, Al) or irreversible sorption of macromolecular organic matter on the resin phase occurred. Whenever occurring, a periodic treatment of the resin with 1 M NaOH solved the problem.

Quite acceptable stability of the resin matrix to oxidation, under the imposed drastic regeneration conditions, was observed, although a slight reduction of the cyclic resin loading (quantified in about 15% of the initial figure) was detected among the cycles run. In this context, the use of alternative resins more resistant to oxidation and organic fouling, i.e., exchangers with higher cross-linking degree and/or with methacrylic matrix, specifically tailored by resin manufacturers, are under investigation.

The main features of the IERECHROM® process can be summarized as follows:

- The detoxified liquid effluent (average Cr concentration: mg/L) is recycled to tannery operations.
- 90% of the influent chromium present in the tannery segregated wastewaters is recovered as $Cr(OH)_3$ free from other metal impurities and organic matter, thus ready for recycling to productive lines.
- Aluminum is partially recovered (50%) as aluminate ion in the first regeneration step, the rest is precipitated as $Al(OH)_3$ and separated from other metals in the other steps of the process.
- Ferric species are quantitatively recovered in the final polishing of the resin.
- Aluminum and ferric chemicals recovered may be recycled in the tannery wastewater treatments as coagulants.
- Organic compounds mostly break through the column (50%), the rest is incidentally co-precipitated with the $Al(OH)_3$ tailings.

(a)

(b)

Figure 9.22 Picture of the mobile installation: (a) general view and (b) view of the control panel.

Economic Considerations

Table 9.7 shows the economic evaluation of the process, carried out as a result of the demonstration campaigns.

Investment and running costs [resin inventory backup, chemicals, operation and maintenance and power (excluding labor)] were quantified, respectively, in 3.3 and 6 Euro/m³ treated water. Based on the average costs of metals control technologies currently applied in Europe, reference figures were 2.5 times lower than metals precipitation and landfilling operations of metals-laden hazardous sludge.

The costs of the resin inventory backup, after degradation and/or fouling, represent only a minor fraction (<5%) of the O&M costs (Table 9.7).

The above figures do not include the economic revenue associated with the minimized environmental impact (no waste is discharged after the operation) and with the commercial value of the recovered metals recycled to the productive lines. A proper economic analysis cannot exclude benefits associated with the above items that are among key features of the proposed innovation. The evaluation of reference costs is, however, strictly associated with the local contingent situation and must be evaluated on a country-to-country basis.

General Considerations on the IERECHROM® Process

A practical alternative to the precipitation methods for metals control has been developed. A conservative technology based on ion exchange allows for quantitative separation and recovery of pure Cr, Al, Fe chemicals and water ready for reuse.

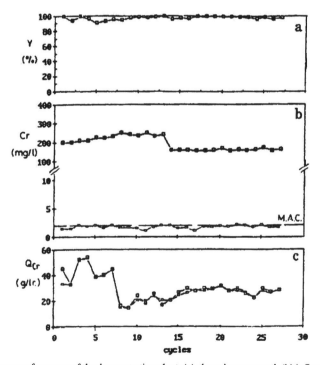

Figure 9.23 Average performance of the demonstration plant: (a) chromium removal, (b) influent (□) aand effluent (■) Cr concentration and (c) chromium exhaustion (□) and regeneration (■) resin loadings.

TABLE 9.7. Economic Evaluation of the IERECHROM® Process Based on the 10 m^3/d Demonstration Plant (December 1998) [45].

Investment Costs (pumps, valves, reactors, reservoirs, filters, control panels)		
Total Investment	Euro	60,000
Amortization (10% in 10 years)	Euro	6,000
Unitary Investment Cost	Euro/m^3	3.3
Operation and Maintenance (O&M) Costs (chemicals, power, resin inventory back-up)		
Chemicals	Euro/m^3	5.5
Resin Inventory Back-up	Euro/m^3	0.25
Power	Euro/m^3	0.25
Unitary O&M Costs	Euro/m^3	6.0
Grand Total (Investment + O&M)	Euro/m^3	9.3

Demonstration campaigns on a 10 m^3/d mobile plant, treating segregated wastewaters from the tannery industry, showed that Cr(III) concentration in the effluents was steadily below the limit imposed by the current European regulations.

Chromium, recovered as chromate from the resin spent regeneration eluates is ready for reuse in the plating industry or in the same tannery industry after reduction to Cr(III).

Application of the IERECHROM® process to the segregated tannery wastewaters leads to formation of detoxified (metals-free) solid residues with minimized (if any) environmental impact, ready for land application as soil conditioners.

CASE HISTORY 1: APPLICATION OF THE IERECHROM® PROCESS IN THE CONVENTIONAL FLOWSHEET OF TANNERY WASTEWATER TREATMENTS

Field experiments were run to test benefits associated with the inclusion of the IERECHROM® process in tannery wastewater operations [45]. Experiments were carried out at tannery wastewater treatment installations that included conventional physicochemical treatments for metals precipitation followed by biological oxidation and final clarification and filtration operations for an overall potentiality exceeding 150 m^3/d [Figure 9.24(a)]. The resulting mixed sludge (physicochemical + biological) was thickened, de-watered and disposed of in controlled landfills.

The inclusion of the IERECHROM® process in the above scheme made physicochemical treatment unnecessary, so a revised scheme was adopted. On these premises, segregated wastewaters were conveyed to the ion exchange operation, while the remaining majority fraction (90% of the overall volume) was reunited with the effluent waters from the ion exchange section and treated according to the simplified scheme shown in Figure 9.24(b). According to this latter scheme, after primary settling, wastewaters were sent directly to biological oxidation (activated sludge) and were finally clarified and filtered. Specifically, the physicochemical selector was converted into an aeration basin for preliminary BOD removal and sulfides oxidation. Phosphates were added to balance demand from biomass, and pH was controlled to 7.5–8. The nonhazardous sludge formed was partially recirculated, and the surplus amounts were thickened, de-watered and applied to land.

Table 9.8 shows comparative results after three months' continuous operation of the two treatment schemes, i.e., conventional and simplified.

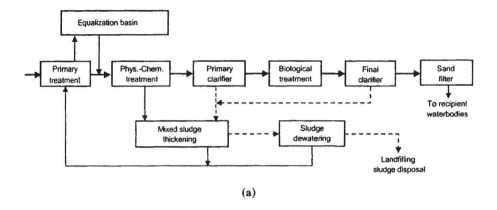

(a)

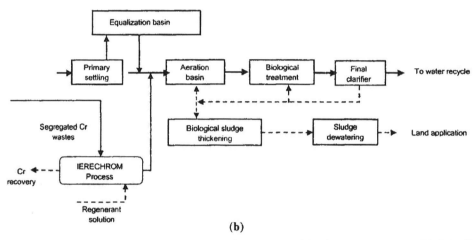

(b)

Figure 9.24 Tannery wastewater treatment schemes: (a) traditional and (b) after inclusiton of IERECHEROM® process.

The overall pollutants removal efficiency is comparable in the two cases, but the following technological improvements were evidenced by the adoption of a simplified scheme:

- elimination of the physicochemical treatment with related chemicals consumption
- drastic reduction of the sludge production (80%)
- minimized Cr residual concentration in the sludge for viable land applications
- recovery of Cr chemicals and coagulants (Al, Fe) for wastewater treatment operations

The economic comparison of the O&M costs after the adoption of the two schemes is reported in Table 9.9. The main costs of conventional wastewater treatment operations (2–3% of the annual tannery budget) refer to manpower, operation, chemicals and controlled landfilling of hazardous sludge. Simplified treatment, with inclusion of the IERECHROM® process, allows for savings in the manpower (0.15 vs. 0.5 Euro/m^3), energy consumption (0.7 vs. 1 Euro/m^3) and transport and sludge disposal (0.3 vs. 1.3 Euro/m^3). Net savings for the overall O&M costs are quantified in 1.5 Euro/m^3 (3.27 vs. 4.85 Euro/m^3).

TABLE 9.8. Comparison of Plant Performance Based on Conventional and Simplified Scheme for Tannery Wastewater Treatment Operations (mg/L).

Parameter	Raw Wastewater*	Conventional Scheme** Physicochemical	Biological	Simplified Scheme† Aeration	Biological
SS**	400–600	90–140	30–40	180–250	60–90
COD	4,000–6,000	400–700	40–70	1,200–2,000	100–200
Cr(III)	100–150	<1	ND	0.1.–0.2	ND
N-NH$_3$	30–50	10–15	3–10	10–20	3–8
N-NO$_3^-$	—	5–10	7–15	7–15	5–10
N-NO$_2^-$	—	0.2–0.4	0.1–0.15	0.4–0.6	0.15–0.25
Surfacants	35–60	10–20	0.15–1.5	10–15	1–2
Sulfite	<1	ND	ND	0.2–0.5	0–0.2
pH	4–7	7.5–8	7–8	7.5–8	7–8

*Raw wastewater is the same for both configurations except for influent Cr concentration which is pretreated by ion exchange in the case of the simplified process.
SS = suspended solids.
**Figure 9.24(a).
†Figure 9.24(b).

TABLE 9.9. Comparative Economic Evaluation of the O&M Costs for the Two Alternative Schemes (December 1998) (Euro/m^3).

Parameter	Conventional Scheme	Simplified Scheme with IERECHROM®
Manpower	0.5	0.15
Chemicals	0.7	1.4
Sludge transport and disposal	1.3	0.3
Analysis	0.15	0.2
Amortization (10%, 10 years)	0.15	0.3
Maintenance	0.4	0.1
Energy consumption	1.0	0.7
Taxes	0.15	0.15
Total	4.85	3.27

Reference costs do not include the economic revenue associated with the commercial value of the metals recovered, as well as the costs-benefits associated with the minimized environmental impact.

CASE HISTORY 2: DIRECT COMPARISON OF THREE ALTERNATIVES FOR THE TANNERY INDUSTRY

Table 9.10 shows a comparative economic evaluation for the applications of three alternative technologies for the environmental problems related to the tannery industry. The application of process 1 (leaching and re-precipitation of metals from sludge) is compared to process 3 (two sections of ion exchange for metals and coagulants recovery from the sludge leachate) and to process 4 (the IERECHROM® process) [9].

With reference to process 1, unitary investment costs for an installation treating, e.g., 20 t/d of wet ($\approx$20% solids) mixed sludge amount to 8 Euro/t (amortization 10% in 10 years), whereas the O&M costs exceed 50 Euro/t dry sludge. The total amount, by including the sludge leaching operation, metals re-precipitation, controlled landfilling of hazardous waste, and land application of detoxified solid residues, amounts to 90 Euro/t dry sludge.

The low technology operations lead to quite high running costs, which are essentially associated with labor and chemicals consumption, whereas the investment costs are mainly related to filtration apparatus.

With reference to process 3, for a full-scale installation treating 80 m^3/d of acidic leachate (corresponding to the volume extracted from 20 t/d of de-watered sludge, 20% solids), unitary investment costs amount to 10.5 Euro/t (same amortization as before), while O&M costs exceed 170 Euro/t. The overall unitary costs for the process, including the acid leaching of sludge, the ion exchange operation and land application of nonhazardous residues amounts to 215 Euro/t dry sludge.

With reference to process 4, costs are expressed on a volumetric basis as operations make reference to segregated liquid effluents. As already substantiated, for the 10 m^3/d

TABLE 9.10. Comparison of Economic Data for Three Alternative Solutions for the Tannery Industry (December 1998).

Item		Process 1*	Process 3**	Process 4†
Investment Costs (pumps, reactors, valves, vessels)				
a) Acid leaching	(Euro)	235,000	235,000	—
b) Leachate treatment	(Euro)	65,000	176,000	
Total investment	(Euro)	300,000	400,000	60,000
Amortization (10%, 10 years)		30,000	40,000	6,000
Unitary investment	(Euro/t)	8	10.5	3.3 (Euro/m^3)
Operation and Maintenance (O&M) (chemicals, energy, labor)				
(a) Acid leaching	(Euro/t)	40	40	—
(b) Leachate re-precipitation	(Euro/t)	11	126	—
Total O&M	(Euro/t)	50	170	6 (Euro/m^3)
Total (Investment + O&M)	(Euro/t)	58	180.3	9.3 (Euro/m^3)
Grand Total	(Euro/t)	90	215	9.3 (Euro/m^3)

*Installation treating 20 t/d of wet mixed sludge.
**Installation treating 80 m^3/d acidic leachate (20 t/d wet sludge).
†Installation treating 10 m^3/d of segregated tannery effluents.
Base of comparison: Cost of the conventional treatment (metals precipitation + landfilling): 180 Euro/t.

IERECHROM® demonstration plant, the O&M costs resulted in the range of 3.3 Euro/m³, with investment costs in the range of 6 Euro/t, for an overall figure exceeding 9.3 Euro/m³ of segregated tannery effluent.

Based on a reference O&M cost for the conventional treatments of tannery wastewaters exceeding 180 Euro/t, the economic comparison of the proposed alternatives allows for the following conclusions to be drawn: process 1 leads to 50% savings (180 vs. 90 Euro/t); costs for process 3 are on the same order with respect to landfilling disposal of mixed hazardous sludge, with the advantage of metals recovery in this latter case; and process 4 allows for savings quantified on the order of 2.5–3 times less than the conventional technology.

As usual, all of the above evaluations do not include revenues from the commercial value of chemicals recovered or the economic returns associated with the minimized environmental impact (no waste discharge).

ACKNOWLEDGEMENTS

The technical assistance provided by Mr. N. Limoni, G. Laera, M. Stramaglia, A. Sasanelli and R. Ciannarella in carrying out the experimental part of the work and fruitful discussions with Prof. G. Macchi, Dr. A. Volpe and Dr. M. Pagano during paper layout are gratefully acknowledged.

REFERENCES

1 Council of the European Communities. Commission Directorate XI, 1999: Best Available Technologies for Treatment of Tannery Wastes. Rep. UK/EIPPCB/tannery.

2 USES, 1980. *Enciclopedia della Chimica,* Florence, Italy, Edizioni Scientifiche. Vol. IV, pp. 287–292.

3 Patterson J. W. 1987. "Metals separation and recovery," in *Metals Speciation Separation and Recovery,* Patterson J. W. and Passino R. eds, Lewis Pub. Inc., Chelsea, MI, USA, pp. 63–93.

4 Rhyner C. R., Schwartz L. J., Wenger R. B. and Kohrell M. G. 1995. *Waste Management and Resource Recovery.* Boca Raton, FL, USA, CRC Press Inc., Lewis Pub., pp. 378–379.

5 Eckenfelder W. W. Jr. 1976. *Industrial Waste Pollution Control.* New York, NY, USA, McGraw-Hill Pub. Co., pp. 51–55.

6 SenGupta A. K. 1995. "Chromate Ion Exchange," in *Ion Exchange Technology. Advances in Pollution Control.* Lancaster, PA, USA, Technomic Pub. Co., Inc., pp. 115–147.

7 Council of the European Communities. 1991. Directives no. 91/271/EEC, 91/156/EEC, 91/689/EEC on Treatment and Discharge of Urban Wastewaters and Effluents from Industrial Sectors.

8 Macchi G., Pagano M., Pettine M., Santori M. and Tiravanti G. 1991. "A Bench Study on Chromium Recovery from Tannery Sludge," *Wat. Res.,* 25(8): 1019–1026.

9 Petruzzelli D., Tiravanti G., Santori M. and Passino R. 1994. "Industrial Waste Management. The Case of the Tanning Industry," in *Chemical Water and Wastewater Treatment III,* Klute R. and Williams H. H. eds., Berlin, Heidelberg, Germany, Springer-Verlag Pub. Co. pp. 269–279.

10 Petruzzelli D., Tiravanti G., Santori M. and Passino R. 1994. "Chromium Removal and Recovery from Tannery Wastes: Laboratory Investigation and Field Experience on a 10 m³/d Demonstration Plant," *Wat. Sci. Technol.,* 30(3): 225–233.

11 Petruzzelli D., Passino R. and Tiravanti G. 1995. " Ion Exchange Process for Chromium Removal and Recovery from Tannery Wastes," *I&EC Research,* 34: 2612–2617.

12 Petruzzelli D., Liberti L., Passino R. and Tiravanti G. 1990. "Specific Resins for Metal Ion Separation. The Cr(III), Fe(III), Al(III) System," in *Recent Development in Ion Exchange,* London, UK, Williams P. A. and Hudson M. J. eds., *Elsevier Appl. Sci.,* pp. 265–275.

13 Petruzzelli D., Alberga L., Passino R., Santori M. and Tiravanti G. 1992. "Exhaustion-Regeneration Behaviour of Carboxylic Resins. The Cr(III)/Fe(III)/Al(III) System," *React. Polym.*, 18:95–105.

14 Tiravanti G., Petruzzelli D., Passino R. and Santori M. 1994. "Metodo per Rimuovere e Recuperare Cromo Trivalente da Acque Reflue," *Italian Patent* no. MI94A 001512.

15 Anderson R. A. 1989. "Essentiality of Chromium in Humans," *Sci. Total Envir.*, 86:75–81.

16 Iyengar G. V. 1989. "Nutritional Chemistry of Chromium Compounds, *Sci. Total Envir.*, 86:69–74.

17 Rinehart W. E. 1989. "Recapitulation," *Sci. Total Envir.*, 86:191–193.

18 Moore J. W. and Ramamoorthy S. 1984. "Chromium," in *Heavy Metals in Natural Waters*, Di Santo R.S. ed., New York, NY, USA, Springer-Verlag, pp. 58–76.

19 Adriano D. C. 1986. "Chromium," in *Trace Elements in the Terrestrial Environment*, New York, NY, USA, Springer Pub. Co., pp. 58–76.

20 Simoncini A. 1982. "Quantità, Caratteristiche e Distribuzione Geografica dei Fanghi di Depurazione delle Concerie Italiane," *Acqua e Aria*, 1:55–57.

21 Silva S. 1989. "Impiego Agricolo dei Residui di Lavorazione delle Pelli," *La Conceria*. Milano, Italia, UNIC, pp. 1–60.

22 Dreiss S. J. 1986. "Chromium Migration through Sludge Treated Soils," *Ground. Wat.*, 24:312–321.

23 Nakayama E., Kuwamoto T., Tsurubo G. and Fujinaga T. 1981. "Chemical Speciation of Chromium in Seawater. Part I. Effects of Manganese Oxide and Reducible Organic Materials on the Redox Processes of Chromium," *Anal. Chim. Acta*, 130:401–404.

24 Early L. E. and Ray D. 1987. "Kinetics of Chromium(III) Oxidation to Chromium(VI) by Reaction with Manganese Dioxide," *Environ. Sci. Technol.*, 21:1187–1193.

25 Jardine P. M., Fendorf S. E., Mayes M. A., Larsen I. L., Brooks S. C. and Bailey W. B. 1999. "Fate and Transport of Hexavalent Chromium in Undisturbed Soil. *Environ. Sci. Technol.*, 33:2939–2944.

26 Metcalf & Eddy Inc. 1991. *Wastewater Engineering. Treatment, Disposal and Reuse.* New York, NY, USA, McGraw-Hill, pp. 445–518.

27 Dette B., Julich R., Buchert M., Bukhave M., Dopfer J., Esquerra J., Gerkens I., Kuppers P., Schneider M., Sorensen Y. S., Stevanato P. and Villa L. 1999. "Waste Prevention and Minimisation," Final Report, European Commission DGXI. Darmstadt, Germany. Institut fur Angewandte Okologie.

28 Westall J. C., Zachary J. L. and Morel F. M. M. 1976. "MINEQL: a Computer Program for the Calculation of the Chemical Equilibrium Composition of Aqueous Systems," Cambridge, MA, USA, Note 18. Department of Civil Engineering, Massachusetts Institute of Technology.

29 Papelis C., Hayes K. F. and Leckie J. O. 1988. "HYDRAQL: A Program for the Computation of Chemical Equilibrium Composition of Aqueous Batch Systems Including Surface Complexation Modeling of Ion Adsorption at the Oxide/Solution Interface," Stanford, CA, USA. Department of Civil Engineering, Stanford University.

30 Allison J. D., Brown D. S. and Novo Gradac K. J. 1991. "MINTEQA2/PRODEFA2, a Geochemical Assessment Model for Environmental Systems," Athens, GA, USA, EPA Rep. no. 600/3-91/021.

31 Baloga M. R. and Early J. E. 1961. "The Kinetics of Oxidation of Cr(III) to Cr(VI) by Hydrogen Peroxide," *J. Am. Chem. Soc.*, 83:4906-4909.

32 Pettine M. and Millero F. J. 1990. "Chromium Speciation in Seawater: The Probable Role of Hydrogen Peroxide," *Lymnol. Oceangr.*, 35:426–432.

33 Hartford W. H. and Copson R. L. 1964. "Chromium Compounds," in *Encyclopedia of Chemical Technology*. Otmer K. ed., New York, NY, USA, Wiley Interscience, 2nd Edition, Vol. 5, pp. 473–514.

34 Hodgkin J. H. and Eibl R. 1986. "Ferric Ion Chelation by Aminophenol Resins," *React. Polym.*, 4:285–292.

35 Helfferich F. 1962. *Ion Exchange.* McGraw-Hill Pub. Co., New York, NY, USA, p. 134.

36 Helfferich F. 1962. *Ion Exchange.* McGraw-Hill Pub. Co., New York, NY, USA, p. 156.

37 Janauer G. E., Gibbons R. E. and Bernier W. E. 1985. "A Systematic Approach to Reactive Ion Exchange," in *Ion Exchange,* Marinsky J. A., Marcus Y. eds., New York, NY, USA, M. Dekker Pub. Co., Vol. 9, Ch. 2, pp. 53–173.

38 Robinson R. A. and Stokes R. H. 1970. *Electrolyte Solutions.* London, UK, Butterworths Pub. Co., p. 233.

39 Dawson M. I., Hobbs P. D. and Dawson D .J. 1983. "Synthesis and Characterization of a Polymeric Water Soluble Chelator for Fe(III)," *J. Polym. Sci. Polym. Lett. Ed.,* 21:381–390.

40 Braun D. and Bouderska M. 1976. "Reversible Crosslinking by Complex Formation. Polymers Containing 2-Hydroxybenzoic Acid Residues," *Eur. Polym. J.,* 12:525–532.

41 Helfferich F. 1965. "Ion Exchange Kinetics. V. Ion Exchange Accompanied by Chemical Reactions," *J. Chem. Phys.,* 69:1178–1189.

42 Eisenman G. 1983. "The Molecular Basis of Ionic Selectivity in Macroscopic Systems," in *Mass Transfer and Kinetics of Ion Exchange,* Liberti L., Helfferich F. G. eds. The Hague, Netherlands, Martinus Nijhoff Pub. NATO-ASI Ser. E71, pp.121–155.

43 Klein G., Sinkovic J. and Vermeulen T. 1982. "Weak Electrolyte Ion Exchange in Advanced Technology Water-Reuse Systems," Washington DC, USA, US Dept. Interior, OWRT Rep. no. RU-82/7, pp. 35–42.

44 Helfferich F. G. 1983. "Ion Exchange Kinetics. Evolution of a Theory," in *Mass Transfer and Kinetics of Ion Exchange,* Liberti L., Helfferich F. G. eds. The Hague, Netherlands, Martinus Nijhoff Pub. NATO-ASI Ser. E71, pp.169–171.

45 Tiravanti G., Petruzzelli D. and Passino R. 1997. "Pretreatment of Tannery Wastewaters by an Ion Exchange Process for Cr(III) Removal and Recovery," *Wat. Sci. Technol.,* 36(2–3):197–207.

CHAPTER 10

HUMASORB™: A Coal-Derived Humic Acid-Based Heavy Metal Sorbent

H. G. SANJAY[1]
AMJAD FATAFTAH[1]
DAMAN WALIA[1]

INTRODUCTION

THE number of hazardous waste sites requiring treatment for soil and groundwater remediation under current federal and state regulations is estimated to be about 217,000 sites in the United States. The sites include those that fall under the National Priorities List (NPL, Superfund), Resource Conservation and Recovery Act (RCRA) Corrective Action, Department of Defense (DOD) and Department of Energy (DOE) installations. The soil and groundwater at these sites are contaminated with various toxic metals (about 50–70% of the sites) and with organic contaminants (40–70% of the sites) [1]. In addition, radioactive contamination is found at 90% of the DOE installations. The DOE estimates that more than 5,700 groundwater plumes have contaminated over 600 billion gallons of water and 50 million cubic meters of soil throughout the DOE complex [2]. Mixed waste containing multiple hazardous and radioactive contaminants is a problem at a number of installations. The types of contaminants present at the sites include the following [1,2]:

- toxic metals such as lead, chromium, arsenic, cadmium, zinc, barium, nickel, copper, beryllium, mercury and others
- organic chemicals such as benzene, toluene and xylenes, chlorinated hydrocarbons such as trichloroethylene (TCE) and perchloroethylene (PCE) and energetic chemicals such as nitroesters and others
- radioactive contaminants such as uranium, plutonium, thorium, cesium, strontium, tritium and others

The remediation of contaminated surface and groundwater is typically attempted with treatments such as precipitation, ion exchange, membrane separation and activated carbon adsorption. At sites having mixed contaminants, two different processes are required to remediate a site, an approach that results in complex and costly processing steps. A typical approach is to remove organics using activated carbon followed by ion exchange to remove

[1]14100 Park Meadow Drive, Suite 210, ARCTECH, Inc., Chantilly, VA 20151, U.S.A., envrtech@arctech.com

metals. The method used most frequently to treat groundwater is the conventional pump-and-treat technology. The groundwater is pumped to the surface and treated using various technologies. However, this method is not very effective in meeting the desired cleanup criteria for sites with various types of contaminants, especially when the aquifers are contaminated with nonaqueous-phase liquids (NAPLs). Pump-and-treat methods are expected to last 30–70 years at a number of sites that contain NAPLs, thus increasing the treatment costs.

The limitation of present treatment approaches can be overcome by the use of HUMASORB™, a humic acid-based product. This chapter describes the performance and applicability of HUMASORB™, a coal-derived humic acid adsorbent for removal of multiple contaminants from groundwater.

An overview of humic substances, humic acid properties, structure and applications are presented in the next section. HUMASORB™ technology, applications and demonstrations are discussed, and the attributes of HUMASORB™ for removal of multiple contaminants from complex waste streams are presented in the following section. The conclusions follow.

HUMIC SUBSTANCES

Humic substances are naturally occurring dark brown to black organic multifunctional polymers with major agricultural and environmental roles. They are one of Earth's richest carbon reservoirs and are complex aromatic macromolecules with various linkages between the aromatic groups. The different compounds involved in linkages include amino acids, amino sugars, peptides, aliphatic acids and other aliphatic compounds. The various functional groups include carboxylic groups (COOH), phenolic, aliphatic and enolic –OH and carbonyl (C=O) structures of various types [3].

Humic substances traditionally are classified into three main fractions on the basis of their aqueous solubility [4]. Fulvic acids ($<M_w>$limit *ca* 5 kDa) are soluble at all pHs and generally exist at low concentrations in natural waters. Fulvic acids are surface active and transport metals, nutrients and pollutants [5]. These properties make fulvic acids hard to isolate, purify and understand. Humins are insoluble at all pHs and are weaker sorbents and metal binders than fulvic acids and humic acids. Humins are further along in the natural progression from live animals and plants toward "dead" coals and carbon [4].

Humic acids, the most important fraction of the humic substance family, are highly functionalized carbon-rich biopolymers that stabilize soils as soil organic matter (SOM). They are anchored by metal binding and attach to clays and minerals, resulting in a decrease of their solubility at a given pH [6]. Humic acids are amorphous and have been described as fractal materials [7]. Humic acids are essential to healthy soils and sediments, as they are the primary water retainers, metal binders and sorbents [8,9]. In addition, the water retention capacity of humic acid provides earth with thermal buffer capacity that prevents catastrophic climates. It is hard to distinguish between low molecular weight humic acids and fulvic acids because useful ways of estimating polymer molar mass (e.g., gel permeation chromatography and viscosity measurements) are frustrated by humic substances' desire to aggregate [4].

Humic Acid's Resources

Humic acids can be extracted from a number of sources like soil, plants, water, peat, sediments, sewage sludge and coal. Coal, being the most abundant and predominant product of plant residue coalification, is a major source of humic acid. All ranks of coal contain humic acid, with the lower rank coals having more humic acid compared to the higher rank coals. Leonardite, an oxidized form of lignite, represents the most easily available and concentrated form of humic acid. Leonardite is a naturally occurring overlay of lignite mines with a concentration of humic acid ranging from 30–90% depending on location.

Coal is the final result of condensation reactions and loss of functionality of small molecular weight substances like humic and fulvic acids. The organic structure of coal and, thus, its characteristics in terms of aromaticity and functional group content depends on the history of coal and is generally related to the rank of coal. Extraction of humic acid from coal involves breaking the condensation linkages through oxidation and increase of functionality, resulting in small fragments and small molecules. This process of extraction is the reverse of the process for coal formation. Thus, coal-derived humic acids are sometimes referred to as regenerated humic acids [9].

Humic Acid's Characterization

Dry, natural or synthetic humic acids are solids with densities in the range 1.5–1.7 gm/ml. Their surface areas, measured by the Brunauer-Emmett-Teller method, with butane as adsorbent, range from 10 ± 3 m^2/gm for natural humic acid to 45 ± 4 m^2/gm for synthetic humic acid. Pure humic acids are dispersed solids with moderately high surface areas, but they are not highly porous. Most of the interior of humic acid particles is made up of covalently linked carbon, hydrogen, nitrogen and oxygen atoms [10].

Elemental analysis is one of the more reliable determinations that can be carried out on humic substances. It has been found that carbon, hydrogen, oxygen, nitrogen, phosphorus and sulfur generally account for 100% of the composition of humic substances on an ash-free basis [8]. The content of pure humic acid and the relative amounts of the functional groups (alcohol, amine, carbonyl, carboxylic acid, phenol, quinone) depend on the source but do not vary greatly. The carboxylic acid and phenolic groups are responsible for the total acidity of humic acid [8,10].

The ultraviolet-visible spectra of humic acids in alkaline solution contain no strong features, although shoulders centered at 260–300 nm are sometimes detected, and the absorbance generally increases with decreasing wavelength. The infrared spectra of HA are useful fingerprints that contain features indicating the presence of oxygen-containing functional groups and aromatic character. Nuclear Magnetic Resonance (NMR) is a valuable fingerprint of the functional groups of HA. The ^{1}H and ^{13}C spectra often are obtained in alkaline NaOD/D$_2$O solutions. ^{13}CPMAS solid-state data show peaks that are often broad, but can be assigned and are useful for characterization [8,10].

Humic Acid's Properties

The properties and characteristics of humic acid include high cation exchange capacity, ability to chelate metals and ability to adsorb organics. Humic acid can help provide slow release of plant macronutrients such as nitrogen, phosphorus and sulfur for agricultural purposes. The ability to form complexes with metal ions and high cation exchange capacity contributes to cation retention. The exchange capacity is useful for retaining plant macro- and micronutrients and preventing leaching.

METAL BINDING

Metals are bound to the carbon skeleton of humic substances through heteroatoms such as nitrogen, oxygen or sulfur. Sulfur is present in low concentrations. Its effect on metal binding, however, is not understood very well. Nitrogen is present in significant concentration and has been shown to have a positive effect on metal binding. According to the evidence in the literature, the most common metal binding occurs via carboxylic and phenolic oxygen.

Humic acid is characterized by high cation exchange capacity compared to leonardite. The exchange capacity of leonardite is 50 meq/100 gm, whereas that of humic acid derived from leonardite is 200–500 meq/100 gm. In comparison, the exchange capacity of many commercial ion exchange resins is approximately 150 meq/100 gm.

ORGANIC COMPOUND BINDING

Humic acid is an association of molecules forming aggregates of elongated bundles of fibers at low pHs and open flexible structures perforated by voids at high pHs. The voids can trap and adsorb organic and inorganic particles if the charges are complementary. Many mechanisms have been postulated to account for the adsorption of organic compounds like Van der Waals attractions, hydrophobic bonding, hydrogen bonding, charge transfer, ion exchange and ligand exchange [11]. Humic acid combines with herbicides by electrostatic bonding (i.e., attraction of a positively charged organic cation to an ionized carboxylic or phenolic group), hydrogen bonding and ligand exchange. In addition, the high concentrations of stable free radicals in humic acid are capable of binding herbicides that can ionize or protonate to the cation form [11].

Van der Waals forces are involved in the adsorption of nonionic and nonpolar compounds. These forces result from short-range dipole-dipole interactions and are additive in nature. The forces between the atoms of the adsorbate and the adsorbent can result in considerable attraction for large molecules. Nonpolar compounds are also adsorbed by the hydrophobic bonding mechanism. This type of bonding is believed to be responsible for the strong adsorption of compounds such as dichlorodiphenyltrichloroethane (DDT) and organochlorine insecticides. Hydrogen bonding is also a dipole-dipole interaction in which the hydrogen atom serves as a bridge between two electronegative atoms. One of the electronegative atoms is held by a covalent bond and the other by electrostatic forces. The adsorption of anionic herbicides on humic acid is attributed to hydrogen bonding. This mechanism may also explain the ability of soils containing humic acid to retain moisture [11].

TOXICITY

Humic acid is less toxic compared to the conventional chelating agents used in agriculture such as ethylenediaminetetraacetic acid (EDTA). The acute oral LD/50 for humic acid is 5.5 gms/kg [12], for EDTA it is 2 gms/kg [13] and, as a reference, it is 10 mg/kg for potassium cyanide [14]. Humic acid is thus three times less toxic than EDTA, and potassium cyanide is 550 times more toxic than humic acid.

Humic Acid's Structure

A complete elucidation of the humic acids' structure is not known yet. This fact led to frontier research on highly purified humic acids and the prospects of understanding humic acids' role in the carbon cycle, biomineralization and other life processes. Because of their polyfunctionality and ability to sorb, bind, fragment, aggregate and be oxidized and reduced, humic acids are much more complicated than nucleic acids, polysaccharides and proteins. Their primary, secondary and higher order structures have been debated almost from the day humic acids were first isolated. The humic acid building blocks proposed earlier were aromatic and "coal-like." Present knowledge indicates a hydrophobic framework of aromatic rings linked by more flexible carbon chains, with alcohol, amine, carboxylic, carbonyl, phenol and quinone functional groups. Humic acids' existence in live plants suggests they have rational primary and higher order structures resulting from biochemically controlled reactions [10].

Recently, a group of researchers from Temple University, Northeastern University and Birmingham University suggested that humic acid is a uniform polymer, consist of repeated units called Temple-Northeastern-Birmingham (TNB) building blocks [15]. The lowest energy conformation of the TNB humic acid building block, which has an empirical formula of $C_{36}H_{30}O_{15}N_2 \cdot xH_2O$ ($x = 0 \approx 15$, water not shown) is shown in Figure 10.1 [15]. This building block was derived after allowance for polysaccharide and protein content from analytical data for humic acids isolated with different methods from many different soil sources and from modeling work [10]. Biosynthesis of humic acid building block from phenylalanine and tryptophan has been rationalized in this model. Figure 10.1 also shows the first hypothetical model structure proposed for humic acid [3].

The cross section of a helical humic acid polymer formed by joining the building blocks of the TNB model is shown in Figure 10.2. It has a central cavity for water, metal and solute

A. Stevenson, 1972

B. TNB, 1998 (Temple, Northeastern and Birmingham)

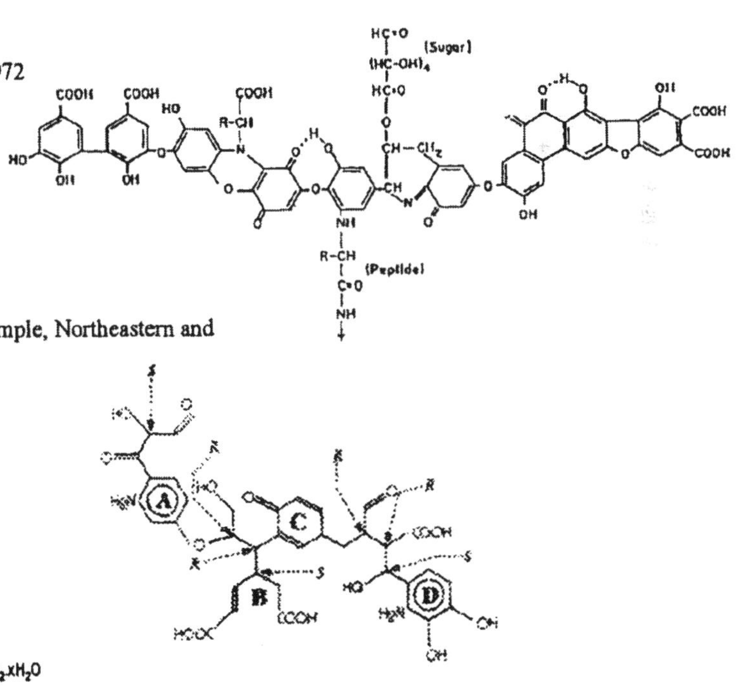

Empirical Formula: $C_{36}H_{30}O_{15}N_2 \cdot xH_2O$

x=0-15

Figure 10.1 Proposed structures of humic acid: (A) From Stevenson, F. J. *Journal of Environmental Quality,* 1(4), 1972; with permission from American Society of Agronomy and (B) from Davies, G. et al. *J. Chem. Soc., Dalton Trans.,* 4047–4060, 1997; with permission from The Royal Society of Chemistry.

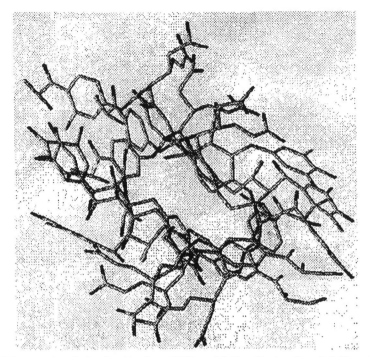

Figure 10.2 Cross section of a humic acid polymer (TNB model) (from Davies, G. et al. *J. Chem. Soc., Dalton Trans.*, 4047–4060, 1997; with permission from The Royal Society of Chemistry).

binding. The synthesis of humic acid is nature's way of retaining water and solutes within and between humic acid molecules, and the humic acid helix normally is filled with water. This water-filled helical model is consistent with many humic acids' properties [10,15].

Humic Acid's Applications

Humic acids' versatile properties can be exploited for many applications like agricultural, environmental and industrial applications.

Agricultural Applications

Agricultural applications include slow release source of the micronutrients for plants and microbial growth, high water-holding capacity, buffering capacity, soil erosion reduction and plant growth stimulation. Humic acid increases the availability of phosphate to the plant by breaking the bond between the phosphate ion and iron or calcium. Phosphate is a stimulator of seed germination and root initiation in plants. In addition, humic acid is very effective for converting iron into suitable forms to protect plants from chlorosis, even in the presence of high concentrations of the phosphate ion. Humic acid contributes to mineralization and immobilization of nitrogen in soil. The complexes formed between ammonium ions and humic acid release nitrogen slowly into the soil. Humic acid serves as a slow release nitrogen carrier in the soil in this respect [16]. In addition, humic acid has a fundamental effect on the physicochemical properties of the soil (water-holding capacity) and is responsible to a large degree for such physicochemical properties as the exchange capacity and buffering properties. These properties are of great importance, not only in controlling the uptake of nutrients by the plant and their retention in the soil, but also in suppressing the deleterious effect of soil acidity [17].

Environmental Applications

The environmental applications of humic acid include metal removal by chelation, removal of organics by adsorption, neutralization of acidic water streams, removal of oxo-anions and reduction of metal species. Ion exchange is the main mechanism for the metal removal by humic acid, in addition to complexation through the heteroatoms—oxygen, nitrogen and sulfur. The cation exchange capacity of humic acid (2–5 meq/g) is much higher than other soil colloids such as vermiculite (1–1.5 meq/g) and montmorillonite (1 meq/g) [18].

Industrial Applications

Humic acids also have many industrial applications. Humic acids and humic acids-containing materials have been used in muds for oil well drilling, as cement additives, as leather dye, as a binder for sand particles and in steel casting. In addition, humic acid has also been used as a pigment in ink, as a component of livestock feed, as a dye for preparation of nylon and in manufacturing and recycling paper [9].

HUMASORB™ TECHNOLOGY

HUMASORB™ is the generic name used for the technology that is the concept of utilizing humic acid-derived products for the mitigation of contaminants from mixed waste streams. The liquid product is termed HUMASORB-L™, and the purified solid humic acid product is termed HUMASORB-S™. However, HUMASORB-S™ is insoluble in water only at lower pH and will dissolve as the pH increases above two and in the presence of monovalent ions such as sodium and potassium. A cross-linked humic acid polymer, HUMASORB-CS™ was developed to overcome this limitation and to lower the solubility at higher pH values.

HUMASORB-CS™ was produced by cross-linking and immobilizing HUMASORB-S™ or HUMASORB-L™ using proprietary methods. A schematic of the approach used to produce various HUMASORB™ products from coal is shown in Figure 10.3. Patents on this material, the process to produce this material, and the mode of application were issued recently [19]. The properties and applications of HUMASORB-CS™ for removal of contaminants are discussed in the following sections.

Solubility of HUMASORB-CS™

The solubility of HUMASORB-CS™ was determined at various pHs and compared with purified humic acid. In these tests, HUMASORB-CS™ (0.5 grams) was mixed with water, and the pH was adjusted to the various pHs by using either sodium hydroxide or hydrochloric acid. The mixture was then placed on a shaker at 300 rpm and 25°C for two hours. After the contact time, the mixture was centrifuged to separate the liquid and solid phases. The liquid-phase pH was measured and analyzed for humic acid to determine solubility. The results, shown in Figure 10.4, indicate that the solubility of HUMASORB-CS™, as deter-

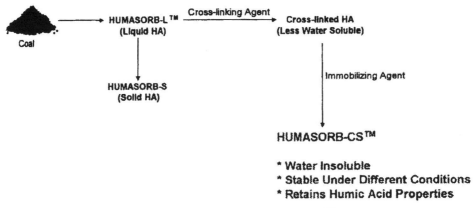

Figure 10.3 Schematic of the approach to produce HUMASORB™ products from coal.

mined under the conditions of this study, is significantly lower compared to that of humic acid. The pH used in Figure 10.4 is that of the liquid phase, measured after the contact time. It was observed that the pH increased to 8 at the end of the experiment for tests with HUMASORB-CS™, when the pH was initially adjusted to 4 and 6.

The different functional groups present in HUMASORB™ before and after cross-linking were estimated using ^{13}C-NMR. The analysis showed that all the functional groups believed to be responsible for contaminant removal are retained after cross-linking.

Stability Tests

HUMASORB-CS™ was subjected to stability tests under various conditions. The objective of these tests was to evaluate the stability of HUMASORB-CS™ using solubility under

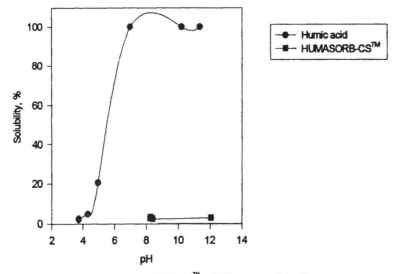

Figure 10.4 HUMASORB-CS™ solubility over a wide pH range.

various test conditions as a criterion. In addition, HUMASORB-CS™ was tested in batch mode to evaluate its effectiveness for removal of target contaminants [chromium (III) was used in the tests as a representative contaminant] after the stability treatment.

METHODS

Stability in Water under Different Conditions

Stability tests were conducted in water at different temperatures and also with water that was spiked with anions such as carbonate and sulfate. In these tests, two grams of HUMASORB-CS™ were taken in centrifuge tubes and 10 ml of tap water or spiked water was added. The mixture was allowed to stand for an extended period after which the solid and liquid phases were separated. The liquid phase was analyzed for humic acid by lowering the pH to below 2. The amount of dry humic acid obtained was used to estimate solubility of HUMASORB-CS™. The following stability tests in water at different temperatures were conducted:

- ambient conditions (1, 4 and 6 months)
- temperature 50°C (1, 2, 3 and 4 weeks)
- temperature 4°C (3, 5.5 and 6.5 months)
- 100 PPM Na_2SO_4 or 100 PPM Na_2CO_3 (1 day, 5 months)
- 100 PPM $CaSO_4$ or 100 PPM $CaCO_3$ (1 day, 5 months)
- 10,000 PPM Na_2SO_4 or 10,000 PPM Na_2CO_3 (3 months)

The solid phase obtained from the tests was then dried in an oven at 50°C and used in batch tests to evaluate removal of a target contaminant [Cr (III)]. Approximately 0.5 grams of the solid were mixed with 25 ml of spiked water containing chromium(III) at 300 rpm and 25°C for two hours. The solid and liquid phases were separated by centrifugation after the contact time, and the liquid phase was analyzed for chromium(III).

RESULTS AND DISCUSSION

The solubility of HUMASORB-CS™ was less than 0.5% in most of the stability tests, indicating that HUMASORB-CS™ is stable under the conditions used in the study. However, HUMASORB-CS™ was soluble in the tests with 10,000 PPM sodium carbonate. HUMASORB-CS™ from the stability tests was used to evaluate chromium(III) removal from a simulated waste stream. More than 90% of chromium(III) was removed as shown in Figures 10.5–10.7. The tests clearly show that HUMASORB-CS™ is not only stable under the conditions evaluated in this study, but also retains its ability to remove contaminants from contaminated waste streams.

Metal Removal

BATCH TESTS

Methods

Batch tests were also conducted with HUMASORB-CS™ at pH 2–2.5 to evaluate its effectiveness for the removal of various metals from simulated waste streams and also to

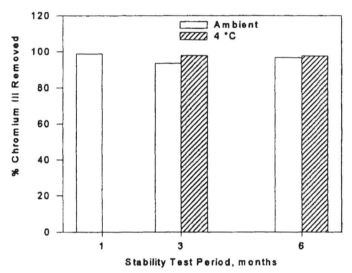

Figure 10.5 Chromium removal by HUMASORB-CS™ after stability tests (ambient and 4°C).

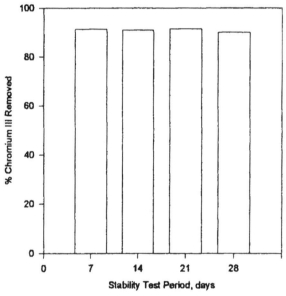

Figure 10.6 Chromium removal by HUMASORB-CS™ after stability tests (50°C).

356

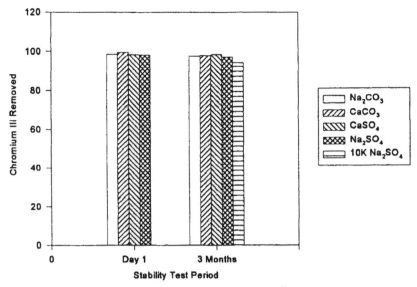

Figure 10.7 Chromium removal by HUMASORB-CS™ after stability tests.

compare with HUMASORB-S™. The tests were conducted using 100 PPM of one metal (As, Cd, Ce, Cs, Cr, Cu, Pb, Hg, Ni, Sr, U, Zn) in the waste stream. In addition, a simulated waste stream containing all 12 metals at 100 PPM each was also contacted with HUMASORB-CS™. The simulated waste solution (25 ml) and HUMASORB-CS™ (one gram) or HUMASORB-S™ (one gram) were shaken at 300 rpm and 25°C for two hours. After the two-hour contact time, the mixture was centrifuged to separate the solid and liquid phases. The liquid phase was analyzed for the metals using inductively couple plasma (ICP) or atomic absorption (AA) spectroscopy.

Results and Discussion

The results from the batch tests are presented in terms of percent removal of metals. The removal of individual metals (shown in Figures 10.8 and 10.9) indicates high removal of a number of metals. The removal was greater than 60% for at least six of the metals under the very low pH (2–2.5) conditions used in the study. A comparison of results from tests with waste streams containing only a single metal using the two forms of HUMASORB™ (Figures 10.8 and 10.9) clearly shows that metal removal is higher for most metals with HUMASORB-CS™. Metal removal from the simulated waste stream containing multiple metals (shown in Figures 10.10 and 10.11) was similar to that observed in tests with individual metals. However, with a waste stream containing multiple metals (Figures 10.10 and 10.11), HUMASORB-S™ is more effective for a few of the metals (copper, lead and mercury). Cesium removal was higher in the test with multiple metals compared to the removal when only cesium was present.

The overall total removal expressed as milliequivalents (meq) of metal removed was similar for both HUMASORB-S™ and HUMASORB-CS™ (approximately 0.25 meq) under the conditions (pH 2–2.5) of this study. It is clear from the batch tests that the proprietary methods used to produce HUMASORB-CS™ improve the solubility characteristics and also retain the ability to remove contaminants.

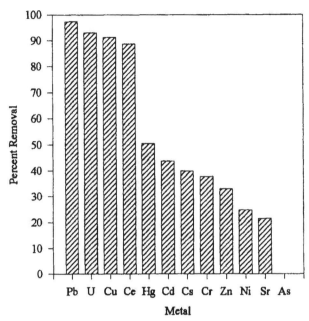

Figure 10.8 Removal of individual metals from simulated waste stream using HUMASORB-CS™.

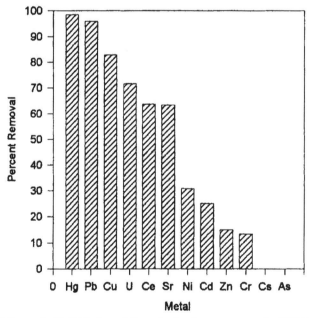

Figure 10.9 Removal of individual metals from simulated waste stream using HUMASORB-S™.

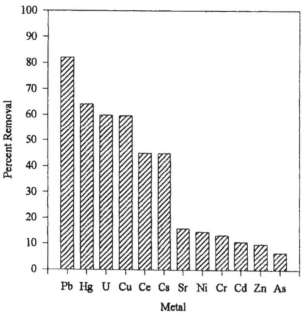

Figure 10.10 Removal of multiple metals from simulated mixed waste stream using HUMASORB-CS™.

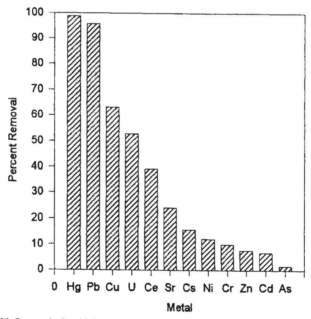

Figure 10.11 Removal of multiple metals from simulated waste stream using HUMASORB-S™.

Isotherms

Batch tests were conducted using simulated waste streams containing individual metals at various concentrations. The objective of the tests was to develop isotherms for different metals to estimate the capacity of HUMASORB-CS™ for metal removal. In these tests, 25-ml of spiked water was treated with one gram of HUMASORB-CS™ for two hours. The concentration of the metal in the spiked water varied from 100 PPM to 10, 000 PPM. The solid and liquid phases were separated using a centrifuge, and the liquid phase was analyzed for the equilibrium concentration of the metal.

The results from these tests were used to fit the Langmuir adsorption model. According to the Langmuir model,

$$\frac{x}{m} = \frac{K_{lan} \times (bC)}{1 + bc}$$

where K_{lan} = amount of solute adsorbed per unit weight of adsorbent to form a monolayer coverage (representing saturation capacity) and b = constant.

The Langmuir equation can be linearized as follows:

$$\frac{m}{x} = \frac{1}{K_{lan}} + \frac{1}{K_{lan} \times b}\left(\frac{1}{C}\right)$$

A plot of the reciprocal equilibrium concentration ($1/C$) versus the reciprocal of amount of metal adsorbed per unit weight of adsorbent (m/x) will be a straight line if the adsorption can be described by the Langmuir isotherm. The intercept of such a plot can be used to estimate the constant K_{lan} (representing saturation capacity) and the slope used to estimate the constant b. The saturation capacities for various metals were estimated from the isotherms and are shown in Figure 10.12. The figure indicates high capacity of HUMASORB™ for metal binding.

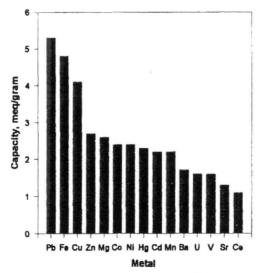

Figure 10.12 Saturation capacity of HUMASORB™ for various metals.

COLUMN TESTS

Methods

Column tests were conducted using glass columns having an internal diameter of 22 mm and an approximate bed height of 20 cm. The columns were packed with 80% sand and 20% HUMASORB-CS™ on a weight basis. Sand was used to lower channeling effects and to allow the use of relatively larger columns. Sand and HUMASORB-CS™ were uniformly mixed and wet-packed into the column. The packed column was visually inspected for uniform distribution of HUMASORB-CS™.

Simulated waste streams were passed through the columns in downflow mode via gravity flow. The flow rate was adjusted using a valve at the column outlet. A constant liquid head was maintained above the bed by using an inverted flask containing the liquid to be passed through the column. This setup eliminated the need for pumps and flow meters while maintaining a relatively constant flow through the column. The flow rate for each test was monitored by measuring the volume of the collected sample against the time used to collect the sample. Column tests were conducted with simulated waste streams at similar rates defined as empty-bed contact time (EBCT). EBCT is the time required for the fluid to pass through the volume occupied by the adsorbent bed. The amount of HUMASORB-CS™ in the column and the bulk density (~1 gram/ml) were used to estimate the volume to estimate EBCT. EBCT and the bed volumes used in these tests are based on the volume occupied by dry HUMASORB-CS™ in the column.

Results and Discussion

The results from the column tests were used to develop breakthrough curves, in which the ratio of the column output to column input concentration is plotted against number of bed volumes passed through the column. It was assumed that the column was saturated when the output concentration was nearly 95% of the input concentration. It was assumed that there was breakthrough when the output concentration was between 2–5% of the input concentration. The flow rates (and thus EBCT) selected were designed to allow for relatively quick breakthrough and saturation of the column for logistical reasons.

The breakthrough curve for removal of lead from a simulated waste stream containing lead and perchloroethylene (PCE) is shown in Figure 10.13. In this experiment, approximately 2,700 bed volumes of contaminated water (lead: 20 PPM) were passed through the column. The contact time based on HUMASORB-CS™ in the column was less than four minutes. As shown in the figure, there was no breakthrough of the contaminant after 2,700 bed volumes. To obtain breakthrough of lead, the same column was used again with the input concentration of lead increased to 200 PPM. Lead breakthrough was observed after an additional 500–600 bed volumes were passed through the column. The column was approximately 70% saturated at the end of the test, at which point 3,600 bed volumes had passed though the column.

A column test was also conducted using a simulated waste stream containing cerium, a surrogate for radioactive plutonium. The concentration of cerium in the column input stream was 200 PPM. The breakthrough curve shown in Figure 10.14 indicates no breakthrough for at least 600 bed volumes; the column was saturated after nearly 1,700 bed volumes.

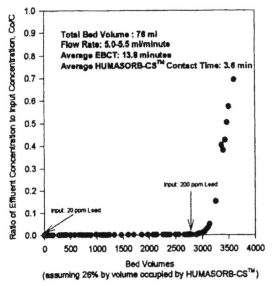

Figure 10.13 Column breakthrough curve for lead.

Reduction of Oxo-Anions

The reduction of different metal species and actinides such as mercury, chromium, vanadium, iron, neptunium and plutonium by humic acid has been reported by a number of investigators [20–25]. Humic acid can·act as a reducing agent and influence the oxidation-reduction of metal species. An unchelatable toxic oxo-anion such as chromium(VI), present as dichromate, is reduced to relatively nontoxic chromium(III). The reduced chromium is then stabilized through chelation by humic acid.

The ability of HUMASORB™ to reduce chromium(VI) to less toxic chromium(III) was

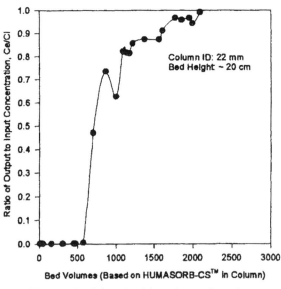

Figure 10.14 Column breakthrough curve for cerium.

evaluated in batch tests. In these tests, 25-ml of wastewater containing 60 PPM of chromium(VI) was contacted with HUMASORB™ for different contact times ranging from one to 16 hours. The reaction mixture was shaken at 300 rpm throughout the test period. The mixture was then centrifuged to separate the solid and liquid phases. The liquid phase was then analyzed for both chromium(VI) and chromium(III). Appropriate control samples were used during the test period.

The results shown in Figure 10.15 show that the chromium(VI) concentration decreases during the test period, but the chromium(III) concentration remains constant at 2–3 PPM. This indicates that chromium(III) is removed immediately from the liquid phase as soon as it is formed due to chromium(VI) reduction. The mechanism for chromium(III) removal is most likely a combination of ion exchange and complexation.

Simulated Barrier Tests

A new approach to treat groundwater is the use of in situ permeable reactive barriers (PRB) to overcome the limitations of above-ground treatment methods such as pump-and-treat. A PRB contains suitable reactive materials and is constructed to intercept the path of the contaminated groundwater plume [26]. The treatment materials in the barrier remove the contaminants as the groundwater passes through it. The residence time for groundwater through the PRB is typically one to two days. The technology to construct and emplace barrier materials is well developed. However, one of the limitations has been the availability of a suitable treatment material that can be used in the barrier. The material must be effective for removal of multiple types of contaminants from these complex streams. In addition, the suitable material must not introduce any contaminants that would be unacceptable into the groundwater. The materials being considered as treatment media in PRBs include Zero-Valent Iron (ZVI), peat, modified zeolites and others [26]. These media can only treat one or two types of contaminants, and the limitations of these media include the potential release of contaminants under changing redox conditions. HUMASORB™ is being considered as an alternative treatment media in PRBs to overcome these limitations and is being evaluated as part of a U.S. Department of Energy (DOE) project [27]. The results from simulated barrier tests conducted in the lab with a simulated wastewater containing multiple contaminants are discussed in this section.

METHODS

The experiments were conducted using two large columns having an internal diameter of two inches and a length of 36 inches. The first column (A) was slurry packed with 100% HUMASORB-CS™ (870 grams) and was subjected to a pressure of 10 psig using nitrogen. The second column (B) was packed with a mixture of 50% sand and 50% HUMASORB-CS™ (750 grams) on a weight basis and was subjected to a pressure of 100 psig. Each column was connected to a pressurized tank containing the simulated contaminated water under the desired pressure, as shown in Figure 10.16. A simulated waste stream containing chromium(VI), cerium (surrogate for plutonium), copper, trichloroethylene (TCE) and tetrachloroethylene (PCE) was prepared and passed at a flow rate of approximately 1.0 ml/min and 0.5 ml/min through columns A and B, respectively. The flow rate was maintained by controlling the output flow with a needle valve (column A) and a metering valve (column B). The flow rates were monitored and measured twice a day. Two samples were collected at the outlet and two at the inlet of each column every ten days for metal and organic analyses, with the samples at the inlet acting as control. The samples were ana-

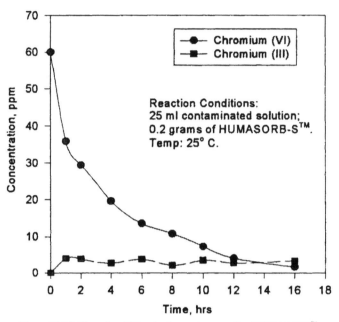

Figure 10.15 Chromium(VI) reduction and removal by HUMASORB™.

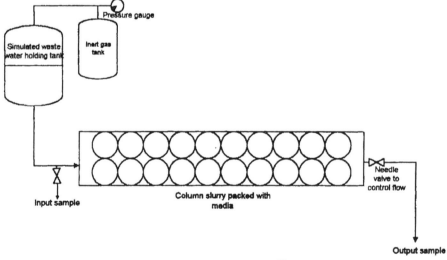

Figure 10.16 Experimental setup to evaluate HUMASORB-CS™ under simulated barrier conditions.

lyzed for organics and metals using calibrated methods by gas chromatography (GC) and inductively coupled plasma (ICP), respectively.

RESULTS AND DISCUSSIONS

The simulated barrier tests were conducted to evaluate the effectiveness of HUMASORB-CS™ for in situ treatment of groundwater. The rate of groundwater flow is very low, and the residence time in a permeable barrier is typically 24–48 hours. The residence time in the simulated barrier tests, based on the amount of HUMASORB-CS™ present in the column was approximately 14 hours for column A and 25 hours for column B.

The breakthrough curves for the metals in columns A and B are shown in Figures 10.17 and 10.18. The simulated water was prepared on a regular basis throughout the nearly two-year study, and the concentration of the contaminants in the input varied during the course of the study. The average input concentrations of the contaminants are also shown in Figures 10.17 and 10.18. In this discussion, it was assumed that there was breakthrough when the output concentration was more than 5% of the input. The breakthrough curves clearly indicate that there is no breakthrough of copper and cerium (surrogate for plutonium). The chromium(VI) concentration in the output is approaching breakthrough after 350 bed volumes but does not increase until at least 700 bed volumes in column A. The chromium(VI) output data is a little scattered, but the output concentration is very close to the breakthrough limit. For column B, the concentration increased above the breakthrough limit after approximately 300 bed volumes but gradually decreased after 350 bed volumes and remained lower than the 5% limit after more than 700 bed volumes. The anomaly in the chromium(VI) breakthrough between 300–350 bed volumes is believed to be primarily because of analytical issues and is also reflected to a certain extent in the breakthrough curves for column A.

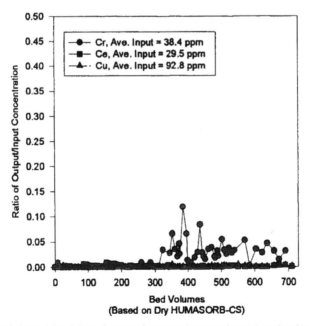

Figure 10.17 Column A breakthrough curves for inorganic contaminants (test duration: 21 months).

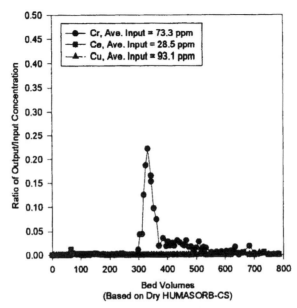

Figure 10.18 Column B breakthrough curves for inorganic contaminants (test duration: 24 months and still running).

The breakthrough curves for PCE and TCE in columns A and B are shown in Figures 10.19 and 10.20. There is no breakthrough of PCE in column A until approximately 400 bed volumes and the column was 30–40% saturated at 700 bed volumes. In column B, PCE breakthrough occurs after 600 bed volumes, and the column was nearly 60% saturated at 700 bed volumes. The higher amount of bed volumes passed through column B before breakthrough is due to the longer residence time compared to that in column A. TCE breakthrough occurs in both columns at approximately 150 bed volumes, and the output concentration continued to increase above the input concentration. This increase in the TCE concentration at the column output is believed to be the result of a number of factors. HUMASORB-CS™ has a higher affinity for PCE compared to TCE, resulting in possible preferential adsorption of PCE. In addition, there is some evidence to suggest the possible degradation of chlorinated organic compounds after adsorption by HUMASORB-CS™ [27]. TCE is a product of PCE degradation, and the steady increase in TCE concentration could be a result of PCE degradation by HUMASORB-CS™.

The data used to generate the breakthrough curves were also used to estimate the capacity of HUMASORB-CS™ for the multiple inorganic and organic contaminants. The capacities were estimated at breakthrough and percent saturation of the column at the time of this report and are tabulated in Table 10.1. The results clearly indicate that HUMASORB-CS™ has a high capacity for the various contaminants and is effective in removing metals, organics and radionuclide surrogates from groundwater using a permeable reactive barrier system.

Pilot Tests at Berkeley Pit

Berkeley Pit is an abandoned open-pit mine located in Butte, Montana, USA, that was designated as a Superfund site by the Environmental Protection Agency (EPA) in 1987. The Berkeley Pit has been filling with acidic, heavy-metal laden water since pumping opera-

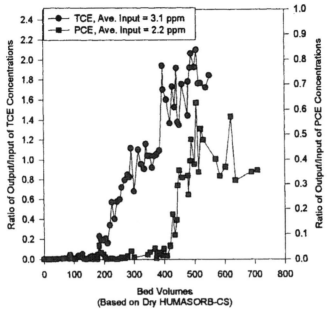

Figure 10.19 Column A breakthrough curves for organic contaminants.

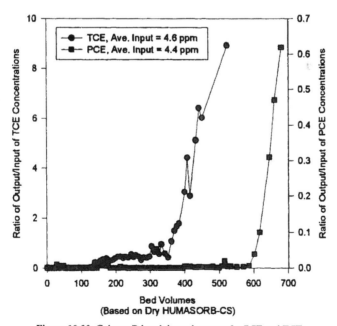

Figure 10.20 Column B breakthrough curves for PCE and TCE.

367

TABLE 10.1.   Estimated Capacity of HUMASORB-CS™ for Various
Inorganic and Organic Contaminants.

| | Estimated Bed Volumes and Capacity at Breakthrough | | | |
| | Column A[1] | | Column B[2] | |
Contaminant	Bed Volumes	Capacity, mg/g	Bed Volumes	Capacity, mg/g
Cu	710^3	65.89	783^3	72.9
Ce	710^3	20.95	783^3	22.3
Cr	352	13.52	318	23.3
PCE	419	0.92	618	2.72
TCE	177	0.55	244	1.12
	Estimated Bed Volumes and Capacity at the Time of the Report			
Cu	710^3	65.89	783^3	72.9
Ce	710^3	20.95	783^3	22.3
Cr	690^4	20.0	680^4	36.6
PCE	690^5	1.22	682^6	2.86
TCE	290^7	0.92	361^7	1.4

[1]pH range of column input: 4.7–6.2; pH range of column output: 8.5–8.9.
[2]pH range of column input: 5.0–6.0; pH range of column output: 8.2–8.8.
[3]No breakthrough.
[4]10% saturation.
[5]30–40% saturation.
[6]60% saturation.
[7]100% saturation.

tions ceased in 1982. Currently, the Pit contains approximately 30 billion gallons with an estimated additional daily in-flow of three to five million gallons. This continues to pose a serious threat to contamination of aquifer and environment in the vicinity of the Butte area. The Pit water has been extensively characterized, with the primary dissolved ions being iron, cadmium, zinc, aluminum, manganese, calcium and copper. Trace amounts of chromium, nickel, lead and arsenic are also present. All metals are present as dissolved solids in ionic form or as dissolved sulfides. The pH of the water is low, approximately 2.6.

The HUMASORB™ process was used to demonstrate resource recovery and remediation of waters from the Berkeley Pit. The HUMASORB™ process for resource recovery and remediation of waters contaminated with multiple metals consists of two stages. In Stage 1, the Berkeley Pit water is treated with HUMASORB-L™ to selectively recover iron, copper and other agricultural micronutrients by formation of humates, which are precipitated as flocs. The precipitated complex can be easily separated in a solid/liquid separation unit. The metals remaining in the water from Stage 1 and all other toxic metals are reduced in Stage 2 using the water-insoluble humic acid-based ion exchange material HUMASORB-CS™. The process utilizes two very diverse properties of humic acid—as an ion exchange chelating material and as a soil amendment product—to make a process that is both effective and yet simple to implement.

BENCH SCALE TESTS

Stage 1

Experiments were conducted using a composite statistical experimental design to optimize the process parameters of pH, HUMASORB-L™ dosage and contact time for Stage 1. The results from Stage 1 tests were analyzed statistically to identify the optimum combination of variables for Stage 1 treatment. The response variables (dependent variables) used in the analysis include the concentrations of iron, cadmium, copper and other metals of interest. The ratio of concentrations of iron and cadmium in Stage 1 was also used as an additional response variable. Regression equations and surface and contour plots were developed using the different response variables. The analysis showed a close fit between the observed and predicted values from regression analysis.

Stage 2

The Stage 2 confirmation tests were conducted at two different flow rates in columns that were one inch in diameter. The total bed height was approximately 42 cm, and the bed volume based on the dry HUMASORB-CS™ in the column was approximately 213 ml. The breakthrough curves for some of the metals are shown in Figure 10.21. The results show breakthrough of some of the metals after 30–40 bed volumes, but for other metals, breakthrough occurs after 50–60 bed volumes have been passed through the column.

The columns used in Stage 2 were also used after the tests to recover the metals by passing 1 N sulfuric acid in countercurrent mode. Approximately 1,000 ml of acid was passed through the columns. The results showed that the metals can be recovered in concentrated form, and the amount of acid required for metal recovery was only 600–700 ml which is less than five bed volumes. The concentrations during metal recovery are shown in Figure 10.22. A comparison of the metals analysis of Berkeley Pit water before and after treatment in the bench-scale tests is shown in Table 10.2.

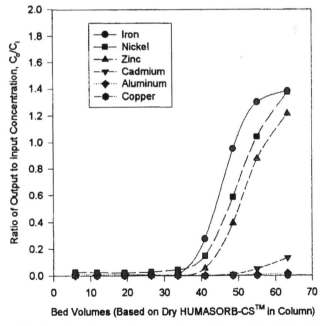

Figure 10.21 Breakthrough curves for some of the metals present in Berkeley Pit water.

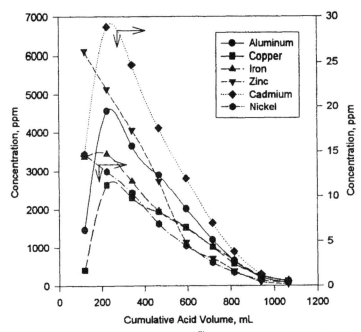

Figure 10.22 Metal recovery from spent HUMASORB™ used to remove metals from Berkely Pit water.

PILOT SCALE TESTS

The HUMASORB™ process was also used in continuous pilot-scale tests for resource recovery and treatment of water from Berkeley Pit. The primary objective in the pilot tests was to produce a micronutrient-enriched commercial soil amendment product. The pilot tests were conducted at a Berkeley Pit water flow rate of 1–3 gallons per minute. In the tests, more than 55% of iron and up to 20% copper were removed in Stage 1 for formulation into a commercial humic acid base soil amendment product. The cadmium removal in Stage 1 was minimized to less than 10%. The removal of a few of the metals in Stage 1 of the pilot tests is shown in Table 10.3. The comparison of the pilot test objectives and HUMASORB-CS™ process performance is shown in Table 10.4.

The results show that the HUMASORB-CS™ process is effective for resource recovery

TABLE 10.2. Concentration of Metals in Berkely Pit Water before and after Treatment with HUMASORB™.

Metals	Raw Water (pH: 2.71)	Treated Water (pH: 5–6)
	PPM	
Aluminum	244	0.428[1]
Cadmium	1.82	0.0016[1]
Copper	201	0.0437[1]
Iron	660	75.1
Nickel	1.02	0.124[1]
Zinc	626	26.2

[1]Below method detection limit.

TABLE 10.3. Metal Removal in Stage 1 of the Pilot Tests with HUMASORB™ Process for Berkely Pit Water Treatment.

Metal	Berkely Pit Water Concentration, ppm (pH: 2.48–2.67)	Stage 1 Treated Water Concentration, ppm (pH: 3.67–3.81)	Percent Removal
	Test #1		
Iron	983	438.75	55.37
Cadmium	2.04	1.98	2.94
Copper	179.5	161.25	10.17
Zinc	561	540.5	3.65
	Test #2		
Iron	963	337.75	64.93
Cadmium	2.04	1.92	5.88
Copper	176	150	14.77
Zinc	550	523	4.91
	Test #3		
Iron	861	449.67	47.77
Cadmium	1.94	2.00	—
Copper	180	165.33	8.15
Zinc	450	537	—
	Test #4		
Iron	998	453.67	54.54
Cadmium	2.05	1.85	9.76
Copper	182	135.67	25.46
Zinc	558	508.33	8.90

and remediation of water contaminated with metals. However, the economics of the process are driven by the benefits of the soil amendment product produced in Stage 1. The following section has a brief description of the results from the product tests.

Soil Amendment Product Tests

The soil amendment product produced by treatment of the Berkeley Pit water was evaluated by the Department of Plant, Soil and Environmental Sciences at Montana State University. The objective of the tests was to evaluate the differences in the uptake of micronutrients, especially iron, derived from the Berkeley Pit water with iron from a commercial source. Columns of six-inch diameter (PVC) pipe, 18 inches high, were used for the tests.

TABLE 10.4. Comparison of Pilot Test Objectives with HUMASORB™ Process Performance.

Metals	Demonstration Objectives	Process Performance
Arsenic	<0.5 ppm in product	0.294–0.481 ppm in product
Cadmium	<0.5 ppm in product	0.280–0.360 ppm in product
Cadmium	>70% removal in Stage 2	67.56–98.93% removal in Stage 2
Copper	>70% removal in Stage 2	85.44–99.96% removal in Stage 2
Iron	>70% removal in Stage 2	73.15–99.97% removal in Stage 2
Zinc	>70% removal in Stage 2	48.02–99.90% removal in Stage 2

The bottom portion of each column was filled to a depth of nine inches with 70 grit sand, and the upper nine-inch portion was filled with soil that had been air-dried, ground and passed through a 2 mm sieve. The test samples used included a combination of commercial fertilizer and the product produced by the HUMASORB-CS™ process.

It was shown conclusively that the micronutrients derived from the Berkeley Pit waters are utilized by the plants. There was no difference between the uptake of nutrients, especially iron, from the Berkeley Pit waters compared to that obtained from a commercial source. In addition, there was an increase in yield of up to 35% for alfalfa from the use of the *actosol*® humic acid product containing iron. Similarly, for spring wheat, there was an increase of up to 20% in the number of tillers/plant.

A preliminary economic analysis was performed using the results from the tests conducted in this study for resource recovery from Berkeley Pit water. The treatment rate was assumed to be 3,500 gpm, which is approximately equal to the amount of water flow in to the Berkeley Pit. At this treatment rate, the water level in the Berkeley Pit will remain constant. A full-scale plant to treat Berkeley Pit water using the HUMASORB™ process will most likely produce a combination of products depending on market needs and requirements. The economic analysis was conducted using two economic scenarios. The net present worth for the proposed full-scale plant with the conservative economic scenario is estimated at $51 million with an internal rate of return of 33% and a payback period of less than three years.

Treatment at Johnston Atoll

A HUMASORB™-based treatment system was used recently at the Johnston Atoll facility of the United States Department of Defense (DoD), for treatment of spent decontamination solution (SDS) to remove lead, mercury and arsenic. SDS was analyzed for residual chemical agent and was found to be below the drinking water standards (DWS) for all agent types. However, much of the waste had concentrations of arsenic, lead and mercury in excess of the limits mandated under the Resource Conservation and Recovery Act (RCRA). The original plan for treatment of the SDS was incineration at the Johnston Atoll Chemical Agent Disposal System (JACADS). However, approximately 24,000 gallons of SDS could not be treated at JACADS because of the current RCRA permit limitations on the processing of such material.

The program manager for chemical demilitarization (PMCD) was asked by the U.S. Environmental Protection Agency (U.S.EPA) to consider treatment of the liquid waste by some alternative technology to reduce the metals concentration to below the regulatory hazardous levels. Technology application studies conducted using the HUMASORB™ process on SDS samples received from Johnston Island showed that the levels of arsenic, lead and mercury could be reduced below the RCRA treatment levels. Based on initial technology application tests, the process was subsequently selected by the PMCD as a viable alternative for removal of the metal contaminants from the SDS waste and was approved for use by U.S.EPA.

A mobile unit based on the HUMASORB™ process was used to treat SDS on-site at Johnston Island. The SDS samples were initially processed through a filter press to remove suspended solids and were then passed through the HUMASORB™ process unit. The concentration of the target metals before and after treatment in the HUMASORB™ process unit is shown in Table 10.5. The concentrations of all the target metals were found to be below the required regulatory mandated levels. An approved independent laboratory completed

TABLE 10.5. Concentration of Metals in SDS before and after Treatment by HUMASORB™ Process.

Sample	Arsenic, ppm		Lead, ppm		Mercury, ppm	
	Before	After	Before	After	Before	After
1	0.410	ND	0.230	0.097	0.260	0.0081
2	190	ND	0.074	ND	6.1	0.011
3	5.3	4.2	1.6	ND	0.250	0.11
4	0.910	ND	0.021	ND	0.340	ND

the analytical activities for regulatory compliance. The technology application tests indicate that the HUMASORB™ process is effective for removal of multiple metals from waste brines produced at chemical agent disposal facilities.

CONCLUSIONS

Humic acid and its interactions with environmental chemicals have been studied extensively. Humic acid has several properties that can be exploited for the treatment and detoxification of hazardous waste streams. The properties that have potential to be exploited for environmental purposes include the ability to chelate metals, reduce oxidized metal species and adsorb organic compounds from wastewater systems. The available literature clearly indicates promise for the use of humic acid-based technology for various applications.

HUMASORB™ is a generic term for humic acid-based products developed by ARCTECH, Inc. for resource recovery and remediation. HUMASORB™ is effective in removing multiple classes of contaminants such as metals, oxyanions, organics and radionuclides from contaminated streams. The cation-exchange capacity of HUMASORB™ can range between 1–5 meq/gram depending on the metal compared to 1–2 meq/gram for many ion exchange resins. Toxic metals present as anions (such as hexavalent chromium) are reduced to a less toxic state and are removed from water. HUMASORB™ is stable in the presence of anions such as sulfate and carbonate and in the presence of microorganisms. HUMASORB™ is also suitable for application as a permeable reactive barrier (PRB) material for in situ treatment of groundwater.

REFERENCES

1 U.S.EPA Report #EPA 542-R-96-005, "Clean Up the Nation's Waste Sites: Markets and Technology Trends," April, 1997.
2 U.S.DOE, Office of Science and Technology, "Technology Summary Reports," August, 1996.
3 Stevenson, F. J. "Reviews and Analyses: Organic Matter Reactions Involving Herbicides in Soil," *Journal of Environmental Quality,* 1(4), 1972.
4 Ziechmann, W. *Humic Substances.* BI Wissenschaftsverlag, Mannheim, 1993.
5 Beckett, R., "The Surface Chemistry of Humic Substances in Aquatic Systems." In: *Surface and Colloid Chemistry in Natural Water and Water Treatment.* R. Beckett, Ed., Plenum, New York, 1990, pp. 3–20.
6 Schulten, H. R. and Schnitzer, M. "Three Dimensional Models for Humic Acids and Soil Organic Matter," *Naturwiss.,* 82, 1995; *Soil Science,* 162, 1997 and references therein.
7 Ghabbour, E. A., Khairy, A. H., Cheney, D. P., et al. "Isolation of Humic Acid from the Brown Alga *Pilayella littoralis,*" *J. Appl. Phycol.,* 6, 1994.

8 MacCarthy, P., Clapp, C. E., Malcolm, R. L. and Bloom., R. R.. *Humic Substances in Soil and Crop Sciences: Selected Readings.* American Society of Agronomy, Madison, Wisconsin, USA, 1990.

9 Senesi, N. and Miano, T. M., *Humic Substances in the Global Environment: Implications for Human Health.* Eds., Elsevier, Amsterdam, 1994, and references therein.

10 Fataftah, A. *Metal-Humic Acid Interactions.* Doctoral Dissertation, Chemistry Department, Northeastern University, Boston, MA, USA, August, 1997, and references therein.

11 Choudhary, G. G. *Toxicological and Environmental Chemistry,* 6, 1983.

12 Keen, D. S. Ed. *Source Book on Humic Acids,* American Colloid Company, Skokie, IL, USA.

13 Aldrich Chemical Co., Inc. MSDS, 1992.

14 Merck Index, 11th Edition, 1989.

15 Davies, G., Fataftah, A., Cherkasskiy, A., et al. "Tight Metal Binding by Humic Acids and Its Role in Biomineralization," *J. Chem. Soc., Dalton Trans.,* 4047–4060, 1997

16 Beames, G. H. "Use of humic substances in plant growth." In: *Source Book on Humic Acids,* Ed. Keen, D.S. American Colloid Company, Skokie, IL, USA.

17 Senn, T. L and Kingman, A. R. *A Review of Humus and Humic Acid Research Series No. 145,* S. C. Agricultural Experiments Station, Clemson, South Carolina, USA, 1973.

18 Tan, K. H. *Principles of Soil Chemistry,* Marcel Dekker, Inc., New York, 1982.

19 Sanjay H. G., Srivastava, K. C. and Walia, D. S. *Adsorbent,* US Patent 5,906,960, 1999; US Patent 6,143,692, 2000.

20 Manahan, S. E. "Interactions of hazardous-waste chemicals with humic substances," In: *Aquatic Humic Substances,* Eds. Suffet, I. H. and MacCarthy, P. American Chemical Society, Washington, DC, USA, 1989.

21 Nash, K., Fried, S., Friedman, A. M. and Sullivan, J. C. "Redox Behavior, Complexing, and Adsorption of Hexavalent Actinides by Humic Acid and Selected Clays," *Environmental Science & Technology,* 15(7), 834, 1981.

22 Alberts, J. J. et al. *Science,* 184, 895, 1974.

23 Szalay, A. and Szilagyi, M. *Geochim Cosmochim, Acta.* 1, 31, 1967.

24 Theis, T. L. and Singer, P. C. *Trace Met. Met-Org. Interact. Nat. Waters. [Symp.],* 273, 1976.

25 Bondietti, E. A. *Transuranium Nuclides Environ., Proc. Symp.,* 273, 1976.

26 Shoemaker, S. H., Greiner, J. F. and Gillham, R. W. "Permeable Reactive Barriers." In *Assessment of Barrier Containment Technologies,* Eds. Rumer, R. R. and Mitchell, J. K. U.S. DOE, U.S.EPA, DuPont Company, NTIS # PB96-180583, 1996.

27 Sanjay, H. G., Fataftah, A., Walia, D. S. and Srivastava, K. C. *Development of Humasorb™ —A Coal Derived Humic Acid for Removal of Metals and Organic Contaminants from Groundwater,* Final Report, U.S. DOE Project # DE-AR21-95-MC32114, 2000.

Index

About the Editor

A RUP K. SENGUPTA is professor and chairman of the department of civil and environmental engineering in Lehigh University. He also holds joint professorship in the chemical engineering department at Lehigh. Dr. SenGupta is currently North American Editor of *Reactive & Functionalized Polymers Journal* and Co-editor of *Ion Exchange and Solvent Extraction Series.* He received a B.S. in chemical engineering in 1972 from Jadavpur University, India and a Ph.D. in environmental engineering in 1984 from the University of Houston. For research in the area of environmental separation, Dr. SenGupta and his graduate students received awards from the American Society of Civil Engineers (ASCE), American Institute of Chemical Engineers (AIChE), American Water Works Association (AWWA) and Lucent Technology.